Opportunities and Challenges for Women Leaders in Environmental Management

Perfecto Gatbonton Aquino, Jr.
The University of Cambodia, Cambodia

Revenio Cabanilla Jalagat, Jr.
Al-Zahra College for Women, Oman

Mercia Selva Malar Justin
Xavier Institute of Management and Entrepreneurship, India

IGI Global
PUBLISHER of TIMELY KNOWLEDGE

A volume in the Practice, Progress, and Proficiency in Sustainability (PPPS) Book Series

Published in the United States of America by
IGI Global
Engineering Science Reference (an imprint of IGI Global)
701 E. Chocolate Avenue
Hershey PA, USA 17033
Tel: 717-533-8845
Fax: 717-533-8661
E-mail: cust@igi-global.com
Web site: http://www.igi-global.com

Library of Congress Cataloging-in-Publication Data

Names: Aquino, Perfecto Gatbonton, Jr., 1962- editor. | Jalagat, Revneio Cabanilla, Jr., 1978- editor. | Justin, Mercia, 1963- editor.
Title: Opportunities and challenges for women leaders in environmental management / edited by: Perfecto Aquino, Revenio Jalagat, Jr., Mercia Justin.
Description: Hershey PA : Engineering Science Reference, [2023] | Includes bibliographical references. | Summary: "This book provides the latest empirical research and theoretical frameworks to analyze the environmental impacts of adopting environmental management practices through the model leadership of top global women leaders in environmental protection and management"-- Provided by publisher.
Identifiers: LCCN 2023037135 (print) | LCCN 2023037136 (ebook) | ISBN 9781668459867 (hardcover) | ISBN 9781668459874 (paperback) | ISBN 9781668459881 (ebook)
Subjects: LCSH: Women environmentalists. | Environmental management. | Environmental management--Vocational guidance. | Women executives. | Leadership in women.
Classification: LCC GE195.9 .O77 2023 (print) | LCC GE195.9 (ebook) | DDC 363.70023--dc23/eng/20231016
LC record available at https://lccn.loc.gov/2023037135
LC ebook record available at https://lccn.loc.gov/2023037136

This book is published in the IGI Global book series Practice, Progress, and Proficiency in Sustainability (PPPS) (ISSN: 2330-3271; eISSN: 2330-328X)

British Cataloguing in Publication Data
A Cataloguing in Publication record for this book is available from the British Library.

For electronic access to this publication, please contact: eresources@igi-global.com.

Practice, Progress, and Proficiency in Sustainability (PPPS) Book Series

Ayman Batisha
International Sustainability Institute, Egypt

ISSN:2330-3271
EISSN:2330-328X

Mission

In a world where traditional business practices are reconsidered and economic activity is performed in a global context, new areas of economic developments are recognized as the key enablers of wealth and income production. This knowledge of information technologies provides infrastructures, systems, and services towards sustainable development.

The **Practices, Progress, and Proficiency in Sustainability (PPPS) Book Series** focuses on the local and global challenges, business opportunities, and societal needs surrounding international collaboration and sustainable development of technology. This series brings together academics, researchers, entrepreneurs, policy makers and government officers aiming to contribute to the progress and proficiency in sustainability.

Coverage

- Knowledge clusters
- Eco-Innovation
- Socio-Economic
- Outsourcing
- Global Business
- Green Technology
- Sustainable Development
- Environmental informatics
- Strategic Management of IT
- Intellectual Capital

IGI Global is currently accepting manuscripts for publication within this series. To submit a proposal for a volume in this series, please contact our Acquisition Editors at Acquisitions@igi-global.com or visit: http://www.igi-global.com/publish/.

The Practice, Progress, and Proficiency in Sustainability (PPPS) Book Series (ISSN 2330-3271) is published by IGI Global, 701 E. Chocolate Avenue, Hershey, PA 17033-1240, USA, www.igi-global.com. This series is composed of titles available for purchase individually; each title is edited to be contextually exclusive from any other title within the series. For pricing and ordering information please visit http://www.igi-global.com/book-series/practice-progress-proficiency-sustainability/73810. Postmaster: Send all address changes to above address.

Titles in this Series

For a list of additional titles in this series, please visit: http://www.igi-global.com/book-series/practice-progress-proficiency-sustainability/73810

Global Perspectives on Climate Change, Social Resilience, and SocialInclusion
Aly Abd ElRazek Galaby (Alexandria University, Egypt) Mohammed Abo ElEnein (American University, UAE) and Hassan Mohamed (Alexandria University, Egypt)
Engineering Science Reference • © 2024 • 380pp • H/C (ISBN: 9781668489635) • US $275.00

Achieving the Sustainable Development Goals Through Infrastructure Development
Cristina Raluca Gh. Popescu (University of Bucharest, Romania & The Bucharest University of Economic Studies, Romania) Poshan Yu (Xi'an Jiaotong-Liverpool University, China & European Business University of Luxembourg, Luxembourg) and Yue Wei (Institution of Public Private Partnerships, Hong Kong)
Engineering Science Reference • © 2023 • 303pp • H/C (ISBN: 9798369307946) • US $230.00

Fostering Cross-Industry Sustainability With Intelligent Technologies
Brojo Kishore Mishra (NIST Institute of Science and Technology (Autonomous), India)
Engineering Science Reference • © 2024 • 400pp • H/C (ISBN: 9798369316382) • US $325.00

Crafting a Sustainable Future Through Education and Sustainable Development
Javier Martínez-Falcó (University of Alicante, Spain) Bartolomé Marco-Lajara (University of Alicante, Spain) Eduardo Sánchez-García (University of Alicante, Spain) and Luis A. Millan-Tudela (University of Alicante, Spain)
Engineering Science Reference • © 2023 • 302pp • H/C (ISBN: 9781668496015) • US $215.00

Sustainable Science and Intelligent Technologies for Societal Development
Brojo Kishore Mishra (NIST Institute of Science and Technology (Autonomous), India)
Engineering Science Reference • © 2023 • 572pp • H/C (ISBN: 9798369311868) • US $325.00

Energy Transition in the African Economy Post 2050
Olayinka Ohunakin (Covenant University, Nigeria)
Engineering Science Reference • © 2023 • 287pp • H/C (ISBN: 9781799886389) • US $215.00

Handbook of Research on Socio-Economic Sustainability in the Post-Pandemic Era
Jozef Oleński (High School Technology and Economics, Jaroslaw, Poland) Jeffrey Sachs (Columbia University, USA) Masayuki Susai (Nagasaki University, Japan) Yannis Tsekouras (University of Macedonia, Greece) and Arjan Gjonça (London School of Economics and Political Science, UK)
Information Science Reference • © 2023 • 401pp • H/C (ISBN: 9781799897606) • US $270.00

IGI Global
PUBLISHER of TIMELY KNOWLEDGE

701 East Chocolate Avenue, Hershey, PA 17033, USA
Tel: 717-533-8845 x100 • Fax: 717-533-8661
E-Mail: cust@igi-global.com • www.igi-global.com

Dedication

This humble work is primordially dedicated to our **Lord Jesus Christ** for giving the us the wisdom and strength until its full completion, without Him we can do nothing. Likewise, the Editors dedicate this work to all women leaders around the world who in one way or the other have contributed their efforts and their skills in saving the Mother Earth from its degradation, this book is specifically dedicated to the following:

To my late mother, Mrs. Josefa Gatbonton Aquino, Philippines;
The late Dr. Ligaya P. Jorge, Graduate School Professor, Royal and Pontifical University of Santo Tomas, Philippines;
Madam Monica Mom Sok, KCS International School, Phnom Penh, Cambodia;
Dr. Pilar Leano Pascual, former DECS-NCR Director and VPAA of Saint Francis of Assisi College System, Las Pinas, Philippines
Mrs. Aurora Mata-Torres, Board Chairman, Saint Anothony Nova School, Caloocan City, Philippines
Mrs. Alma Jalagat, Sultanate of Oman

EDITORIAL ADVISORY BOARD

Table of Contents

Detailed Table of Contents

Environmental management has emerged as a critical field of study and practice, aiming to address the challenges posed by environmental degradation and climate change. Women have played an increasingly prominent role in this domain, contributing significantly to sustainable development and environmental conservation. This research explores the opportunities and challenges faced by women leaders in environmental management, aiming to highlight their contributions, identify barriers, and proposes strategies to promote gender equality within the field. The study draws upon academic literature, case studies, and expert opinions to present a comprehensive analysis of the topic.

Women leaders in environmental management face unique opportunities and challenges in both developed and developing countries. In developed countries, opportunities include progressive policies, well-established institutions, and greater access to resources. Women can access quality education and training programs, legal frameworks promoting gender equality, and professional networks. However, in developed countries, women face challenges such as deep-rooted gender biases, stereotypes, and the 'glass ceiling' phenomenon, which limits women's representation in top decision-making positions. In developing countries, women leaders can engage with local communities and integrate traditional knowledge and sustainable practices into environmental management strategies. Limited resources and infrastructure in developing countries foster innovation and creativity, allowing them to develop contextualized solutions that address environmental challenges efficiently. However, challenges in developing countries include traditional gender roles, cultural norms, and resource constraints.

Women leaders who are at the forefront of environmental conservation efforts, have been instrumental in mitigating the effects of climate change, protecting biodiversity, and promoting sustainable development. The EMEA Region has seen an increasing number of women leaders in various industries and they play a crucial role in shaping public opinion on environmental issues. In this Chapter, their specific contributions are examined, including awareness, grassroots mobilization, and policy and advocacy. The challenges and opportunities they face, which include gender bias, risk and safety, collaboration and partnerships, and the role of media are also identified. Women can be effective leaders in environmental management through: collaboration and teamwork; communication; problem solving and innovation; attention to detail; as well as empathy and inclusivity. Women leaders promote sustainable development, protect the environment and their work in the media inspire and empower other women to take on leadership roles and contribute towards sustainable development.

This chapter examines the opportunities and challenges faced by Indian women leaders in the field of environmental management. As India undergoes rapid development and faces growing environmental concerns, the need for effective leadership in environmental management becomes increasingly crucial. The chapter highlights the progress made in promoting women's involvement in this field, such as increasing awareness, government policies, education and skill development, networking opportunities, and global collaborations. However, it also identifies the challenges that hinder women's progress, including gender biases, limited representation, work-life balance, resource access, and cultural constraints.

This book chapter explores the significant role of women leaders in environmental management and their impact on sustainable development. Historically, women have been underrepresented in leadership positions in the environmental sector, but in recent years, their contributions and achievements have gained recognition worldwide. The chapter begins by discussing the barriers and challenges that women have faced in accessing leadership roles in environmental management. It examines social, cultural, and institutional factors that have limited women's participation and advancement in this field. Despite these obstacles, the chapter highlights the success stories of pioneering women leaders who have defied societal norms and shattered glass ceilings to make a difference in environmental conservation and sustainability. The chapter then delves into the unique perspectives and approaches that women bring to environmental management.

Women and gender-diverse leaders play vital roles in society, ranging from maintaining households to protecting the environment. They are not homogenous but contribute significantly towards building a sustainable world. Being historically at the forefront of social and environmental movements, they have been advocating for change and are an integral part in promoting sustainability. This chapter explores the contributions of women and gender-diverse leaders in building a sustainable world. Women's unique perspective and skills will be highlighted.

Preface

In the vast tapestry of human existence, there has always been a silent force shaping the very foundation of our planet – women. From time immemorial, they have quietly and powerfully orchestrated a symphony of care for the environment. Since the 1972 International Conference on Human Environment in Sweden, the question of who holds the responsibility for safeguarding and managing the Earth's precious natural resources has echoed through countless gatherings and forums.

In a world where everyday actions like tossing waste into the sea, factories disgorging pollutants, and vehicles belching carbon emissions have become routine, the urgent need for responsible stewardship of our planet is undeniable. Through the ages, women, whose roles were once confined to the hearth and home, have transcended these boundaries to play pivotal roles in preserving and rejuvenating our environment.

This book, *Opportunities and Challenges for Women Leaders in Environmental Management*, is a testament to the remarkable contributions of women across the globe who have emerged as staunch environmentalists, making extraordinary efforts to combat the relentless degradation and potential destruction of our shared home. From restoring polluted rivers to ensuring the availability of safe drinking water, women have risen as dynamic custodians of the Earth's God-given gifts.

Within these pages, we aim to shed light on the roles and strategies of women who have dedicated themselves to environmental preservation. Through data, empirical studies, and inspiring narratives, we reveal the vital roles women play in activism, advocacy, research, and communication, showcasing their profound impact on both international and domestic environmental conservation.

Our intention is clear: to provide the latest empirical research and theoretical frameworks that illuminate the environmental impact of women's leadership in environmental protection and management. This book is not merely a collection of words; it is a call to action. We aspire to enlighten world leaders and global citizens about the significance and advantages of responsible custodianship of the environment.

The intended audience for this volume is as diverse as the women whose stories it shares. Environmentalists, students of business and environmental management, professors, academicians, business leaders, government agencies, research scholars, librarians, and non-governmental organizations devoted to environmental conservation and management will find invaluable insights. Women's organizations, too, will discover a wellspring of inspiration and knowledge within these pages.

The book encompasses an array of topics, from environmental management systems to energy conservation, water use, waste management, and the preservation of ecological parks and public sites. It serves as a comprehensive guide for those who seek a deeper understanding of environmental challenges and the transformative power of women's leadership.

ORGANIZATION OF BOOK

Chapter 1

This chapter delves into the Philippines' pioneering gender-sensitive policies aimed at equalizing men and women's roles in leadership positions across sectors. It explores the evolution of these policies from the '80s to the '90s, particularly focusing on women's participation in natural resource and environmental management. The study uncovers that while women have made strides in regional and national leadership, there are fewer opportunities at the grassroots level. The chapter employs documentary evidence to investigate the state of women's leadership in environmental management, revealing challenges such as limited access to education, male-dominated decision-making, poverty, and discrimination. It also presents recommendations for enhancing training and leadership at the grassroots level.

Chapter 2

This chapter sheds light on the roles of educated and professional women as leaders in their homes and workplaces, emphasizing their contributions to environmental management. These women possess a deep understanding of environmental issues and are well-placed to control environmental degradation. The chapter explores opportunities for women to engage in various levels of environmental management and the challenges they face as leaders. It reviews relevant literature, highlighting how women lead environmental management initiatives at home, in the workplace, and society.

Chapter 3

This chapter recognizes the unique perspectives, skills, and talents that women bring to environmental management. It emphasizes their participation in various disciplines and the valuable contributions they make to sustainability and conservation. The chapter underscores the importance of women's involvement in environmental decision-making processes, where they often prioritize long-term sustainability, social equality, and community well-being. It reviews research on women's environmental leadership, provides policy recommendations for increasing women's representation in environmental decision-making, and identifies areas for further research.

Chapter 4

This chapter takes us on a journey through India, a country renowned for its rich biodiversity and environmental challenges. It highlights the emergence of women leaders who have played pivotal roles in shaping environmental policies and implementing sustainable practices. These women have actively participated in governance processes related to environmental issues, resulting in the implementation of various environmental initiatives. The chapter showcases their contributions, the challenges they face, and the impact of their leadership on ecological issues.

Chapter 5

This chapter delves into the critical field of environmental management, where women have assumed increasingly prominent roles. It explores the opportunities and challenges faced by women leaders in this domain, aiming to highlight their contributions, identify barriers, and propose strategies for promoting gender equality within the field. The study draws upon academic literature, case studies, and expert opinions to provide a comprehensive analysis of the topic.

Chapter 6

This chapter provides a global overview, comparing the opportunities and challenges faced by women leaders in both developed and developing countries. It underscores the advantages of progressive policies and well-established institutions in developed countries, as well as the unique strengths of women in developing countries, who integrate traditional knowledge and sustainable practices into environmental management.

Chapter 7

This chapter focuses on the EMEA Region, where an increasing number of women leaders are making significant contributions to environmental conservation. It examines their roles in raising awareness, grassroots mobilization, policy advocacy, and other aspects. The chapter also addresses the challenges and opportunities they encounter, including gender bias, risk, collaboration, and the role of media.

Chapter 8

This chapter delves into the shift from the 'green agenda' to the 'brown agenda,' emphasizing the importance of gender equality in the context of environmental issues. It explores the evolving dynamics of environmental discourse and the need for greater gender inclusivity in environmental management.

Chapter 9

This chapter delves into the socio-political perceptions and challenges that women leaders face in environmental management. It explores whether these challenges differ from those in other spheres, as well as the role and success of women leaders in raising awareness and educating society about environmental concerns.

Chapter 10

This chapter explores the vital role of fisher women in preserving the coastal environment. It highlights the challenges they face, including their indirect involvement in coastal fishing and the impact on their livelihoods. It also discusses how fishing activities have shaped their roles and opportunities in coastal environmental management.

Chapter 11

This chapter examines the opportunities and challenges faced by Indian women leaders in environmental management. It underscores the progress made in promoting women's involvement in this field and identifies the challenges that hinder their progress, including gender biases, limited representation, work-life balance, resource access, and cultural constraints.

Chapter 12

This chapter explores the significant role of women leaders in environmental management and their impact on sustainable development. It discusses the barriers and challenges women face in accessing leadership roles in environmental management and highlights success stories of pioneering women leaders who have defied societal norms.

Chapter 13

This chapter recognizes the significant contributions of women and gender-diverse leaders in building a sustainable world. It highlights their unique perspectives, skills, and advocacy for change, addressing issues related to social and environmental sustainability.

As editors of this volume, we are humbled by the stories, research, and wisdom shared by the contributors. We believe that the stories within these pages will inspire action, spark change, and encourage a shared commitment to the protection and preservation of our planet for future generations.

Perfecto Aquino
The University of Cambodia, Cambodia

Revenio Jalagat
Al-Zahra College for Women, Oman

Mercia Justin
Xavier Institute of Management and Entrepreneurship, India

Chapter 1
Women Leadership in Environment Management:
The Philippines' Challenge at the Grassroots Level

Revenio Cabanilla Jalagat Jr.
https://orcid.org/0000-0002-8878-3825
Al-Zahra College for Women, Oman

Perfecto Gatbonton Aquino Jr.
The University of Cambodia, Cambodia

Mercia Selva Malar Justin
https://orcid.org/0000-0002-9721-1503
Xavier Institute for Management and Entrepreneurship, India

Karima Sayari
Al-Zahra College for Women, Oman

ABSTRACT

The Philippines is among the pioneers of gender-sensitive policies that aim to neutralize the role of men and women in leadership opportunities in almost all sectors. From the '80s to the '90s, a bunch of policies were implemented, ensuring the women's participation in natural resource and environmental management; however, numerous populations of women have earned leadership in the regional and national levels but fewer opportunities at the grassroots level. In line with this, the study sought to investigate the state of women's leadership focusing on environmental management. Documentary evidence will be collected and interpreted to address the study objectives. Major findings revealed that at the municipal and barangay level, women have struggled to get education and knowledge to lead, are confined to male-dominated decisions, experienced poverty, and are often discriminated. Recommendations were laid that concerned government agencies and other stakeholders emphasized training and leadership enhancements at the grassroots and sustained until women leaders are raised.

DOI: 10.4018/978-1-6684-5986-7.ch001

INTRODUCTION

Philippines is one of the strong catalysts of women leadership and management in the society and all sectors. The active role of women in the society becomes vibrant since Philippines has been regarded as the leader of the international women's right platform that enables the local women's populace movement that fights for gender equality and other related concerns (UN Women, 2014). Moreover, the Philippines has received an accolade of being the first nation is ASEAN region that approves the Elimination of all Forms of Discrimination Against Women. The Government Service Insurance System (GSIS, 2021) has emphasized the implementation and sustainability of the Republic Act no. 9710 known as the Magna Carta of Women which focused on the Gender and Development Program (GAD) which aspires to obtain higher level of gender equality where women would be developed as active agents of the country's development. Accordingly, the mechanisms adopted by GAD actively supports the institutionalization of gender issues along the major country's national government environmental management agencies such as the Department of Agriculture (DA), Department of Environment and Natural Resources (DENR), and Department of Agrarian Reform (DAR). The implementation of GAD is pioneered by the DENR with their program called United Nations Collaborative Program on Reducing Emissions from Deforestation and Forest Degradation (UN-REDD) with preference to the role of women in the natural resource management sector (Philippine Commission on Women, 2023).

In 2010, an act was implemented called the Philippine Disaster Risk Reduction and Management Act (RA 10121) through the National Disaster Risk Reduction and Management Plan (NDRRMC) from 2011-2028 and to be run by a council composed of the government and non-government agencies. This national plan was cascaded into the local city/municipality, and the barangay and village level. One of the recipients of the implementation NDRRMC is a Barangay in Tacloban City called Barangay San Jose, a susceptible location for major typhoons. In a barangay setting, the NDRRMC committee is ruled by officials and employees of the barangay including the barangay health workers (BHWs), barangay nutrition scholars (BNSs), and the barangay service point officers (BSPOs) wherein most of these workers are women (The Japan International Cooperation Agency, 2016). In addition, the barangay's initiative of forming the committee is in line with the Republic Act 10121 which declares that "disaster risk reduction and management measures should be gender-responsive, sensitive to indigenous knowledge, and respectful of human rights, focusing on the importance of the protection of women (p. 15)." With the aim of promoting women rights and equality, the budget allocation of the NDRRMC for women or GAD activities reached to five percent. Gender policies are integrated in the activities implemented by the NDRRMC.

On the other hand, the Department of Natural Resources (DENR) has established strong mechanisms to raise the voice of women's participation in the area of resource policy and management since 1989 which includes for instance the formation of the Technical Working Committee on Women (TWCW) for the purpose of serving the needs of women and ensure the implementation of DENR programs and policies that protects the welfare of its women workforce (PCW, 2009). In relation to promoting the Philippine Strategy for Sustainable Development (PPGD), women are empowered to make decisions and fully participate in activities that promote sustainable development. Nevertheless, even with the many policies that advances women interest and leadership, women are still struggling and are facing obstacles because of the country's male dominant culture (Philippine Commission on Women, 2016). Results from interviews from informants with higher socio-economic status revealed that, women encountered difficulties to lead in higher ranks in the environmental industry unless they have higher education status such as master's or PhD when compared to men with less educational qualification but occupying higher positions through experience. One of the hardest difficulties encountered by women is experienced in the poor provinces and municipalities nationwide because of poverty, increasing population, social and economic factors (PCW, 2009). Women's role in remote areas is crucial in controlling reproductive health, overcoming limitation of educational opportunities, maintain resource sustainability.

Based on the above-mentioned scenario, this paper tries to investigate the state of women's leadership at the grassroots level focusing on environmental management. It emphasized their challenges and struggles in leading and upholding environment management and presenting realizable solutions and recommendations to alleviate the incurrence of such difficulties in the future.

WOMEN ENVIRONMENT LEADERSHIP AT THE GRASSROOTS LEVEL

Philippines has been one of the frontrunners of the gender sensitivity laws and policies that emphasized the role of women in the national level and it has concretized with the creation of the Philippine Commission on Women (International Fund for Agricultural Development, 2014). However, the women's contribution in both the national and local levels in terms of the rural workforce and its impact in rural companies are still low with limited opportunities of women leadership. The country has a program centered on Women situated in rural areas known as "Rural Women's Leadership Programme" (RWLP) that strives to boost the abilities of the women in rural communities at all levels and their professional organizations and improve the quality of policy advocacy and negotiations. In line with this policy was the creation of women organization called, "Women Organizing for Change in Agriculture and Natural Resource Management" (WOCAN) that employed the system in cooperation with the Farmers Association for Sustainable Rural Development (AFA) and its Philippines member "PAKISAMA", a union of 28 establishments nationwide operating as fishermen, small-scale farmers, and indigenous individuals. The purpose of the creation PAKISAMA is to enhance the capabilities of the Filipino farmers to join in agrarian reform and development in rural activities which includes the improvement of financial, operational, and organizational systems. It also endeavors to establish a sound development framework with its member associations.

PAKISAMA has its own group of women situated in rural areas called "LAKAMBINI" who is responsible for implementing the Gender and Development Programme. It also successfully started a sequence of gender-awareness seminars for women in provincial and national levels in coordination with RWLP to ensure the effective operation of LAKAMBINI. Among the provinces where the program was instituted are Camarines Norte, Albay, Sorsogon, Camarines Sur, and Aurora in Luzon; Iloilo and Capiz in Visayas; and Lanao del Norte, Bukidnon, Agusan del Norte and Sur, and North Cotabato in Mindanao (International Fund for Agricultural Development, 2014). While the program achieved a certain level of success, it has experienced difficulties and one of these difficulties dwelt on overcoming stereotypes and cultural norms. The solicited norms that men are more capable to work than women and in line with the perceived thought that the country is a male-dominated workforce. It has become a growing trend women must have the approval of the husband and direct family before deciding to become leaders in organizations. Moreover, the notion that women struggled to create a balance between their work and family are prevalent to most populace nationwide and at the grassroots level, most of the women prefer to work at their homes as housewife rather than looking for jobs and livelihood (WOCAN, 2021). An excerpt from an interview related to women struggles in leading narrated that, *"a woman attended an Entrepreneurship training for one week and at once she came home at 9:00 PM because of the distant location. The husband did not open the door saying that the wife gave more attention to women's organization than family matters which had caused disagreement" (p. 42).*

The lack of support of the family members especially husbands in villages and rural communities has become a struggle for women. In other words, they are not empowered to lead because family has become a major barrier. Moreover, the culture and environment of men as decision makers deprived women the self-confidence to excel, hence; it has become a major barrier (WOCAN, 2021). Women have experienced feeling of insecurities and incompetence and are stuck with the mindset of not seeing themselves as leaders or not having the potentials to lead or to become a leader.

GENDER GAP ASSESSMENT: WOMEN IN ACTION

According to USAIF-LEAF (2014), the country Philippines undeniably attained higher scores in gender equality index worldwide highlighting the contribution of women in the society. The result also accounts the higher percentage of educated women than men from secondary education to higher degrees. Women have represented the parliamentary seats accounting 20% of the total political servants which can interpreted that, the country has a vivacious community and a strong women movement that contributes to the Convention on the Elimination of All Forms of Discrimination Against Women (CEDAW) not only in the Philippines but also in the ASEAN Region. In terms of national development, women participation is high. In fact, in the recent Philippine Plan for Gender-Responsiveness (1995-2025) the emphasis is to ensure the massive participation of women in national development. Since 1990, the Philippine government had enacted different laws for women protection, laws against discrimination, and the empowerment of the indigenous people to have equal rights among Filipino citizens by virtue of the Indigenous People's Rights Act of 1997 requiring the active participation of indigenous women in decision-making. Moreover, cascaded from the 1987 Constitution, the passage of laws encouraging women's participation in politics and governance were supported in response to the call of gender equality and in line with the national development plan that focused on the empowerment and advancement of women. More important, in 2009, a law protecting women against discrimination was passed under Republic Act 9710 known as the "Magna Carta of Women" that mainly tackles on anti-discrimination and gender equality anchored primarily on CEDAW regulations.

At the grassroots or rural level, it has been the tradition that men use the natural resources to earn a living while women consider utilizing those resources as source of medicine, food, drinking water, fuel, and others. Gender biases can be obviously observed considering that most women are confined in their own homes, changes took place as time passes by as considerable number of women accepted various roles in community life, economic sphere, and even in traditional indigenous roles in governance and in managing the natural resources. As example, the women actively participate in the maintenance of the ecosystems while in agriculture, about 56.2 percent of the workforce are women. In fishery industry, women have engaged leadership in a range of production activities such as fish processing, preservation and packing, fish and shrimp collection, fish marketing, fish gear maintenance, etc. Though these activities are already prevalent in rural areas, the women contribution are still less recognized and mainstreamed because of the limitations in the access to productive resources, technology, services and infrastructure support, and capitalization (USAIF-LEAF, 2014).

Furthermore, because of the diversity in culture and geographical landscape, most of the women situated in rural and urban areas faced discrimination due to the inherited historical patriarchal system and cultural expectations that deprived women to be engaged environmental decision-making situations. To reiterate, the male dominance culture has enforced the women's role as domestic caretakers, less opportunities for education, and little participation in decision-makings involving environment management. Most especially in rural areas and remote places, knowledge and capabilities of women are undermined, lack technical trainings for livelihoods, and other related activities. Akter et al. (2017) stressed that, women in the Philippines in general have been burned out because of more household and farming responsibilities in rural areas which led to poor health conditions and well-being. Women engaged in farming activities experienced sickness, fatigue, weakened bodies, muscle pains as results of prolonged exposure to pesticides and insecticides. Also, women farmers and fisherfolks in some municipalities and cities were not given due attention as Local Government Units (LGUs) and regional offices have not acknowledged the legitimacy and capability women in farming activities and livelihoods (Ani & Correa, 2016). Information from CEDAW (2015) states that women are still struggling in representing and participating in organizations of indigenous communities, agriculture, and fishery. And many women in the agriculture sector in rural areas have not been registered in fisheries and agricultural databases such as for example the RSBSA and FishR. Women are practically left out in the activities usually dominated by men.

EDUCATION AND TRAINING: A BARRIER TO RURAL ENVIRONMENT LEADERSHIP

Weinstein (2010) revealed in his study that, approximately 80% of Filipinos socially classified under poverty status are situated in rural places in the country and these are in the mountainous areas and rice fields. Accordingly, these locations lack the resources especially in the access of education and knowledge with the lack of schools and classrooms. Even if public schools are free, many families can hardly bring their children to schools for multifarious reasons. Apart from this is the inseparable poor economic conditions due to poverty and unemployment. Most members of the family contribute income as out-of-school youth being exposed to different kinds of income generating activities. In agriculture for example, families from rural areas are expected to work in the field with children wherein most of the time children schooling being compromised just to earn a living resulting to repetition of the children in their grade levels and for others being forced to stop their education for a while. Women are focused on sustaining family needs and poverty situation that hinder them in participating agriculture-related trainings while also considering that their husbands are busy at the farm. As a matter of fact, about 26% of women can access trainings related to agriculture and pest control management. On the other hand, husbands are not supportive with their wives on attending education and trainings but rather discouraging them by telling them to stay in the house to do the household chores and taking care of the children.

Furthermore, women in rural communities especially in the provinces of Mindanao have inadequate exposure to decision making activities and have limited opportunities for education because of the exposure to gendered stereotypes that hinder them to learn and be involved with managing natural resources (Personal Communication, 2016). Because of poverty situations in many households, families have no option but to send their sons to schools at the detriments of daughters thinking into consideration that, the sons will manage future families and are the ones who will earn for a living for the family.

CULTURAL BARRIER: MALE DOMINATED CULTURE

Reports from WOCAN (2021), another challenge that women leaders in rural areas experience is the cultural challenge and stereotypes. Women must overcome insecurities, unconfident, and perceived incompetence. In other words, discriminatory practices in socio-cultural aspects are evident as hindrances to women's inclusion while the representation of men in the society at all levels remains strong. In the aspect of local resource management organizations, recent studies have described the women's participation in the resource management as passive or nominal in the environmental programs such as community forestry groups and Village Forest Joint Management (USAIF-LEAF, 2014). The dominance of men practically prevented the women to excel in leadership because of the general cultural norms that men decide fairly, justly, and firm compared to the changeable decision making of women. The perceived less exposure of the women in rural and remote areas resulted into their lack of the confidence to lead. Furthermore, due to the women's pressure to adhere husbands' decisions in almost everything, they became verbally abused, beaten, domestic violence, victim-blaming, shame, and other unacceptable behaviors from husbands (Valdez et al., 2022). Such violent activities demoralized women and they have lost their self-esteem knowing that at any moment they can subject to abuse within their family circle. Although many laws and policies favor and protect women, the implementation is still a big concern to reckon with.

POVERTY: A HINDRANCE TO WOMEN LEADERSHIP IN RURAL COMMUNITY?

Women have big role to play in the society and the Philippine Commission on Women (PCW) is the government arm that advocates women empowerment and any issues and concerns related to women's general interest and welfare. It caters to legislative agenda, leadership and political involvement, governance on gender responsiveness, economic empowerment, and violence against women. However, even with the plans and programs exist, difficulties and challenges encountered among women are continuously experienced. Due to poverty is one of the main reasons in impeding women leadership at the grassroots level. Rural settlers and indigenous women continuously experienced scarcity of resources and deprived basic needs. According to the survey conducted by PCW, about 11 women die because of pregnancy complications and many have experienced lack of access to productive health (Philippine Commission on Women, 2023). This lack of access to health is evidence that in the rural areas, poverty has caused women to hardly go to private centers to access health and while most of them avail of the services in rural public medical centers, the volume of people wanting for medical services are also challenging as the scarcity of medical workers becomes prevalent in many places. Accordingly, reports from PCW (2009) exclaimed that one of the greatest hardships and difficulties experienced in impoverished and underdeveloped provinces and municipalities are accounted to poverty and rapid population growth. Women play a vital role in controlling reproductive health and minimizing family size with proper awareness of family planning methods.

According to the Asian Development Bank (2009) report, the country is still suffering from primary education, poverty, low access to medical and reproductive health, and the aftermath brought by the global economic crises. The government has done their share in educating the women in family planning and birth control in congruence with alleviating poverty. The classification of poor and the marginalized people are found in rural areas working in the agricultural sector such as the farmers and fishers. In addition, the major causes of poverty in the Philippines are attributed to but not limited to the following: low to moderate economic growth for the past 40 years; low growth elasticity of poverty reduction; weakness in employment generation and the quality of jobs generated; failure to fully develop the agriculture sector; high inflation during crisis periods; high levels of population growth; high and persistent levels of inequality (incomes and assets), which dampen the positive impacts of economic expansion; and recurrent shocks and exposure to risks such as economic crisis, conflicts, natural disasters, and "environmental poverty (Asian Development Bank, 2009, p. 2).

WOMEN DISCRIMINATION IN LEADERSHIP: MUNICIPAL AND BARANGAY LEVEL

Discrimination among women in the leadership and higher positions in the government, private, and the localities are eminent due to the previously mentioned cultural norms. At the grassroots level, the role of NGOs and women organizations are vital to the success of mainstreaming gender issues particularly on the role of women in local government and communities. To combat this trend of discrimination, the Philippine government has created the so called Barangay Bayan Governance Consortium (BBGC) to ensure that women in organized barangay and communities through initiating the program centered on gender mainstreaming in local governance through the participatory resource appraisal, the interest and welfare of women participation has been taken due consideration by strengthening the barangays to prepare and keep a barangay development plan that primarily served as appraising the agency's resources (Villarin & Ramos, 2009). This attempt of preparing a clear-cut plan of resources appraisal resulted into an awareness of women on the importance of women's voice in the leadership and an overall development to their respective communities. Survey has shown that, almost 80% of rural and remote residents affirmed that the needs of their community should be responded on time and women may have the opportunity to attend in different community-oriented activities as well as expressing the needs and demands to the local government. The rate of discrimination has gradually declined as people are becoming that the feeling of being discriminated will be replaced by the women's ability to support and lead side by side with men.

TRAINING AND EDUCATION APPROACHES: WOMEN LEADERSHIP EMPOWERMENT

The government has a pivotal role in ensuring that the right amount of education and training is imparted to localities, especially in rural and remote places. Concerned government agencies who can contribute to the development and empowerment of women towards leadership maybe emphasized. According to The Global Gender Office of IUCN (2015), the government along with civil society must mechanized for a long-term plans and initiatives that will breakdown the common cultural norms of men over women in leadership in decision makings. Empowering women to lead at the grassroots level by the government's initiation of activities and programs that allows women to perform a leading role that would be beneficial not only to themselves but the husband and the entire family. As example, women in barangays and villages require experiences in meetings, trainings, and access to resources such as educational and financial trainings. More exposures of the women in associations and groups for a cause specifically in environment management like energy conservation, responsible and eco-friendly farming, tree planting activities in coordination with government agencies such as the Department of Agrarian Reform (DAR) and Department of Environment and Natural Resources (DENR) to name a few.

Another possible strategy and mechanisms to ensure that women get the needed education suitable to leadership development is to educate to the rural communities the importance of sending not only sons to school but also daughters thus, promoting equality and equal access to basic education. Implementing this strategy enable a higher literacy rate among women in rural places and encourage them to be involved in decision making with confidence. The IFAD (2014) has laid out the mechanisms to institute education and knowledge to rural women through the PAKISAMA policies and plans to build women leadership skills and evaluate performance. Programs instituted through PAKISAMA will go through organizational assessment to ensure that women are given due importance in the decision makings by also conducting sensitivity analysis to all implemented projects and programs. Moreover, to bolster the knowledge management of women in PAKISAMA, a presentation was conducted that focuses on promoting the women farmers as agents of change and highlighting the leadership enabling features of women leaders, challenges and enabling factors to overcome them. Research as support was conducted that emphasized the equal role and responsibilities of men and women in farming wherein its key findings from the leadership stories from 45 women and 15 men leaders centered on encouraging and motivating women leaders to rise and exercise the equal rights to men. It is empowering women to excel and be at par with men leaders in all aspects including business, government service, employees, and others.

Hence, education and training cannot be underestimated and the more the women in rural communities have access to education and training, the more empowerment and confidence the people will encountered. PAKISAM and LAKAMBINI conducted a training on gender equality with 239 women and 88 men billed as "Training of Trainers". The aim is to raise women leaders who will become leaders from the rural communities to the national level. The training was implemented over two years from the national level down to the provincial levels including municipalities, barangays, and villages. The program is expected to attain a certain level of success and leaders will multiply. These trainings include but not limited to the following:

Training provision in relevant skills: Advocacy, facilitation, and negotiation. It is of utmost importance as this is the point of origin towards gender-sensitivity training before behavioral and leadership training takes place. Institutional changes and its processes will not be achieved without starting with the basic processes. Lots of the people including women situated in rural communities have not fully understood the concept gender sensitivity and associate it western culture. The need of disseminating the information to local/rural communities to educate them about gender with sensitivity and culturally favorable way.

The content of the training is also very important taking into consideration the context of recipients where most of the women have not gone formal education or limited formal education. As an example, informal trainings were provided with simplicity on accounting and financial management to women.

Delivery of the training or the methods of presenting it is as important as the content and the training itself. In other words, the delivery should make sure that it is aligned with the cultural characteristics of the participants. Traditionally, storytelling is one of the effective tools in conveying information to them. Alternative strategies may include one-on-one discussion, group mentoring, and coaching. However, to overcome stereotypes, other concrete models can be introduced to motivate women workers and build self-confidence.

Provision of trainings in localized context with creative methodologies such as providing materials that caters to values and emotions, active sharing of experiences, reflection, and interactive approach.

Selection of trainers that has adequate knowledge at the grassroots level on gender equality, handle sensitivity issues and maximized women learning.

BREAKING THE CULTURAL BARRIERS: MALE DOMINATED ENVIRONMENT

To reiterate, Philippines has embraced for a long time a patriarchal society wherein men have the only say in the household which has become a trend in family structures and in larger social institutions. As a tradition, men are breadwinners as a source of pride and reputation, are expected to rule the household wherein decisions emanated. Women are expected to submit to the husbands with self-sacrificing attitude and perform any tasks with the approval of the latter. The need for trainings and education that transforms the traditional mindset of the people in rural areas, and this can be done when such training shall be given to men and women that may provide them more awareness to relate the present situation and realities which is apart from the traditions and customs old cultural norms. Intensifying the provision of education to remote areas and allocated by the government a budget for education that is reflective of the need of the rural communities. Even the difficulty of transport to these areas are experienced which may limit the access of information by the government concerned agencies and other stakeholders to assess the basic needs of the populace. The lack of confidence by people especially the women can be lessened when these women were given more opportunities to be associated with initiatives and environment management platforms at the grassroots level. The local government agencies should devise strategies to coordinate with these potential women for leadership.

ADDRESSING POVERTY TO LEAD: RURAL WOMEN LEADERS

Undeniably, poverty is one of the major challenges that the country is recurringly faced. As previously explained, poverty is more prevalent in rural areas in the agricultural sector basically the farmers and fishers. Research has proven the association of poverty condition to educational attainment as cited in Asian Development Bank (2009) referring to poverty as mainly a rural phenomenon although urban poor populace is also increasing. Accordingly, the survey also stated that two-third of the poor households nationwide refer to the citizens that have undergone elementary school education or below and the family size average six or more members where the provinces of Mindanao are considered the most affected with poverty situation. To address these gaps, the government may devise or revise an effective strategy to combat poverty from the national level down to the municipal barangay level that incorporates social protection. The local government with the aid of the national government should mechanize a strategy that ensure the effective delivery of basic social services and poverty alleviation plans at the local and barangay level aimed at poverty reduction which includes providing education and training for them. Moreover, strengthening research at the local level for more accurate and reliable information about the living conditions of the underprivileged and marginalized families for proper targeting and addressing their needs. Considering that the problem of poverty is multidimensional, addressing its resolution requires participation of multi-government agencies, sectors, and other stakeholders. Integration of multicompanies and agencies is instrumental in collaboratively addressing the problems of poverty. Poverty can become chronic if left unresearched or inaccurately attended. Hence, it is a challenge of the different government and private sectors to alleviate the cases of poverty, its causes, constraints, and creating actional plans and programs that contribute to its resolution. United efforts by multi-stakeholders are the key to combat poverty at certain levels.

BATTLING WOMEN DISCRIMINATION IN RURAL LEADERSHIP

The revealing cases of women discrimination in the country cannot be underestimated as women in rural areas are faced with persistent discrimination like for instance, deprivations in the access of productive resources, economic opportunities, and services. The United Nations (2023) stated through a Philippine representative that, the Philippine government has passed laws and strategies that enhance women in rural communities for access to informal materials, equipment, trainings, machines, facilities, market assistance, etc. In addition, the government also allocated five percent of the agency's budgets on gender and development to cater to the needs of women population in equality to men. The law that promotes the Violence Against Women is empowering the women to exercise their rights for protection against violence and abuse. The vulnerability of women to abuses can emanate from their own families especially in rural areas wherein husbands usually have the jobs turning the wives into household chores unable to make decisions for the family, an at times an object to domestic violence. Hence, organizing women into associations, charitable works, and other avenues that women can associate with each other for a cause in the rural communities can be considered as a breakthrough from their usual stay at home lifestyle; educating them of their rights, promoting rights to have pleasure and safety, take part in enhancement programs and trainings, enhance self-seeking behaviors, and adherence to the Magna Carta for Women.

Concerned agencies and stakeholders may develop plans and programs that eliminate gender biases, non-discriminatory employment, equal opportunities for trainings and education, improved access to information, more opportunities for women leadership roles, acknowledging women achievement in rural communities and villages. The enhancement of the community-based reporting should be fully implemented to create an active response mechanism against discrimination against women. Building a strong foundation of women in empowerment and providing more opportunities to improve in leadership and management coupled with education and training whether formally or informally could be a very satisfying experience to them and help them move forward for future undertakings.

PAKISAMA has suggested on the ways and strategies to become an effective woman leader with gender sensitivity the organizations of farmers. Firstly, a woman leader should have an adequate knowledge on how to decide based on consultation and consensus, and careful analysis anchored on local values. Secondly, effectively communicate to the constituents with access to available resources. Thirdly, can communicate in the wider public in relation to gender issues and be able to read and understand documentations and presentations on financial statements. And effectively facilitate meetings, resolve conflicts, and build consensus and empowerment. Devise schemes and strategies that support women to attend activities outside the homes just like association meetings, charitable institutions, and activities. Supporting women's continuous learning in the communities through agricultural extension work activities and offering support to women who would embark in income-generating activities to encourage and enhance women's leadership characteristics such as confidence, skills, self-esteem, and others.

CONCLUSION AND RECOMMENDATIONS

This chapter dealt with investigating the state of women's leadership at the grassroots level which is centered on environmental management. It also tackles on challenges and difficulties encountered by women leaders in the locality at the municipal, barangay, villages, and rural communities such as the lack of education and knowledge creation, cultural norms or male-dominated decision making, poverty situation, and discrimination against women. Evidence was gathered to arrive at a clear picture of the real condition of women leadership at the grassroots and key findings indicated that, women leaders and women populace in the barangays and villages have struggled to have formal education due to the family's preference of sending sons to schools coupled with the lack of financial resources and mobility. Women are deprived the opportunities to study, and the scarcity of public schools and teachers have increased the knowledge and learning issues. In the aspect of cultural norms and male-dominated communities, women are generally deprived to engage into activities outside their homes like attending into seminars and trainings related to environment management, a need to convince the husbands to agree with the wives for the said activities. Moreover, the lack of exposures of women into leadership programs and trainings that enhance confidence and skills to lead. In terms of poverty, women are suffering from extreme poverty situations, the lack of access to health in the rural areas, poverty has caused women to hardly go to private centers to access health and while most of them avail of the services in rural public medical centers, the volume of people wanting for medical services are also challenging as the scarcity of medical workers becomes prevalent in many places.

Furthermore, discrimination against women had also prevented women to excel and lead because of the increasing cases of violence against women and especially in the rural areas, it is a cultural norm that men are the ones who find jobs while women stayed in their homes which can result into domestic violence and depriving women to make decisions, self-esteem, and confidence to lead.

Anchored on these results, some recommendations are suggested to address the challenges faced by women in environment leadership. Concerned government agencies and stakeholders play a vital role in addressing issues on the lack of education and training, cultural barriers, poverty among women, and women discrimination in rural leadership. Training and education should be provided like women in barangays and villages require experiences in meetings, trainings, and access to resources such as educational and financial trainings. Intensify the provision of education to remote areas and allocated by the government a budget for education that is reflective of the need of the rural communities. The lack of confidence by people especially the women can be lessened when these women were given more opportunities to be associated with initiatives and environment management platforms at the grassroots level. The local government agencies should devise strategies to coordinate with these potential women for leadership. Considering that the problem of poverty is multidimensional, addressing its resolution requires participation of multi-government agencies, sectors, and other stakeholders. Integration of multicompanies and agencies is instrumental in collaboratively addressing the problems of poverty. And, in battling discrimination issues against women, concerned agencies and stakeholders may develop plans and programs that eliminate gender biases, non-discriminatory employment, equal opportunities for trainings and education, improved access to information, more opportunities for women leadership roles, acknowledging women achievement in rural communities and villages. The enhancement of the community-based reporting should be fully implemented to create an active response mechanism against discrimination against women.

REFERENCES

Akter, S., Rutsaert, P., Luis, J., Htwe, N. M., San, S. S., Rahajo, B., & Pustika, A. (2017). Women's empowerment and gender equity in Agriculture: A different perspective from Southeast Asia. *Food Policy*, *69*, 270–279. doi:10.1016/j.foodpol.2017.05.003

Ani, P. A., & Correa, A. B. (2016). *Agricultural Extension Policies in the Philippines: Towards Enhancing the Delivery of Technological Services*. FFC. https://ap.fftc.org.tw/article/1092

Asian Development Bank. (2009). *Poverty in the Philippines: Causes, Constraints, and Opportunities*. ADB. https://www.adb.org/publications/poverty-philippines-causes-constraints-and-opportunities

Government Service Insurance System. (2021). *The Gender and Development (GAD) Program*. GSIS. https://www.gsis.gov.ph/gad/

International Fund for Agricultural Development. (2014). *2014 Annual Report Highlights*. IFAD. https://beta.ifad.org/documents/38714170/39625423/Highlights+AR+2014_E.pdf/74576162-ae01-48a8-8db9-acd53a250682

Islands, M. (2016). *Consideration of reports submitted by state parties under Article 18 of the convention*. UN. https://digitallibrary.un.org/record/849531?ln=en

Personal Communication. (2016). *Filipino Women in International Migration Situation Analysis, Policy Context and International Mechanisms*. UN. https://www.unwomen.org/sites/default/files/Headquarters/Attachments/Sections/Library/Publications/2016/Filipino-Women-International-Migration.pdf

Philippine Commission on Women. (2009). *Annual Accomplishment Report*. PCW. https://pcw.gov.ph/assets/files/2020/05/2009_pcw_accomplishment_report.pdf

Philippine Commission on Women. (2016). *A Handbook on the Application of the Enhanced Gender Mainstreaming Evaluation Framework*. PCW. https://library.pcw.gov.ph/wp-content/uploads/2020/12/PCW-A-Handbook-on-the-Application-of-the-Enhanced-Gender-Mainstreaming-Evaluation-Framework-2016.pdf?fbclid=IwAR2ehD2vaFFqQxbjGy4MIEyst7P7PAK83lRWaxROyyLpgQ74maFTqCtl19E

Philippine Commission on Women. (2023). *Citizen's Charter*. https://pcw.gov.ph/citizens-charter/

The Japan International Cooperation Agency. (2016). *Transforming Gender Relations in Disaster Risk Reduction*. https://www.jica.go.jp/Resource/jica-ri/publication/booksandreports/l75nbg00000697z9-att/JICA_Philippines_Sri_Lanka.pdf

WeinsteinJ. (2010). *The Problem of Rural Education in the Philippines*. https://joshweinstein.wordpress.com/2010/03/02/the-problem-of-education-in-the-philippines/

WOCAN. (2021). *RWL Project Philippines Report*. https://www.wocan.org/resource/rwl-project-philippines-report/

United Nations. (2023, March 10). Delegates Highlight Rural Women's, Girls' National, Economic Contributions, But Point to Gaps in Education, Food Security, as Commission Session Continues. https://press.un.org/en/2023/wom2225.doc.htm

Women, U. N. (2014). *United Nations Framework Convention on Climate Change*. https://www.unwomen.org/en/how-we-work/intergovernmental-support/climate-change-and-the-environment/united-nations-framework-convention-on-climate-change

USAIF-LEAF. (2014). *Women's Inclusion in REDD+ in The Philippines Lessons from Good Practices in Forest, and Other Natural Resources Management Sectors*. https://redd.unfccc.int/uploads/2234_1_philippines_gender_and_redd_report_sm.pdf

Valdez, I. K. M., Arevalo, M.P.N., Robredo, J.P.G., Gacad, S.L.S., Villaceran, M.A.J., Libang, G.R., Dela Paz, E.P., Rubin, K.S.T., & Eala, M.A.B. (2022). Violence against women in the Philippines: Barriers to seeking support. *The Lancet Regional Health - Western Pacific, 23*, 1-2.

Villarin, T., & Ramos, S. (2009). *Transformative Power: Case Studies on Political Participation of Community Women Leaders* (1st ed.). Institute of Politics and Governance and Friedrich Ebert Stiftung.

Weinstein, J. (2010). *The Problem of Rural Education in the Philippines*. Josh Weinstein. https://joshweinstein.wordpress.com/2010/03/02/the-problem-of-education-in-the-philippines/

Chapter 2
Opportunities and Challenges of Educated Professional Women Leaders in Environment Management

Mercia Selva Malar Justin
https://orcid.org/0000-0002-9721-1503
Xavier Institute for Management and Entrepreneurship, India

Revenio Cabanilla Jalagat Jr.
https://orcid.org/0000-0002-8878-3825
Al-Zahra College for Women, Oman

Perfecto Gatbonton Aquino Jr.
The University of Cambodia, Cambodia

ABSTRACT

The chapter intends to present the role of educated and professional women as leaders of home, how they contribute to environmental management, the opportunities, and the challenges they meet in environmental management. Educated and professional women have greater awareness and knowledge about the environmental management. In managing houses and at their workplaces they have a greater role to play. Therefore, they have greater opportunity to control environmental degradation and contribute positively to environment management. It also explores the opportunities for women to involve in various forms and levels of environmental management and the challenges women face as leaders to initiate environmental management programs. A literature review will cover the research articles that focus on the opportunities and challenges of women leaders in environmental protection and expose how women lead environmental management initiatives at home, workplace, and society. Findings revealed that workplace and society women must assume greater role to support environment management.

DOI: 10.4018/978-1-6684-5986-7.ch002

INTRODUCTION

The chapter intends to present the role of educated and professional women as leaders of home, education, corporate, society/NGO/Government and Environmentalists and how they contribute to environmental management, as well as the opportunities and the challenges they meet in environmental management. Educated and professional women have greater awareness and knowledge about the environmental management. In managing homes and at their workplaces they have a greater role to play with the knowledge and skill sets they develop. Therefore, they have greater opportunity to control environmental degradation and contribute positively to environment management. The opportunities can be well utilized to carry out environmental management activities by educated professional women. Yet, the opportunities may not always be converted to real possible actions as there could barriers that prevent them from translating opportunities to realities. The chapter explores the opportunities available for women to involve in various forms and levels of environmental management. It also brings out insights on the challenges women face as leaders who would intend and initiate environmental management programs.

The awareness level on environmental degradation and destruction is high among educated professional women leaders. They are consistently in touch with reading and listening material that appeals for environmental protection. They are also constantly in touch with NGOs that work for and advocate environmental protection. The environmental management curriculum prescribed in graduate level also makes them clearly aware of their responsibility towards environment management. The knowledge and exposure to the need for environmental management does not always translate into environmental protection and preservation. There are several challenges women as leaders of households and as professional leaders encounter in executing the environmental management for nurturing and protecting nature. The chapter explores the various forms and nature of challenges educated, professional women leaders face in being environmentally responsible.

METHODOLOGY

The chapter is written based on literature review and primary data to fulfil the various objectives of the chapter. Literature review covered the research articles that focus on the opportunities and challenges of women leaders in environmental protection in the domains of home, corporate and public life and in the professions of teaching and environmentalism. Primary data was be collected from educated professional women and men using questionnaire that will bring the required data to study the opportunities and challenges for women leaders in environmental management. The primary data required for the analysis was collected from educated professional women and men across India as to how they have exercised their leadership in various domains to bring about a sustainable environment. Further, case studies of three educators cum environmental protectionists is included in the chapter.

The chapter explains how women lead environmental management initiatives at home, work place and society. Managing kitchen waste and waste water at home can be a significant way women can lead environment protection at home front. At work place women leaders can ensure that proper waste disposal, energy conservation, car-pooling, use of renewable energy sources, etc. can be adopted to protect environment. In the society women can be great change agents by volunteering in CSR initiatives and with NGOs to bring about a great transformation in the attitude and behavior of common people in conserving and protecting environment. Thus, in the three common domains and two specific professions where women play a crucial role they can effectively ensure that environment is well managed. Further, when women take up crucial and powerful roles, they can exercise great initiatives as environment protectors. Apart from the primary data and secondary data analysis, even case studies have be cited to make the significant role of women in environment protection. The chapter brings a very wide perspective of how Indian women are currently playing their leadership roles to protect environment covering their regular domains of functioning and their extraordinary roles as powerful leaders in organizations and nations. The chapter will be organized as four parts: Introduction, Literature Review, Analysis and Discussion, Case Studies and Conclusion.

REVIEW OF LITERATURE

Women and Environmental Protection

Shinbrot et. al. (2019) in their study titled 'Unlocking Women's Sustainability Leadership Potential: Perceptions of Contributions and Challenges for Women in Sustainable Development', presented the fact that almost exclusively male list of past winners of the Sustainable Development Leadership award given by the World Sustainable Development Summit exemplifies the issue that women leadership in sustainability initiatives is almost missing. Looking specifically at women leaders in sustainable development, the research identified challenges women faced as leaders from the viewpoint of sustainable development activists and explored their perceptions of the unique contributions of women as leaders in sustainable development. Over the decades, research has demonstrated how women are uniquely positioned as stewards of natural resources (Cavendish, 2000; Quisumbing et al., 2001; Fernandez, 2008; Nightingale, 2011). FAO (2011) stated that over 40 percent agricultural workers are women. Women play a crucial role in economic development as well. Women make collaboration possible as leaders (Eagly & Carli), Women face more and more barriers to become leaders is proved again and again. Barriers include unequal access to economic opportunities (UN, 2013), legal barriers to asset ownership (Deere et al., 2013), lack of representation in politics (Agarwal, 2010), barriers in access to education (UN, 2013), trenchant stereotypes (Fiske, 2010; Koenig et al., 2011), expected gender norms (Eagly & Karau, 2002), as well as self-expectations and devaluation of successes (Haynes & Heilman, 2013) that hinder leadership aspirations. Glass et al. (2015) Do Women Leaders Promote Sustainability? Analyzing the Effect of Corporate Governance Composition on Environmental Performance investigated the impact women leaders had on the corporate environmental strategies of organizations. A dataset of all Fortune 500 CEOs and boards of directors for a ten-year period, was examined for several aspects of gender in leadership on environmental strategy.

Shambel (2012) presented how in Ethiopia women have demonstrated unique knowledge and expertise in leading strategies to combat the effects of climate change, through the management of natural resources and environment. The women especially at the grassroots level were part of the study, though they were not well recognized. Further, their efforts were met with lot of challenges. It was conducted in South Wollo, Qalu district, Ethiopia aiming to see gender impact of climate change on women's life and women's role in environmental management. It also examined natural resource protection to avert the risks of climate change and to scrutinize major impediments. The research employed qualitative orientation through basic qualitative survey techniques. Focus group discussion, individual interview, and participatory observation were used as basic data gathering tools. These tools were accompanied by a guide checklist for primary data and published and unpublished documents for secondary information. The study found that women have active body of knowledge in protecting the environment and natural resources despite various challenges emerging from political, economic, institutional and other arenas.

Nugent and Shandra (2009) found that there is a large focus at the organizational level on the importance of a link between women and the environment, yet little empirical research explores this. They examined how women's status in economic, political, educational, and health spheres affect state environmental protection efforts in the form of protected land area. Hypotheses derived from ecofeminism and empirical literature of gendered differences in individual-level attitudes and behaviors were tested using multiple regression models with a cross-national sample. It was found that no support for broader ecofeminist claims that the overall oppression of women and environmental degradation are linked by a common source. Yet it was observed that strong support for the idea that increasing women's political status in particular through representation in national government has a positive effect on state environmental protection efforts. There was no evidence found that connection to a world polity has a significant effect on nation-state designation of protected land area.

Jewitt (2000) examined how the women and their positive environmental link assumption is a problem and the pitfalls that could arise by translating the idea into development policy making. The study challenged the idea of women as environmental guardians, following the work of Bina Agarwal and Cecile Jackson, as well as others. It highlighted the fluidity and localized nature of inter- and intra-community variations and work allocation and significant gender variations in decision making and control over environmental resources. The study emphasized on the impact of patrilineal inheritance systems and patrilocal residence patterns on women's familiarity with and their control over local environments.

Women as Leaders at Home

Saraçli et al. (2014) found that though environmental problems have led to increased environmental concerns, mothers were more concerned and worried about environmental issues than fathers. Using Structural Equation Modeling the study took mothers' education into consideration, to assess the environmental sensitivities if university students and the causal relationships between their environmental protection commitment and environmental friendly consumption behaviors. Pilot study was conducted on 60 randomly chosen students and the final version of the tool was refined eliminating unclear questions. Later it was administered to 520 students of Eskişehir Osmangazi University, Turkey randomly. The study found that environmental sensitivity impacted the environmental protection commitment positively – higher the sensitivity, higher the commitment. The students' genders, their families' place of domicile and particularly the parents' educational levels were significant factors in determining differences in the development of students' environmental sensitivities and behaviors. Above all, students with higher education qualified mothers promised to be active environmentalists by joining related organizations. Women students were more inclined towards environmentalism as much as students whose mothers were university graduates. Higher the level of education, higher the level of commitment towards environmental protection was a great fact that the study revealed.

Wuyep et al. (2014) examined women's involvement in environmental management in Plateau state. They generated primary data of women from six local government areas. Majority of the women were involved in farming and contributed significantly to land/soil conservation. Most of them have planted trees over the last five years while. Majority were involved in involved in clearing their surroundings daily in terms of sweeping, clearing drainages and refuse disposal. The significant problems faced by the women included lack of waste disposal equipment, poor drainage systems and lack of awareness among the general public. The opportunities to conserve and protect environment for educated home makers and mothers were found in the following aspects: Kitchen waste management, grey water management, planting trees at home and at properties, Gadgets of low carbon emission, Electric vehicle to minimize carbon emission and purchasing eco-friendly clothes/ cosmetics/ toiletries. Educated women home makers found ways and means to treat both solid waste management from the kitchen, processed food consumption, garden, linen and stationery, etc. through their reading, interaction with other educated women and environmental protection NGOs. They tried to recycle and use the grey water, ensured that they harvest rain water, etc. They actively involved in planting trees and other conservation initiatives at home and elsewhere. They tried to invest in gadgets that eliminate or minimize carbon emission, consume minimum energy or use renewable energy. They were willing and interested to consume ecofriendly products in food, toiletries, cosmetics, dress, etc.

For educated mothers and home- makers the challenges to protect environment arises from: their work – life imbalance, commitment to career, non-cooperative family members & maids. The imbalance in work-like could take away all the time one could allot for improving and planning for an ecofriendly home. Career focused women can neglect their environment commitment. Even for a committed environment focused and zealous home-maker or mother it becomes a challenge when the house maid and the family members don't follow the norms, principles, guidelines, processes and procedures to be followed to protect environment through the household functioning.

Women Leaders at Educational Institutions

Quigley et al. (2017) opined that Environmental Education Research (EER) rarely includes women's perspectives. It indicates that in environmental education research, an entire knowledge source is largely ignored. The study employed a methodology called Participatory Rural Appraisal, a methodology new to the field of EER, of Kenyan teachers from the Maasai Mara region to understand the tensions around environmental views. The study documented gendered tensions such as the burdens of responsibility, the power imbalance disadvantaged women feel regarding solving environmental issues, and the blame that is directed at both men and women as a result of these burdens and power imbalances. Yachina et al. (2018) observed that currently, the contradictions have reached global proportions and are becoming increasingly difficult to resolve. The study aimed to examine the process of formation and development of the regional system of continuous environmental education for teachers. The study involved 60 university students, 150 students of general education schools and 10 pre-school teachers. The study adopted a complex of research methods which included the following: a comprehensive, systematic, personal, integrative approach to the formation of ecological culture, the study of training educational programs for environmental education of preschool children, students of general education institutions and university students; a sociological survey of schoolchildren and students university. The study found the contradiction that prevailed between the need for a regional system based on the regional requirements and the insufficiently developed regional system of continuous environmental education in the teacher's training.

Women teachers who are concerned about environment and are keen to contribute to the protection of environment can contribute to environment protection in various ways – the opportunities are plenty. The teacher can include environment protection thoughts, concerns, opportunities, needs, quotes, etc. as part of the general discussions in the class, thereby nurturing environmental protection concern and spirit in her students. There are curriculum designed for environmental protection right from school education to higher education. The teacher can teach the curriculum with zeal and enthusiasm. The teacher can connect the students to NGOs working on environment protection as volunteers. The teacher can organize workshops, training programs, seminars, conferences, etc. focusing on environment protection. The teacher can initiate a social forestry program in the campus or in the village/city where the education institution is located. All these opportunities can be made into challenges if the school administration is not supportive, students are not cooperative, parents are challenging the teacher's initiatives, lack of financial resources, too much of teaching workload and others.

Women Leaders in Corporates

Ricci et al. (2023) conducted a Scopus search to take stock of the empirical works on gender diversity and CEP, then we analyzed relevant articles using the SRL methodology combined with an in-depth content analysis, to identify the main research topics and the major existing gaps. They found that the existing research has focused on four main thematic areas: (1) Board gender diversity, female leaders and environmental performance; (2) Board gender diversity, female leaders and environmental disclosure; (3) Board gender diversity, female leaders and environmental innovation; and (4) Board gender diversity and female leaders as moderating factors between corporate environmental performance and corporate financial performance. It was found that the literature tended to focus on the direct effect of gender diversity on CEP, but relatively few studies investigated the contingencies factors that would affect this association. They suggested that the future research should explore the boundary conditions of this linkage at different levels of analysis (i.e., at organizational, institutional, and individual levels) so as to reconcile previous inconsistent findings. It was also found that research has not thoroughly analyzed the mediating mechanisms of the effect of gender diversity on CEP. Hence it is important that studies should fill this void as well. The authors also found that in terms of empirical strategies, the vast majority of studies employ quantitative research methods and assume a linear relationship between gender diversity and environmental performance. Therefore it is was suggested that research should also explore the potential curvilinear effects of gender diversity and employ more qualitative and mixed-methods approaches.

Liu's (2018) ako study examined the relationship between board gender diversity and corporate environmental violations. Drawing on gender socialization and diversity theories, greater female board representation and female chief executive officers (CEO) are expected to reduce the frequency of corporate environmental violations. Empirical evidence in this study shows that firms with greater board gender diversity are less often sued for environmental infringements. In contrast, CEO gender is linked to reduced environmental litigation only in firms with low female board representation. The study explored the relationship between board gender diversity and improved corporate environmental policies as a mechanism to explain the reduced litigation frequency. These findings provide important insights on women on board to investors, managers, and policymakers into the role of female leadership in public companies.

Glass et al. (2016) investigated the impact women leaders have on the corporate environmental strategies of organizations. Using a dataset of all Fortune 500 CEOs and boards of directors for a ten-year period, the study examined several aspects of gender in leadership on environmental strategy. Specifically, we test the impact of women CEOs, the proportion of women on the BOD, the number of interlinks women board members hold, and the interactive and cumulative effects of women CEOs and gender diverse boards. It was found that firms characterized by gender diverse leadership teams are more effective than other firms at pursuing environmentally friendly strategies.

Wei et al. (2021) opined that environmental quality strongly depends on human behavior patterns. They suggested that many environmental challenges are rooted in human actions, and thus, it is believed that these problems can be reduced through the promotion of pro-environmental behaviors (PB). The moderating role of female leaders in the proposed relationship was found to be more significant than that of male leaders. The study has provided insight to policymakers on how to restructure their CSR preferences, priorities on the environment, and gender differences

Li et al. (2017) tested the relationship between gender diversity on boards and firms' environmental policy. They predicted that gender diversity on boards of directors should have a positive relationship with firms' environmental policy. Further the authors suggested that firm character in terms of pollution creation likelihood moderates the relationship between gender diversity on boards and firms' environmental policy. They analyzed data from 865 publicly listed firms in the United States and found direct and significant empirical evidence for their predictions. The study inferred the importance of gender diversity for the development of good firm environmental policy as well as for the improvement of corporate governance. Also, the study concluded that the more likely firms in a given industry are to cause environmental pollution, the more salient will be the beneficial effect of gender diversity on boards on firms' environmental policy in the industry

Jouber (2022) investigated the impact of top management team (TMT)'s gender diversity on corporate social performance (CSP). The study found inconsistent results in literature by testing the moderator effects of chief executive officer (CEO) managerial ability and corporate governance (CG) on such impact. The study applied dynamic panel estimator on an international sample of 8640 firms for a period of five years from 2013 to 2017. The author found reliable evidence that the critical mass of at least three women leaders has a positive impact on the firm's CSP. Obtained results suggested, that the deterrence effects of CEO managerial ability and CG tools (board independence, board gender diversity, the presence of a corporate social responsibility committee and family control) on the women leaders' contribution to the firm's CSP level. These results remain consistent with alternative measures for women leaders and CEO managerial ability. The differences seem to disappear when women achieve the CEO position, the chairperson position or both positions. The study also revealed that women leaders' effect on CSP seems dependent (do not) on the country (industry) which a firm belongs to.

The study highlighted the importance of fostering the achievement of a critical mass of women leaders and the combination of CEO managerial ability – educational/professional backgrounds – and CG attributes to improve the firm's CSP. If investors wish to increase CSP, they should ask for more gender diversified TMTs. Furthermore, this study supported regulators in their efforts to increase senior women's quotas by providing empirical evidence of better social outcomes under leader gender diversity. The study becomes useful for companies in setting the criteria to identify CEOs who can support their strategic decisions.

Women in corporate have opportunities to conserve and protect environment by creating policies and procedures for protecting environment, getting involved in CSR activities of the corporate in protecting the environment, becoming supporters/partners of environmental protection, establish recognitions/awards for environmental protection, work towards recognitions/awards through corporate programs/resources. Women as CEO or Board Chairman or Board members will be able to influence and bring in environmental protection focused policies and procedures. Women leaders can directly engage in CSR activities as role models and as inspiration to their teams. They can fund environment initiatives both within the corporate and at external bodies like NGOs, Community or government. As there are environment oriented awards by highly recognized international and national organizations they can focus and work to achieve those awards. They can constitute awards and recognitions for employees and suppliers who are environment friendly to promote environment protection.

Women in corporate may face the challenges in protecting the environment such as not given responsible positions to make decisions that favor environment protection, dismissed from being heard, condemned as irrelevant, unaccepted as detrimental to profit, not considered/ recognized/rewarded. Women are not always given the responsible roles in corporate. Glass ceiling is real and women have to struggle through to reach that level. Even qualified and potential women face the discrimination as they support environment conservation and protection. The ideas and plans of women leaders in corporate for environmental protection gets dismissed and mocked, not taken seriously and accused as detrimental to profit making motive of the firm. Women leaders may face the challenge of neglect and cornering, even being not recognized and rewarded for the good work they carry out as they promote environment protection.

WOMEN AS LEADERS IN SOCIETY/NGOS/GOVERNMENT

Opportunities to Conserve and Protect Environment

Labaris (2009) provided a general review and a case study of women's involvement in environmental management in Nasarawa State. Primary data were generated from a questionnaire survey of women from five local government areas. Majority of the women were involved in farming and contributed significantly to land/soil conservation. Most of them had planted trees over the last five years while. Majority were involved in involved in clearing their surroundings daily in terms of sweeping, clearing drainages and refuse disposal. The significant problems faced by the women included lack of waste disposal equipment, poor drainage systems and lack of awareness among the general public.

Hillman (2020) Environmental politics is becoming an increasingly important field because of severe issues like climate change. Research has argued that women care more for the environment. Meanwhile, female politicians in national parliaments are reported to have stronger environmental concerns and be more supportive of environmental protection than their male colleagues, but it is still unsure if they exhibit these characteristics in their decision-making. The purpose of this study is to capture the effect of women's political presence on environmental outputs which has not previously been tested much empirically from a cross- country perspective. A quantitative method with a statistical analysis is used on cross-sectional, time-series data from 34 European countries to test if an increase in the proportion of seats held by women in national parliaments have a positive effect on climate change legislation, joining environmental agreements and decreasing greenhouse gas emissions. The result of the analysis finds little support to the suggested hypothesis of a positive effect, except for joining environmental agreements that showed significance. The correlating trend of gender equality in the parliaments and improving environmental politics might instead be due to modernization. However, it cannot be ruled out by this analysis that an increase of the number of women in national parliament does not have a positive effect on environmental politics. Further research is needed to determine how the environment is prioritized by female politicians in order to study the effect of their presence.

Resurrección (2013) observed that since the 1980s, the discourse that women are intrinsically closer to nature, are hardest hit by environmental degradation, and have special knowledge of natural resource systems has influenced development policy circles and intervention programmes globally. The paper explored why and how women–environment linkages remain seductive and influential. They forwarded three arguments for this: first, for gender to muster entry into climate politics, women's identities are projected as fixed, centered, and uniform — and tied to nature; second, the discourse of climate change vulnerability has proven to be a strategic entry point for feminist advocacy; and finally, inertia associated with past environmental projects has reinstated the women–environment discourse in contemporary climate change discussions and possibly, future interventions.

Women as part of the society have several opportunities to protect environment such as: advocacy and propagating environment protection, mobilizing women power, activism for environment protection and striving for awards and recognitions. Women can advocate and propagate the message of conservation of environment to all their neighbors, villagers, housing society/association members, etc. They can collect other women for the society's environment protection through planting trees, rain water harvesting mechanisms, control carbon emission in the locality by burning stubble, making organic manure with earth worms, organizing butterfly parks, running an apiary, etc.

Women as part of the society may face challenges to protect environment: opposition from the public/government in the context of development, safety and security for life. If the initiative is not viewed well by the public/government or if it is considered unacceptable to the social norms, they will oppose the women indulging into it. Some women have faced life threats and death when they tried to protect the environment.

Women as Conservational Leaders: Case Studies

Environmentalists have greater opportunities to protect and conserve the environment because of their expertise and the facilities and resources given to them for nurturing and protecting environment. Their professional education and training makes them more competent and equipped to work in this space.

Dilokwanich et al. (2015) observed that Thai women have learned how to extent their roles from a caretaker of children and a household to natural resources and environmental protection and management in local and inter-regional communities. Thai woman environmental leaders who are taking a successful role of environmental guardian in their communities. This research attempted to explore why they took leadership role in environment, how they work so successful as an environmental guardian, and what their next move is. The study engaged with the 28 Thai woman leaders who received the award of Thai Environmental Conservation Mother from the Faculty of Environment and Resource Studies, Mahidol University between 2004 and 2012. They were in-depth interviewed and collected data were preceded by content analysis. The study concluded that most leaders saved their communities' environment and natural resources from the intervention of new development activities; most of them had their parents as a good role model in environmental management who provide knowledge of morals and environmental ethics as a good basic of leadership while some shared their husband's responsibility in the same matter. Further, it was found that, teamwork was their working style with the assistance of public participation to hold teamwork and collaboration of the community. It was observed that all leaders had systematic working with talents of patience, gentleness and sensitivity. The leaders by working along with the network broadened their new information and knowledge between practitioners. Also above 50% of the leaders were preparing their successors from the younger generation who can continue environmental conservation of the community, while others could not do it. This pushed some leaders find alternative succession process.

It was further found that in consistent with the findings of Drew Hutton and Libby Connors, the environmental movement has a different meaning for almost every person interviewed. The research used the term 'Australian environment movement', and focused on the work of leaders in organisations that: had originated in Australia; or had been embedded in Australia for a long time and have a specific Australian identity; or had a clearly stated objective of changing political and industry policies and practices on environmental protection and management; or had a membership base of either individuals or organisations (the latter being peak bodies).

It excluded women in organisations whose principal activities are education, scientific investigation, recreation, networking and information exchange or planning and organisations which have a very recent presence in Australia, but whose bases lie overseas.

Many of the women leaders who have been part of this project have moved between organisations that fit the four criteria above and those that do not. While there are many other women who meet these criteria for leadership, the intention of this research was to gather a representative sample of environmental leadership across the decades since Federation

Jones and Solomon's (2019) study identified gender-related challenges women conservation leaders experienced in their careers and supports helping them in advance. The authors conducted and analyzed semi-structured interviews with 56 women leaders in conservation organizations across the United States. It was found that 100 percent interviewees reported experiencing or witnessing a gender-related workplace challenge in at least one of six categories, and the vast majority reported encountering four or more of these challenges: salary inequality and difficulty negotiating, formal exclusion, informal exclusion, harassment and inadequate organizational response, assumptions of inadequacy, and assumptions of wrongness. Participants also experienced two categories of supports: structural supports and supportive relationships. Women's experiences varied based on age, race and ethnicity, and leadership position. Our results indicate more effort is needed to identify effective strategies for making conservation a more inclusive, empowering, and appealing profession in which to work.

Opportunities to conserve and protect environment for women environmentalists gives them and establishes gender equity. It further gives them an opportunity to be included both in formal and informal contexts and situations. They get the required nurture and encouragement as part of the society and formal training and development. They can establish their adequacy through fact based analysis and data. Challenges to protect environment for women environmentalists arise from gender discrimination, inadequate organizational response and assumptions about the adequacy of women

Analysis and Presentation of the Primary Data

An opinion survey was conducted to gather primary data. The opinion survey was carried out with a questionnaire administered over all academic, corporate, social, NGO, contacts in mail and in WhatsApp, during April 2023. The questionnaire was fully responded by 500 out of 2000 target recipients of the questionnaire. The respondent profiles is as below:

Respondents were from organizations of budget up to USD 7.2 billion Organization

Table 1. Gender

Gender	Numbers	Percentage
Male	380	76
Female	120	24

As seen from Table:1, majority of the respondents were women.

Table 2. Educational level

Education Level	Numbers	Percentage
Under Graduate	50	10
Post Graduate	280	56
Professional	50	10
Doctoral	120	24

As seen from Table 2, the majority of the respondents were Post-Graduates. Professionals (10) and Doctoral (24) degree holders were also part of the survey. All respondents were degree holders or of higher-level education.

Table 3. Education level

Leadership Role	Numbers	Percentage
Educator	290	58
NGO	60	12
Corporate	50	10
Religious	20	4
Community	20	4
Home	10	2
Others	50	10

According to Table No.3, significant majority of respondents are from education sector as teachers, professors, principals and directors. NGO leaders and corporate leaders were part of the survey. Exclusive religious, community and home leaders were insignificant.

92 % respondents agreed that women are able to protect the environment better and 96% respondent agreed that Leadership position helps protect environment. 94% of the respondents' organizations were taking environmental protection initiatives of varied range.

100 percent of the respondents agreed that women leaders are capable of and in advantageous position to protect environment. 92 % respondents felt that educated women also need orientation, reminder and

motivation to engage in environment protection. Frequency of the orientation, reminder or motivation as required by respondents - 40% quarterly, 30% half-yearly, 22% annually and 8% monthly.

92% respondents opined that educated women too need training and orientation in environment protection and conservation. Women leadership in environment protection and conservation bring better results believed 96% respondents. 44% of the respondents were involved with NGOs in volunteering for environment protection. Only 10% respondents were involved with CSR initiatives as only 10% respondents were with corporate. All the respondents were involved in encouraging and motivating others to be part of environment protection activities.

The environment protection training initiatives that would be beneficial as suggested by respondents were as follows:

Table 4. Environment protection training needs

Training	Numbers	Percentage
Environment Consciousness	200	40
Tree Planting	70	14
Solid Waste Management	70	14
Waste Water Recycling	70	14
Waste to Wealth	70	14
Butterfly Garden	10	2
Bee Keeping	5	1
Bamboo Garden	5	1

As seen from Table 4, the respondents wanted environment protection training in different areas – the significant one being 'Environment Consciousness'. Tree Planting, Solid Waste Management, Waste Water Management and Waste to Wealth were all equally preferred and wanted by 14 percent respondents. Butterfly Garden, Bee keeping and Bamboo Garden were desired by very few respondents.

Table 5. Environment Protection Initiatives already being done by the respondents

Environment Protection Initiatives	Responses	Percentage
Water recycling	225	45
Water Conservation	200	40
Rain water harvesting	200	40
Reduce and reuse plastic waste	225	45
Generate and use renewable energy	125	25
Proper disposal of e-waste	100	20
Planting trees	375	75
Bamboo garden, herbal garden etc.	75	15
Create awareness	425	85
Compost Pit	300	60
Lake Conservation Initiative	125	25
Recycling Products	100	20

Regarding the environment conservation initiatives being done by respondents each respondent had more than one initiative in operation. Creating awareness was the most popular initiative carried out by the respondents (85%). Tree planting (75%) was the next most happening initiative. Creating a compost pit to compost the vegetable and garden waste (60%) was the next predominant effort. Recycling water (45%) and reduce & recycle of plastic waste (45%) was also being followed by many. Water conservation and rain water harvesting were yet other initiatives followed by many (40%)

Table 6. Opportunities seen in protecting the environment

Opportunities	Responses	Percentage
Contribution to environment protection	450	90
Healthy living	375	75
Better improved planet in future	300	60
Be an ambassador of conservation	350	70
Contribute to Sustainability	400	80
Protect Humanity	300	60
Protect Planet earth	250	60
A great career	300	50
Career growth and advancement	200	40
Recognition/Reward/ Award	100	20

As seen from Table 6, the opportunities that flow out of environment protection are many – contribution to environment protection (90%) is seen as the most predominant opportunity. Contribution to sustainability (80%) is the next important opportunity seen. Healthy living (75%) and the possibility of being an ambassador for the conservation (70%) are other significant opportunities mentioned. Recognition (20%), career growth (40%) and career opportunities (50%) are not seen as great opportunities in protecting the environment.

Table 7. Challenges in taking environmental protection initiatives

Challenges	Responses	Percentage
Funds	400	80
Non-Cooperation/ No Support	350	70
Career Focus/ Lack of Time	325	65
Unconcerned Colleagues/ Society/ Leaders	375	75
No broader outlook/ Selfish ambitions	425	85
Not motivated to common good	400	80
Inconsistent efforts	350	70
Lack of awareness	300	60
Lack of support from Government	200	40
Lack of support from Management	150	30

As seen from the Table 7, the challenges faced by the respondents were many – Selfish ambitions, Funds, No motivation towards common goal were being the serious challenges (80% - 85%). Non-cooperation, unconcerned people around and inconsistent efforts were the next level of challenges (70% - 75%). Lack of awareness, lack of government support and lack of management support were other less important challenges faced.

Case Studies of Three Indian Educated Women

Dr. Parvathy Warrier is a retired professor and has been a lifetime member of OISCA-International, South India Chapter. OISCA –International is an environment protection NGO based and head quartered in Japan and has its efforts spread across the globe. Her challenges have been to involve more women into the NGO's initiatives. She has effectively guided and lead the local chapter in Palakkad and had emerged as a South India Chapter President.

Dr. Mansee Bal Bhargava is a transdisciplinary learner of the built environment from sustainability and livability perspectives. She has ventured into entrepreneurial initiatives along with research and education. Her entrepreneurial ventures focus has been on large scale developments with inclination towards ecological restoration. Her education has been from Cambridge and Erasmus University. She has won several global scholarships GATE-India, Cambridge Commonwealth, Nuffic, RIHN, Fulbright-Nehru to name a few. She has dedicated her life to conserve and protect environment through research and public speaking. She is on a life mission to direct, guide and support to find their life purpose in the space of environment conservation and protection. The challenges she faces are not being on a regular job as organizations don't give the freedom to move according to the need for public speaking and research. Further, finding the right partners and right talent is a challenge.

Dr. Mercia Selva Malar Justin is a professor of finance in Xavier Institute of Management and Entrepreneurship, Bangalore. She started her intiatives towards environment since 2002 as part of the Junior Chamber International's community initiatives. That year she won the National New Jaycee Runner Up award. Her initiatives in partnership with Siruthuzhi and OISCA International started from then on. Using these two organizations she has raised awareness on environmental concerns among her colleagues and students. She was invited to the International Board meeting held at Tokyo, Japan of OISCA International during the year 2018. She organized two international Youth Forums for OISCA International with UN Sustainable Development Goals as the theme. She has organized tree planting drives and planted more than 500 trees as part of her family celebrations. The challenges she faces are: not everyone in Business Schools value the efforts and supports the initiatives, environment conservation initiatives are treated with step motherly consideration and less important to organizations and to the corporate.

CONCLUSION

"A healthy ecology is the basis for a healthy economy."—Claudine Schneider, former U.S. Representative from Rhode Island, in The Green Lifestyle Handbook

India's female literacy rate is 53.7%. If all the literate women contribute their best efforts as leaders in environment management, the nation can definitely march towards a great future with conserved, protected and preserved environment. Women are at the helm of affairs of their households and the impact expected can be highly remarkable. In workplace and society women must assume greater role and come forward to support environment management. The most critical challenge women leaders will face is gender discrimination and disempowering attitude of men in the society. Discarding the need for environment protection and conservation is yet another strong challenge. Exhibiting extraordinary courage and strength to bring about a remarkable change in the environment is required of women leaders.

REFERENCES

Christia, L. M., & Vijayanand, A. (2023). Women Leadership in Environmental Conservation in View of Laudato SI. *Journal of Pharmaceutical Negative Results*, *14*, 226–230.

Dewi, K. H. (2017). Protecting Environment and Securing Power: Female Leaders' Political Initiative in the Age of Climate Change in Indonesia and Japan. *International Journal of Indonesian Studies*, *1*, 122–160.

Dilokwanich, S., Kaewsawang, S., Savatdipap, S., Iochawna, J., & Limganjanawat, S. (2015). Lessons Learnt of Thai Women Environmental Leaders. *Environment Asia*, *8*, 18–25.

Einarsen, S., Raknes, B. R. I., & Matthiesen, S. B. (1994). Bullying and harassment at work and their relationships to work environment quality: An exploratory study. *European Journal of Work and Organizational Psychology*, *4*(4), 381–401. doi:10.1080/13594329408410497

Elix, J., & Lambert, J. (2014). 16. Passionate defenders, accidental leaders: Women in the Australian environment movement. *Diversity in Leadership, 301*.

Forster, M. E., Palmer, F., & Barnett, S. (2016). Karanga mai ra: Stories of Māori women as leaders. *Leadership, 12*(3), 324–345. doi:10.1177/1742715015608681

Freudenberg, N., & Steinsapir, C. (2014). Not in our backyards: The grassroots environmental movement. In R. E. Dunlap & A. G. Mertig (Eds.), *American Environmentalism* (pp. 27–37). Taylor & Francis.

Glass, C., Cook, A., & Ingersoll, A. R. (2016). Do women leaders promote sustainability? Analyzing the effect of corporate governance composition on environmental performance. *Business Strategy and the Environment, 25*(7), 495–511. doi:10.1002/bse.1879

Hillman, R. (2020). *Are Women Leaders of Nature? -a quantitative study on the effect of women's political representation on environmental politics in European countries*. Gupea. https://gupea.ub.gu.se/handle/2077/63686

Jagnani, M., & Mahadevan, M. (2021). *Women Leaders Improve Environmental Outcomes: Evidence from Crop Fires in India*. IZA. https://conference.iza.org/conference_files/Gender_2021/mahadevan_m31449.pdf

Jewitt, S. (2000). Mothering earth? Gender and environmental protection in the Jharkhand, India. *The Journal of Peasant Studies, 27*(2), 94–131. doi:10.1080/03066150008438733

Jones, M. S., & Solomon, J. (2019). Challenges and supports for women conservation leaders. *Conservation Science and Practice, 1*(6), e36. doi:10.1111/csp2.36

Jouber, H. (2022). Women leaders and corporate social performance: Do critical mass, CEO managerial ability and corporate governance matter? *Management Decision, 60*(5), 1185–1217. doi:10.1108/MD-07-2020-0953

Labaris, A. (2009). Women involvement in environmental protection and management: A case of Nasarawa state. *Journal of Sustainable Development in Africa, 10*, 179–191.

Li, J., Zhao, F., Chen, S., Jiang, W., Liu, T., & Shi, S. (2017). Gender diversity on boards and firms' environmental policy. *Business Strategy and the Environment, 26*(3), 306–315. doi:10.1002/bse.1918

Liu, C. (2018). Are women greener? Corporate gender diversity and environmental violations. *Journal of Corporate Finance, 52*, 118–142. doi:10.1016/j.jcorpfin.2018.08.004

Merchant, C. (1981). Earthcare: Women and the environment. *Environment, 23*(5), 6–40. doi:10.1080/00139157.1981.9933143

Niu, Y., & Sims, C. (2022). Women Leaders in the Non-Governmental Organization (NGO) Sector. In J. Ke & G. N. McLean (Eds.), *Chinese Women in Leadership* (pp. 203–226). Springer International Publishing.

Nugent, C., & Shandra, J. M. (2009). State environmental protection efforts, women's status, and world polity: A cross-national analysis. *Organization & Environment, 22*(2), 208–229. doi:10.1177/1086026609338166

Ornelas, R. (2011). Water and the Indigenous Women's Leadership Project. *Journal of Research on Women and Gender, 2*, 111–134.

Pohlmann, L. M. (2013). *Women leaders of environmental advocacy organizations: A qualitative study.* University of Southern Maine.

Prindeville, D. M. (2003). Latina and American Indian women leaders, political representation, and environmental policy. *Journal of Latino/Latin. American Studies (Lawrence, Kan.), 1*, 55–76.

Resurrección, B. P. (2013). Persistent women and environment linkages in climate change and sustainable development agendas. *Women's Studies International Forum, 40*, 33–43. doi:10.1016/j.wsif.2013.03.011

Ricci, F., Scafarto, V., & della Corte, G. (2023). Board Gender Diversity, Female Leaders and Corporate Environmental Responsibility: A Structured Literature Review. When the Crisis Becomes an Opportunity: *The Role of Women in the Post-Covid Organization,* 411-432.

Saraçli, S., Yilmaz, V., & Arslan, T. (2014). The Effects of Mothers' Educational Levels on University Students' Environmental Protection Commitments and Environmental Behaviors. *Eurasian Journal of Educational Research, 14*(55), 177–199. doi:10.14689/ejer.2014.55.11

Shambel, G. (2012). The gender dimension of climate change and women's role in environmental protection and natural resource management in South Wollo, Ethiopia. *Research on Humanities and Social Sciences, 2*, 32–40.

Wei, S., Sial, M. S., Zhou, W., Badulescu, A., & Badulescu, D. (2021). Improving the environmental footprint through employees: A case of female leaders from the perspective of CSR. *International Journal of Environmental Research and Public Health, 18*(24), 13082. doi:10.3390/ijerph182413082 PMID:34948692

Wuyep, S., Dung, V., Buhari, A., Madaki, D., & Bitrus, B. (2014). Women participation in environmental protection and management: Lessons from plateau state, Nigeria. *American Journal of Environmental Protection, 2*, 32–36. doi:10.12691/env-2-2-1

Yachina, N. P., Khuziakhmetov, A. N., & Gabdrakhmanova, R. G. (2018). Formation and development of the regional system of continuous environmental education of a teacher. *Ekoloji, 27*, 1315–1322.

Chapter 3
The Role of Women Leadership in Environmental Management

Kiran Shrimant Kakade
https://orcid.org/0000-0003-0556-3390
Lala Lajpatrai Institute of Management, University of Mumbai, India

Anjali M. Kulkarni
Lala Lajpatrai Institute of Management, University of Mumbai, India

Kasturi R. Naik
https://orcid.org/0000-0003-0335-7534
Sparsh Global Business School, India

ABSTRACT

Women provide environmental management with unique perspectives, skills, and talents that help improve sustainability and conservation. This chapter examines women's unique position in environmental management, emphasising the various disciplines in which they participate and the valuable contributions they make. Women must participate in environmental decision-making processes for long-lasting outcomes. They frequently value long-term environmental sustainability, social equality, and neighborhood well-being highly. Women are better equipped than men to manage complex environmental issues holistically, considering interrelated social, economic, and ecological factors. This chapter reviews the research on women's environmental leadership in environmental management, makes policy recommendations for possible government initiatives to increase the representation of women in environmental decision-making, and identifies areas for additional data collection and research.

DOI: 10.4018/978-1-6684-5986-7.ch003

INTRODUCTION: WOMEN AND THE GREEN FRONTIER—A STUDY OF FEMALE LEADERSHIP IN ENVIRONMENTAL PRACTICES

The interaction between human civilization and the natural environment has profoundly shaped societies, particularly in third-world countries. Yet, managing environmental challenges, deeply intertwined with cultural nuances, has become complex. Governments spearhead environmental preservation but require collaboration across all society strata, with a particular emphasis on women. Historically, despite their significant contributions to environmental protection, societal biases have often overshadowed women's roles. With women's actions directly influencing environmental sustainability, international organizations frequently highlight them as pivotal players. By addressing existing challenges, women can harness this potential, effectively driving environmentally friendly behaviors and decisions.

What is Women's Leadership?

Women's leadership refers to the act of women taking on roles of leadership in various sectors and capacities, influencing decisions, policies, and practices, shaping visions, and inspiring others to achieve shared goals. It encompasses not only the presence of women in top-tier positions but also the distinct perspectives, styles, and values they bring to the table. Often characterized by collaboration, inclusivity, and holistic thinking, women's leadership frequently emphasizes sustainable growth, social justice, and long-term solutions.

Why is Women's Leadership Important?

Diverse Perspectives: Leadership that is diversified, encompassing both male and female leaders, is more likely to generate innovative solutions to complex problems. Women bring a unique set of experiences and insights that can foster out-of-the-box thinking.

Holistic Approach: Studies have suggested that women are more likely to take a holistic and inclusive approach to problem-solving. This is particularly crucial for environmental issues that require a multidimensional perspective (Eagly & Carli, 2003).

Enhanced Collaboration: Women leaders often prioritize collaboration, consensus-building, and teamwork. Such a collaborative approach can be vital in environmental practices where multi-stakeholder cooperation is the key.

Ethical Decision Making: Women leaders have been noted for their ethical decision-making approach, which can be instrumental in environmental conservation where ethical considerations are paramount (Kidder, 2001).

Recognition of Women's Leadership in Environmental Practices

While strides have been made in recognizing the efforts of women in leadership roles, there is still a significant gap, especially in fields traditionally dominated by men, such as environmental management. Nevertheless, the narrative is gradually changing. Women across the globe, and notably in India, are now at the forefront of environmental movements, conservation projects, and sustainable policy-making. Their efforts, though sometimes under-recognized, have created lasting impacts. For instance, Dr. Vandana Shiva, an Indian scholar and environmental activist, has been a staunch advocate for biodiversity, organic farming, and the rights of farmers. Her leadership has been pivotal in challenging the practices of multinational corporations and emphasizing the importance of indigenous knowledge and practices (Shiva, 1993).

Similarly, Medha Patkar's leadership in the Narmada Bachao Andolan shed light on the environmental and societal impacts of large dam projects. Through her relentless efforts, she has spotlighted the need for a more inclusive approach to development, where both environmental and social factors are taken into account (Roy, 1999). Despite these successes, it's crucial to note that many women leaders often operate without the spotlight, making significant changes at the grassroots level. Their contributions, though less visible in mainstream media, are equally, if not more, significant in shaping sustainable futures.

Women's leadership in environmental practices is an asset that has yet to be fully tapped. While the contribution of female leaders like Dr. Shiva and Medha Patkar is undeniable, countless unsung heroines are making daily strides in environmental conservation. Recognizing, supporting, and amplifying their efforts is not just a matter of gender equality but a necessity for a sustainable future.

Historical Context

Tracing back to the Western Industrial Revolution at the end of the 19th century, the genesis of leadership theories emerges. As the study of leadership burgeoned during this period, the conventional belief system linked leadership qualities predominantly with men. However, the past few decades have seen a progressive rise of women in varied sectors, leading to a shift in leadership theories and practices. Today, there's an urgent call for gender diversity across business spheres, with women leaders often exemplifying employee empowerment and nurturing interpersonal workplace relations.

LITERATURE REVIEW

As women climb corporate hierarchies, examining their leadership becomes crucial. As noted by Basu & Mukherjee, women's ascent to management positions has been a gradual journey. Various studies, like those by Cockburn, suggest that women's characteristics, molded by early experiences and societal interactions, foster traits such as empathy and cooperation. Such traits, as observed by Eagly, Johannesen- Schmidt, and Van Egen, make women well-suited for contemporary leadership roles, emphasizing collaboration and adaptability.

Table 1. Representing literature review

Year	Paper Title	Author(s)	Key Findings
2014	Feminine Perspective on Green Leadership	Jane Smith	Found that women leaders in environmental NGOs often prioritize collaborative and inclusive decision-making.
2015	The Eco-feminism and Environmental Advocacy	Laura Johnson & Mark Lee	Highlighted the intersection of feminism and environmentalism in driving sustainable practices.
2016	From Grassroots to Global: Women Leading in Conversation	Anika Patel	Showcased stories of women from different backgrounds making significant impacts in conservation efforts.
2017	Women in Green Tech: Breaking Barriers	Aliya Khan	Analyzed the increasing role of women in green technological innovations and sustainable startups.
2018	Female Leadership: A Sustainable Future in Business	Michael Luo & Lily Chen	Demonstrated a correlation between female leadership and the adoption of sustainable business practices.
2019	The Influence of Gender on Environmental Policy Making	Samuel O'Reilly	Found that policies designed by women leaders were often more comprehensive and sustainability-focused.
2020	The Empowering Women in the Era of Climate Change	Rebecca Torres	Discussed strategies for leveraging women's unique skills and perspectives in addressing climate change.
2021	The Role of Women CEOs in Promoting Green Corporate Agenda	Carlos Gomez & Zoe White	Provided evidence that firms led by women CEOs were more likely to have robust environmental sustainability strategies.

WOMEN LEADERS IN ENVIRONMENTAL MANAGEMENT IN INDIA AND THE ROLE OF WOMEN: A LITERATURE REVIEW

India, with its rich and diverse culture, has witnessed the rise of influential women leaders who have driven significant transformations in various sectors, particularly in environmental management. The role of women in this sphere cannot be understated, and literature consistently underscores their indispensable contributions.

Women Leaders in Environmental Management

Dr. Vandana Shiva, a globally acclaimed environmental activist, spearheads the Navdanya initiative which champions organic farming and biodiversity conservation. Her seminal work "Earth Democracy: Justice, Sustainability, and Peace" outlines her perspectives on conserving biodiversity, traditional knowledge, and cultural diversity as part of global justice (Shiva, 2005). Another luminary, Medha Patkar, took the lead with the Narmada Bachao Andolan. Her dedication to grassroots activism exemplifies women's capacity to mobilize masses against environmentally destructive projects (Roy, 1999). Sunita Narain's leadership at the Centre for Science and Environment (CSE) consistently emphasizes sustainable development. CSE's studies under her guidance, especially regarding air pollution, have been instrumental in shaping environmental policies in India (Narain & Ghosh, 2017).

Role of Women

Apart from leadership positions, the pivotal role of women in environmental management is evident across various layers of society. Literature suggests that women, due to their traditional roles in communities, often have a deeper, firsthand understanding of environmental challenges. Agarwal (2000) argues that women's roles in forest resource management are particularly significant. Their daily interactions with these ecosystems place them in a unique position to suggest sustainable solutions. In urban settings, women have been vocal about water management issues, with researchers like Dr. Priyanka Jamwal making strides at institutes like ATREE, and their findings playing a role in policy formation (Jamwal & Nambi, 2018).

Furthermore, the literature establishes that women's participation in environmental decision-making processes ensures a more comprehensive approach to sustainability. Their inclusion promotes biodiversity conservation, resource management, and, more importantly, brings to the fore traditionally overlooked perspectives (Leach, 2007).

Factors Influencing Women's Leadership in Environmental Management

The role of women in environmental management has been increasingly recognized over the last few decades. Women often bear the primary responsibility for managing natural resources at the household level, making their involvement in broader environmental initiatives both logical and necessary. However, several factors influence their participation and leadership in environmental management. Analyzing these factors can provide insight into how best to facilitate and amplify the impact of women in this crucial sector.

Cultural and Social Norms: In many societies, especially in developing countries, cultural and social norms restrict women's participation in public spaces and decision-making processes (Ghosh & Rao, 2019). While women might play crucial roles in the daily management of resources like water and firewood, their voices may be underrepresented in larger discussions about environmental policies and practices.

Access to Education: Education is a significant enabler for women's leadership. An educated woman is more likely to be informed about environmental issues, aware of her rights, and capable of advocating for sustainable practices (Sharma, 2015). However, disparities in educational opportunities for women, particularly in rural areas, can limit their potential leadership roles in environmental management.

Economic Empowerment: Women's economic independence can serve as a catalyst for their active participation in environmental decision-making. When women control financial resources, they often invest in sustainable and eco-friendly technologies and practices, leading to long-term environmental benefits (Kumar & Patel, 2017).

Supportive Legislation and Policies: Governments and institutions play a pivotal role in facilitating women's leadership in environmental management. Policies promoting gender equality, representation quotas for women in decision-making bodies, and initiatives targeting capacity-building for women can significantly influence their leadership roles (Malik & Verma, 2018).

Community and Grassroots Movements: Grassroots movements, such as the Chipko Movement in India, have historically empowered women to take up leadership roles in environmental causes. Such movements not only amplify the voice of women but also highlight their intrinsic connection to and knowledge of local ecosystems (Das & Jain, 2021).

Access to Resources: For women to take leadership roles, they must have access to resources – both in terms of tangible assets like land and water and intangible resources like information, training, and networks. When women have control over land, for instance, they are more likely to adopt sustainable agricultural practices (Srivastava, 2020).

Global and National Networks: Participation in global environmental conventions and forums can influence women's leadership at the national and local levels. Exposure to global best practices, networking opportunities, and platforms to share their experiences can empower women leaders in environmental management.

While women have immense potential to contribute to environmental management, several factors can influence their leadership roles in this domain. Addressing these factors - be it through policy interventions, societal shifts, or capacity-building measures - is crucial to harnessing the full potential of women's leadership for a sustainable future.

The Link Between Women, Development, and Environment

Environmental degradation, costing nearly 5.7% of global GDP according to the World Bank, underscores the need for sustainable growth. Historically, women have been at the forefront of environmental preservation. Saikia (2021) emphasizes the intrinsic connection women share with nature, prompting them to champion environmental causes. The interplay between gender and environmental sustainability is gaining recognition, with a focus on increasing women's representation in environmental decision-making.

Theoretical Framework

Modern environmental management hinges on several foundational aspects. Effective management encompasses planning, organization, leadership, and monitoring. Today's challenges necessitate a revamped management approach, integrating ethics and human resource considerations. Environmental management, at its core, seeks a harmonious interplay between biotic and abiotic elements, with human intervention playing a pivotal role. The ecofeminist perspective underscores women's intrinsic connection with nature, emphasizing their pivotal role in environmental stewardship. Women's leadership in environmental management is both historical and crucial. To navigate contemporary environmental challenges, harnessing women's potential becomes indispensable. Encouraging their increased representation and participation in environmental decision-making processes can pave the way for sustainable solutions, benefiting societies globally.

The discourse surrounding the role of women in development and the environment has evolved over time. The "women in development" paradigm emerged as a significant school of thought, emphasizing the specific needs of women both as individual entities and collective groups. This phase of reflection was crucial for conceptualizing developmental projects and programs (Rico, 1993). A corollary to this movement was the "women and the environment" perspective, which sought to understand and underscore the unique relationships and dependencies women have with their environment. The principles stemming from this paradigm were notably evident in the materials from the NGO Forum during the 1985 World Conference to Review and Appraise the Achievements of the United Nations Decade for Women.

Women's Involvement in Decision Making and Environmental Programs

As Koohi et al. (2016) rightly pointed out, women's role as principal stakeholders in environmental projects and programs has become indispensable. Their involvement in local-level engagements has markedly influenced the realization of government development agendas. This decentralized participation aligns with the tenets of Indian democracy, wherein civil society, public officials, and policymakers collaborate.

However, historically there has been a trend of segregating women's issues and addressing them specifically. Although certain issues particular to women can be independently addressed, such an approach has been shown to be insufficient. According to Rico (1998), failure to recognize the interplay between men's and women's social conditions, as well as between the gender system and development policies, has hampered strides toward equity in daily life and public policies.

Evidence of Women's Influence

A shift towards more environmentally conscious governance and business practices when women are in leadership roles has been observed. As per recent research by Östros (2023), companies led by women display an enhanced commitment to implementing eco-friendly initiatives. Such women-led enterprises also tend to invest more in sustainable practices, such as renewable energy, recycling, and improved energy efficiency.

Women play an instrumental role in influencing consumption patterns due to their pivotal position in household management. Their intimate interactions with resources like water and forests make them acutely conscious of the environment's state. As primary consumers of these resources, women's roles are directly impacted by the environment, often making them more environmentally conscious than their male counterparts. Their immersion in agricultural and livestock responsibilities further binds them to nature, enhancing this connection.

In India, women form a significant one-third of the workforce, making substantial contributions to the nation's economy and their families' sustenance. Women have emerged as catalysts for change in grassroots programs focusing on economic empowerment, environmental conservation, and sanitation. Programs adopting a gendered lens have been developed, leveraging the Self Help Group model. These groups have successfully tackled environmental challenges like vermicomposting, tree plantation, and joint forest management. The involvement of grassroots workers, such as Anganwadis and ASHA, has been instrumental in these endeavors (Kumari, 2019).

To pave the way for a sustainable future, it is imperative to recognize and harness the distinct role women play in environmental conservation. Historically, women have been at the forefront of environmental initiatives, advocating for reforms and sustainable living. Recognizing their distinctive contributions and understanding the underlying motivations is vital for future conservation efforts.

The Role of Women in Environmental and Sustainability Initiatives

Women have played a pivotal role in environment and sustainability domains, influencing various sectors through their unique knowledge, leadership abilities, and activism. This comprehensive analysis is organized chronologically and cohesively to capture the multifaceted contributions of women to sustainability and environmental protection.

Empowerment and Education

The bedrock of women's involvement in environmental and sustainability issues is empowerment and education. Increased access to education arms women with the knowledge and skills required to spearhead environmental initiatives. An educated woman is not just an active participant but can also influence decision-making processes, advocate for progressive environmental policies, and engage her community. Furthermore, empowerment provides women with opportunities to venture into careers in environmental sciences, policy-making, and other sustainability-related fields. The influence of education and empowerment transcends the academic realm, equipping women with the tools to assert their rights, gain financial independence, and make critical decisions that impact both their immediate environment and the world at large.

Traditional Knowledge and Wisdom

Historically, women have been custodians of an extensive reservoir of traditional knowledge and practices related to the environment. From sustainable farming methods to conservation practices, women's wisdom has often been rooted deeply in the customs of their communities. By integrating this traditional knowledge into contemporary sustainability endeavors, women provide invaluable insights that champion the preservation of ecosystems and judicious use of resources.

Sustainable Agriculture and Food Security

Globally, women are instrumental in agriculture and ensuring food security. They have consistently championed sustainable farming methods, integrating practices such as organic agriculture, agroforestry, and permaculture. These techniques not only bolster food security but also curtail environmental degradation.

Conservation and Biodiversity

Beyond agriculture, women are dedicated environmental stewards, taking on roles in habitat restoration, wildlife conservation, and natural resource protection. Their intimate association with nature, coupled with their active participation in eco-tourism and sustainable livelihood projects, underscores their commitment to biodiversity conservation and community development.

Climate Change Mitigation and Adaptation

While climate change affects vulnerable demographics, women have transformed from merely affected victims to being pivotal players in mitigation and adaptation strategies. Women-driven projects in areas like renewable energy, sustainable transportation, and climate-resilient agriculture have made significant strides in combating the adverse effects of climate change.

Community Engagement and Grassroots Activism

Women have demonstrated unparalleled skills in community engagement and grassroots activism. Their leadership skills in organizing communities, advocacy for various causes, and development of grassroots projects are transformative. Women also excel in building communities, volunteering, engaging in intersectional activism, and influencing policy and governance. Their contributions are not only diverse but also crucial for nurturing empowered, inclusive communities.

Corporate Sustainability and Leadership

In the corporate realm, women leaders are becoming increasingly influential, pushing for sustainable business practices. The presence of women in leadership positions has often translated to improved environmental and social performance in organizations. Their emphasis on sustainable business strategies, resource efficiency, and green innovation contributes significantly to industry resilience and sustainability. From grassroots activism to corporate leadership, women have been integral in shaping the sustainability and environmental protection discourse. Their unique blend of traditional knowledge, academic prowess, and leadership skills positions them as pivotal contributors to global sustainability efforts.

Women Leaders in Environment Management in India

India, often revered as the mother of the earth, has long nurtured its inhabitants. However, with the relentless exploitation of natural resources, the ecological balance is perpetually at stake. This paper examines the vital role of women in environmental management in India, spanning from grassroots movements to high-level policy interventions.

Elevating Environmental Management: Women Leaders Making a Difference in India

The power of women leaders stretches beyond corporate arenas. In fields such as environmental management, their unparalleled vision and capabilities lead to profound transformations. Real-life success stories, beyond the corporate domain, provide tangible evidence for the claim that women's leadership can notably uplift various sectors, including environmental management.

1. Dr. Vandana Shiva - Navdanya:
   Esteemed environmental activist and scholar, Dr. Vandana Shiva founded Navdanya, an initiative championing organic farming and biodiversity conservation. Thanks to her leadership, Navdanya has played a pivotal role in preserving over 3,000 rice varieties (Shiva, 2005).
2. Medha Patkar - Narmada Bachao Andolan:
   Medha Patkar's spearheading of the Narmada Bachao Andolan is a testament to the potency of grassroots activism in India. She valiantly opposed large dam projects that posed environmental and human rights concerns, igniting discussions at national and international levels (Roy, 1999).
3. Sunita Narain - Centre for Science and Environment (CSE):
   Under the direction of Sunita Narain, CSE's initiatives have consistently emphasized the need for sustainable development in India. Influential studies, like the landmark report on Delhi's air pollution, have steered governmental policies towards environmental prioritization (Narain & Ghosh, 2017).
4. Dr. Priyanka Jamwal - Ashoka Trust for Research in Ecology and the Environment (ATREE):
   Dr. Jamwal's contributions at ATREE are centered on India's pressing water challenges, with her findings playing a role in urban water management policy formation (Jamwal & Nambi, 2018).
5. Ritu Kumar - Sustainable Fashion:
   Ritu Kumar, renowned for her fashion designing, has been instrumental in championing sustainable fashion practices. Her emphasis on traditional craftsmanship combined with eco-friendly materials has set precedents in the fashion industry (Mishra, 2014).

In summary, the accomplishments of these women in environmental management manifest the boundless potential that arises from integrating women's insights and leadership. Such narratives affirm the need for further tapping into this potential for sustainable futures.

Historical Context: Early Women-Led Environmental Movements

1731: Bishnoi Movement - Often regarded as the world's first environmental movement, Amrita Bai and the Bishnoi women of Jodhpur's Khejarali village played a pioneering role. They protested against deforestation by embracing the Khejri trees, vital to their community.

1917: Champaran Movement - Led by Mahatma Gandhi against indigo planting imposed by British landlords. Women, mainly from urban settings, supported this.

1930: Salt Satyagraha - Women supported Gandhiji's movement against the British Salt Tax Monopoly.

1972: Chipko Movement - Initiated in the Uttarakhand region by Sundarlal Bahuguna, notable female environmentalists like Bachni Devi and Gaura Devi were instrumental.

1978: Silent Valley Movement - The campaign in Kerala's Silent Valley, with notable contributions from poet Sugatha kumari, aimed at preserving the evergreen forest against a hydroelectric project.

1982: Navdanya Movement - Established by environmental advocate Vandana Shiva, this movement emphasized biodiversity and included women in its eco-feminist agenda.

1983: Aapiko Movement - This campaign in Karnataka's Uttara Kannada district was against commercial forest policies, which threatened the Sirsi forest.

1989: Narmada Bachao Andolan - Led by Medha Patkar, this movement opposed the construction of the Sardar Sarovar Dam on Gujarat's Narmada River.

CONTEMPORARY WOMEN ENVIRONMENTAL LEADERS AND ACTIVISTS IN INDIA

Vandana Shiva

Background: Vandana Shiva, trained as a physicist, later shifted her focus towards interdisciplinary research in science, technology, and environmental policy.

Contribution: Founder of the Navdanya movement, she has ardently advocated for the rights of people in India to have access to food security and biodiversity. She's renowned for her fierce opposition against genetically modified crops and for championing organic farming. Dr. Shiva has also written extensively on topics related to biopiracy, bioethics, and intellectual property rights.

Awards & Recognition: She has received numerous awards for her work, including the Right Livelihood Award, often dubbed the "Alternative Nobel Prize".

Sunita Narain

Background: An environmentalist and political activist, she's also the Director-General of the Centre for Science and Environment (CSE).

Contribution: Sunita Narain has been a prominent advocate for sustainable development and for addressing the local needs within the global climate change debate. She has been a major figure in influencing Indian environmental policy.

Awards & Recognition: Awarded the Padma Shri in 2005, she was also listed in the Time 100 influential people list in 2016.

Jamuna Tadu

Background: Fondly referred to as the "Lady Tarzan" of Jharkhand, Jamuna comes from a modest background in Jharkhand.

Contribution: Under her leadership and with the support of 100 Adivasi women, she's been a force against illegal logging by protecting forests in Jharkhand from the timber mafia. This movement not only protected the environment but also secured the livelihoods of many local tribes.

Radha Bhatt

Background: An Indian social activist and Gandhian.

Contribution: She initiated the Nadi Bachao Abhiyan in 2008 which was a campaign against hydroelectric projects that were causing harm to the Ganga River and its tributaries. She has been at the forefront of sustainable development and environmental conservation.

WOMEN AND ENVIRONMENTAL PROTECTION POLICIES AND PROGRAMMES IN INDIA

Environmental Gender Index (EGI)

Overview: The EGI ranks countries based on the integration of gender perspectives into their environmental policies.

India's Position: Out of 72 countries, India is positioned at the 46th spot, indicating there's still a considerable way to go to fully integrate gender perspectives into environmental policies.

Joint Forest Management (JFM) and National Forest Policy (1988):

Overview: A collaboration between local communities and the forest department to jointly manage and conserve forest resources.

Key Feature: Mandates that 33% of the members of VanaSamarakshana Samitis be women, thereby promoting a gendered approach to cooperative forest management.

2016 National Policies for Women:

Overview: A comprehensive policy aiming at the overall development of women in India.

Key Feature: The policy prioritizes and emphasizes the equitable control of women over natural resources and their environment, acknowledging their significant role in environmental management.

Forest Rights Act:

Overview: Recognizes and vests forest rights to forest dwelling Scheduled Tribes and other traditional forest dwellers.

Key Feature: This act emphasizes the importance of women's rights in forest governance, ensuring they have a voice and stake in the management and conservation of forests.

Awareness Programs:

Overview: Various initiatives have been launched by the government and NGOs to promote environmental awareness among citizens.

Key Feature: The incorporation of environmental education into the school curricula and regular broadcasts on environmental issues through mediums like the National Television and All India Radio play pivotal roles in raising awareness and fostering a culture of environmental responsibility.

WOMEN IN CORPORATE LEADERSHIP: A PARADIGM SHIFT

Over the past two decades, the perception and treatment of women in corporate settings have undergone significant transformation. According to a 2015 McKinsey estimate, achieving gender equality could amplify global GDP by $12 trillion by 2025. Moreover, Harvard Business School's study showed firms with 10% more female partners experienced an average 1.5% annual increase in total fund returns. The 2017 Sustainable Development Goals underscored gender equality as pivotal. Studies like "Better Leadership, Better World" by WomenRising2030 and "Leadership for Sustainability is Female" from Malmo University highlight that women leaders can effectively steer sustainability projects. Their unique competencies in innovation, long-term thinking, and social inclusion position them to spearhead economic opportunities aligned with the SDGs. The trajectory of women in environmental leadership in India, from grassroots movements to boardrooms, accentuates their indispensable role. As guardians of both nature and the next generation, women's influence on environmental strategies and policies holds promise for a sustainable future.

Women's Leadership and Environmental Decision-Making: A Comprehensive Analysis

Historically, women's engagement in environmental advocacy has been transformative. The recognition of Wangari Maathai, the founder of the Green Belt Movement, with the 2004 Nobel Peace Prize marked a significant milestone in acknowledging women's role in environmental movements. Furthermore, in 2002, female environment ministers along with 28 female heads of environmental NGOs established the Network of Women Environment Ministers, showcasing the collective power and representation of women in the domain.

The Current Scenario

There is a growing recognition of the benefits of women's leadership in environmental strategies. Recent findings suggest that enterprises led by women show a propensity towards green initiatives. For instance, the European Investment Bank reports that 47% of firms with female ownership employ over 50% of women, in contrast to only 26% of male-owned businesses. This not only emphasizes women's inclination towards gender parity but also underlines their contribution to fostering sustainable growth.

Moreover, evidence indicates that women are more inclined to establish green startups. With the shift towards a greener and more digitized economy, this gender dynamic implies expansive opportunities for women. Given that women investors show a higher inclination towards emphasizing environmental, social, and governance concerns, their growing relevance cannot be understated. Such trends highlight the significance of promoting female leaders across diverse sectors.

STRATEGIES TO ENHANCE WOMEN'S LEADERSHIP IN ENVIRONMENTAL CONTEXTS

Addressing Discrimination: At the forefront of any initiative should be regulatory actions that combat gender discrimination, societal norms, and behaviors like harassment and gender stereotyping.

Empowerment Initiatives: Governments can take proactive steps such as offering training, mentoring, and coaching to bolster women in environmental leadership roles. Collaboration with businesses, educational institutions, and vocational entities can further amplify the effects of these programs.

Promoting Female Ownership: Actively supporting projects that advocate for female environmental leadership and green business ownership in the civil sector is crucial.

Gender-sensitive Procurement: Applying a gender perspective in public procurement, especially for green infrastructure projects, can champion female-owned businesses.

Protective Measures: Ensuring the protection of female environmental activists, especially in instances of harassment, and promoting the formation of civil society organizations focused on sustainable objectives.

Indicators to Gauge Women's Environmental Leadership:

Understanding women's influence on sustainability mandates comprehensive research. Some key indicators include:

The ratio of women in senior leadership roles in green private sector companies.

The number of female heads and delegates in international environmental negotiations such as the UNFCCC COPs.

The gender distribution in national-level environmental forums.

The Flywheel Effect and Leadership Dynamics:

Women who ascend to leadership roles serve as beacons for their peers. This creates a flywheel effect, progressively altering decision-making dynamics. Comprehensive studies emphasize the need for a critical mass of women in leadership roles to foster active listening, collaboration, and deeper discussion of intricate issues.

The Way Forward

A major takeaway from the largest-ever all-female Antarctic expedition in 2019 was the palpable association between women's leadership and favorable environmental outcomes. Countries that support women in significant positions tend to show a heightened commitment to environmental protection, often ratified by stringent climate change legislations, establishing protected areas, and international treaty endorsements.

Furthermore, countries that uphold higher political and social statuses for women have reported reduced carbon footprints. This can be attributed to the leadership styles of women which emphasize long-term strategies, inclusivity, and collaboration.

However, despite the positive trajectory, challenges persist. Women still face barriers such as lack of access to property, challenging economic circumstances, time constraints, and socio-cultural norms that often stereotype their roles. To ensure a sustainable future, it is imperative to address these challenges and promote inclusive decision-making at all levels.

In sum, women's leadership is not just beneficial for gender parity; it's paramount for the planet. The path to a sustainable future requires the collective effort of both genders, advocating for change driven by data, science, and inclusivity.

WOMEN'S CONTRIBUTION TO CORPORATE SUSTAINABILITY AND ENVIRONMENTAL MANAGEMENT

Women play an indispensable role in promoting environmental sustainability and corporate responsibility. With a unique perspective and dedication, they have demonstrated an ability to enact change, foster collaboration, and drive innovation. This paper examines the myriad ways women have positively impacted environmental conservation, sustainability, and corporate growth.

THE ROLE OF WOMEN IN PROMOTING CORPORATE SUSTAINABILITY

Prioritizing Ethical Practices

Women in leadership roles have consistently showcased the importance of embedding ethical practices within organizational structures. These practices not only promote transparency but also emphasize social and environmental responsibility. Helena Helmersson's implementation of H&M Group's "Circular and Climate Positive" strategy is a testament to this.

Incorporating Sustainability in Business Strategies

Embedding sustainability into the core of business strategies is pivotal. Women leaders, such as Beth Ford of Land O'Lakes, have championed the integration of environmental considerations into organizational growth plans and risk assessments.

Driving Sustainable Innovation

Women entrepreneurs have often been at the forefront of sustainability-focused innovation, developing products and technologies that prioritize eco-friendliness. Lisa Jackson's role at Apple, emphasizing renewable energy and recycled materials, exemplifies this trend.

Stakeholder Engagement and Collaboration

Collaborative efforts, particularly those led by women like Christiana Figueres, have shown that pooling resources and knowledge can lead to substantial progress in addressing environmental challenges.

Fostering Internal Engagement

By creating a culture of sustainability within corporations, women leaders ensure that every stakeholder, from top management to junior employees, is aligned with the organization's sustainable objectives.

Influencing Supply Chain Sustainability

The influence of women is also palpable in the realm of supply chains. Leaders like Anita Roddick have set industry standards by prioritizing ethical sourcing and reshaping global supply chains.

Advocating for Policy and Regulatory Changes

Advocacy efforts, especially those by figures like Mary Robinson, underscore the importance of policies that support sustainability, emphasizing the need for businesses to function within environmentally conscious regulatory frameworks.

Community Involvement in Environmental Management

Local community involvement, particularly by women, has been pivotal in driving environmental conservation initiatives. Grassroots organizations, by engaging with communities, ensure that local perspectives are considered in larger developmental programs. However, the bureaucratization of local bodies has posed challenges. Future initiatives, such as Swach Bharat and Clean Ganga, further underline the importance of community engagement in development. Policymakers must thus prioritize the training and incentivization of grassroots workers to ensure sustained progress.

The role of women in environmental conservation and corporate sustainability cannot be understated. Their unique insights, collaborative spirit, and unwavering commitment have driven significant change. Ensuring their participation and leadership is integral to achieving broader sustainability goals. As we look towards a sustainable future, it becomes imperative to recognize and champion the contributions of women, both at corporate levels and within communities.

Indicators to Measure Women Leadership in Environmental Management

The role of women in leadership, particularly in environmental management, is gaining prominence as the world confronts various environmental challenges. To understand and quantify the impact of women's leadership in this sphere, several indicators can be utilized:

Representation in Decision-making Positions: Examine the proportion of women in leadership roles within environmental bodies, agencies, and organizations. This could be within government environmental agencies, international environmental NGOs, or even local community environmental protection groups (Denton, 2002).

Women-led Environmental Projects: Count the number of projects spearheaded by women that have yielded positive environmental outcomes. This can be reflected in terms of policy changes, habitat restoration, waste reduction, or other tangible metrics (Aguilar, 2009).

Funding Allocation: Observe the amount of funding and resources allocated to projects led by women or initiatives championed by women leaders in the environmental sector (Leach, 2007).

Policy Advocacy: Measure the number of policies or regulations proposed, modified, or enforced due to the advocacy and influence of women leaders (MacGregor, 2010).

Capacity Building and Training: Assess the initiatives led by women that focus on educating, training, and building capacity in environmental management. This includes workshops, training programs, and educational campaigns aimed at building sustainable communities (Aguilar & Perrot-Maître, 2008).

Community Engagement: Gauge the level of community engagement and participation in projects and initiatives helmed by women. This can include public meetings, consultations, or other forms of public participation (Resurrección & Elmhirst, 2008).

Achievements and Awards: Identify the accolades and recognitions received by women in the environmental field. This can be in the form of international awards, publications, or other acknowledgments that recognize their contributions (Dankelman, 2002).

Research and Publications: Count the number of research papers, articles, and publications authored by women in environmental management that contribute to the field's knowledge and practices (Arora-Jonsson, 2011).

ANALYSIS AND DISCUSSION

The notion of "Women and the Green Frontier" encapsulates the increasing involvement and influence of women in the environmental sector. Women's leadership in this realm is not merely about equal gender representation but also about introducing diverse approaches, insights, and solutions to pressing environmental challenges. Historically, women have been at the forefront of many environmental movements, from grassroots activism to policymaking (Denton, 2002). Their intimate knowledge of local ecosystems, often stemming from traditional roles in agriculture and household resource management, equips them with unique insights into sustainable practices (Aguilar, 2009).

Women leaders in the environmental sector have introduced innovative solutions that address the intersectionality of gender, socio-economic disparities, and environmental concerns. For instance, policies advocated by women often consider the indirect consequences on vulnerable populations, ensuring a holistic and inclusive approach to environmental management (MacGregor, 2010). Moreover, women-led initiatives in environmental practices often prioritize community engagement and empowerment. These initiatives recognize that sustainable environmental practices require the collective effort of communities and are most successful when there's local ownership (Resurrección & Elmhirst, 2008). However, challenges persist. Women leaders in environmental practices still face barriers such as limited access to resources, inadequate representation in decision-making forums, and even societal perceptions that sometimes undermine their authority (Leach, 2007). Addressing these barriers requires not only institutional changes but also shifts in societal perceptions regarding women's roles in leadership and environmental stewardship. As the world grapples with unprecedented environmental challenges, the leadership of women in this sector becomes not just essential but pivotal. Recognizing, supporting, and amplifying their roles can accelerate the journey towards a more sustainable future.

CONCLUSION

In the context of "Women and the Green Frontier: A Study of Female Leadership in Environmental Practices", the evaluation of field data underscores the pivotal role women play in driving environmentally sustainable practices, policies, and initiatives. Numerous studies (Sharma, 2015; Kumar & Patel, 2017) have highlighted the positive correlation between women's involvement in decision-making and sustainable environmental outcomes, especially in rural and indigenous communities of India. The field data from various regions of the country reflects that women, often seen as primary resource managers in their households and communities, demonstrate a heightened sense of responsibility towards resource conservation and sustainable practices (Ghosh & Rao, 2019). Their direct involvement in grassroots activities, be it forest conservation, water management, or organic farming, results in outcomes that are not only environmentally sound but also socially equitable.

Moreover, women-led environmental movements, like the Chipko Movement and Narmada Bachao Andolan, have historically shown that women leaders can galvanize community action, influence policy changes, and ensure that environmental considerations are mainstreamed into development initiatives (Malik & Verma, 2018). However, despite these positive trends, challenges persist. Structural inequalities, limited access to resources, and ingrained socio-cultural norms often sideline women from decision-making processes at higher levels, even when these decisions directly affect the environment and resources they manage (Srivastava, 2020). Policy frameworks must be revised to ensure not just representation but active participation of women in all tiers of environmental decision-making. In light of the above, policy recommendations focus on the urgent need for capacity-building initiatives tailored for women, amplifying their voices in decision-making fora and ensuring that they are integral stakeholders in formulating and implementing environmental policies. Policies must be rooted in the understanding that empowering women is not just a gender mandate but also an environmental imperative (Das & Jain, 2021).

Conclusively, as we chart the course for a sustainable future, it is evident that harnessing the leadership and expertise of women in environmental management is not only desirable but absolutely essential. As the data suggests and as countless on-ground initiatives have shown, when women lead, the environment, communities, and economies thrive.

REFERENCES

Agarwal, B. (2000). Conceptualizing environmental collective action: Why gender matters. *Cambridge Journal of Economics*, *24*(3), 283–310. https://www.jstor.org/stable/23601169. doi:10.1093/cje/24.3.283

Basu, M. (n.d.). Impact of Environmental Factors on Women Leadership. In M. Basu (Ed.), *Emerging Trends and Innovations in Modern Management* (pp. 365–376). Inspira Research Association.

Das, A., & Jain, R. (2021). The influence of women's leadership on community-based conservation efforts in India. *Conservation & Society*, *19*(2), 158–170.

Ghosh, S., & Rao, D. (2019). Women as key drivers of sustainable practices: An empirical study of rural India. *Sustainability*, *11*(3), 689.

Gomez, C., & White, Z. (2021). The Role of Women CEOs in Promoting Green Corporate Agenda. *Corporate Sustainability Review*, *18*(8), 110–125.

Jamwal, P., & Nambi, A. (2018). Wastewater treatment for carbon mitigation. *Current Science*, *114*(5), 1031.

Johnson, L., & Lee, M. (2015). Eco-feminism and Environmental Advocacy. *Green Women Studies*, *8*(2), 30–47.

Khan, A. (2017). Women in Green Tech: Breaking Barriers. *Sustainable Tech Review*, *10*(1), 12–25.

Koohi, E., Shobeiri, S.M., & Koohi, E., & Meiboudi. (2016). Women's participation in environmental management and development Promotion Culture. *The Open Access Journal of Resistive Economics*, *4*(1), 23–35.

Kumar, A., & Patel, N. (2017). Gender and green governance in India. *Indian Journal of Gender Studies*, *24*(3), 345–367.

Kumari, S. (2019). Role of Women in Environmental Management. *International Research Journal of Management Sociology & Humanity*, *10*(1), 432–436. doi:10.32804/IRJMSH

Leach, M. (2007). Earth mother myths and other ecofeminist fables: How a strategic notion rose and fell. *Development and Change*, *38*(1), 67–85. doi:10.1111/j.1467-7660.2007.00403.x

Levin, M. (2021, November 24). *Why female leadership is crucial to tackling climate change and other crises*. Climate Change Champions. https://climatechampions.unfccc.int/why-female-leadership-is-crucial-to-tackling-climate-change-and-other-crises/

Luo, M., & Chen, L. (2018). Female Leadership: A Sustainable Future in Business. *Journal of Business Ethics*, *20*(6), 123–140.

Malik, K., & Verma, R. (2018). Women-led environmental movements in India: Exploring the role of gender. *Environment and Natural Resources Research*, *8*(4), 65–76.

Mediawire. (2021, Sep 27). Women Leaders: The better half for building a better planet. *Economic Times*. https://economictimes.indiatimes.com/industry/csr/initiatives/women-leaders-the-better-half-for-building-a-better-planet/articleshow/86563194.cms?from=mdr

Mishra, V. (2014). Handmade in India: The craft of sustainable fashion. *Fashion Theory*, *18*(4), 465–484.

Narain, S., & Ghosh, A. (2017). Why Should I Not Be Angry? State of India's Environment 2017. In *Figures*. Centre for Science and Environment.

O'Reilly, S. (2019). The Influence of Gender on Environmental Policy Making. *Policy and Sustainability*, *9*(3), 15–29.

Östros, B. B. (2023, March 8). *Business case for women's leadership*. EIB. https://www.eib.org/en/stories/climate-women-profits

Patel, A. (2016). From Grassroots to Global: Women Leading in Conservation. *International Journal of Conservation*, *15*(4), 78–95.

Rico, M. N. (1998). Gender, the Environment and the Sustainability. In M. N. Rico (Ed.), *Gender, the Environment and the Sustaibility of Development* (pp. 15–20). United Nation.

Roy, A. (1999). The greater common good. *India International Centre Quarterly, 26*(1), 25–36.

Saikia, U. (2021). Role Of Women In Environmental Movement In India. *Ilkogretim Online - Elementary Education Online, 20*(6), 2473-2478. doi:10.17051/ilkonline.2021.06.228

Sharma, P. (2015). Women's role in environmental conservation in India. *Journal of Sustainable Development, 8*(2), 250–255.

Shiva, V. (2005). *Earth democracy: Justice, sustainability, and peace*. South End Press. doi:10.5040/9781350219755

Sigita Strumskyte, S. R. (2022). *Women's leadership in environmental action*. Organisation for Economic Co-operation and Development.

Smith, J. (2014). Feminine Perspectives on Green Leadership. *Journal of Environmental Leadership, 12*(3), 45–60.

Srivastava, N. (2020). Gender dynamics in forest management: A case study from Uttarakhand, India. *Forest Policy and Economics, 112*, 102097.

Torres, R. (2020). Empowering Women in the Era of Climate Change. *Climate Change and Gender, 7*(5), 60–77.

Chapter 4
Indian Environmental Management:
Empowering Communities Through Shaping Policies by Women Leaders

Sumi A. M.
Institute of Management in Kerala, India

K. S. Chandrasekar
Institute of Management in Kerala, India

ABSTRACT

India, a country known for its rich biodiversity and environmental challenges, has witnessed the emergence of numerous women leaders in environmental management. These leaders have played pivotal roles in shaping policies, implementing sustainable practices, and promoting environmental conservation nationwide. Women leaders in India have actively participated in policy formulation and governance processes related to the environment. Their contributions have resulted in implementing several environmental initiatives, including the protection of forests, the conservation of wildlife, and the promotion of sustainable land and resource management practices. They have established ecological education programs and encouraged youth and communities to adopt eco-friendly practices. Despite this, women leaders in environmental management in India face various challenges. This chapter provides an overview of the role of women leaders in environmental management in India. It highlights their contributions, the challenges faced, and the impact of their leadership on various ecological issues.

INTRODUCTION

Indian women have been actively engaged in various environmental protection activities, making significant contributions to sustainability and conservation efforts. They have been at the forefront of initiatives to address environmental challenges and promote a greener future.

DOI: 10.4018/978-1-6684-5986-7.ch004

Women often serve as key actors in sustainable resource management in rural communities, contributing significantly to activities related to water and forest management (Agrawal & Gibson, 1999).

One notable area of focus for Indian women has been sustainable agriculture and organic farming. Many women have taken up cultivating organic crops, promoting eco-friendly farming practices, and preserving traditional seed varieties. They have established organic farming cooperatives, conducted training programs for farmers, and raised awareness about the harmful effects of chemical-intensive agriculture.

Women's participation in community-based natural resource management programs has been associated with improved resource conservation outcomes and increased community resilience (Agarwal, 2001). Indian women have also played a crucial role in raising awareness about waste management and recycling. They have initiated waste segregation programs, established community recycling centres, and promoted composting and upcycling practices. Through their efforts, they have reduced the amount of waste going to landfills and fostered a culture of environmental responsibility.

Furthermore, Indian women are actively involved in preserving forests and biodiversity. They have organised tree-planting drives, fought against deforestation, and advocated for protecting endangered species. Women-led organisations have taken the lead in establishing community-based conservation initiatives, demonstrating the vital role of women in safeguarding ecosystems.

Water conservation is another area where Indian women have made significant contributions. They have been instrumental in implementing rainwater harvesting projects, promoting water conservation practices at the household level, and advocating for sustainable water management. Their efforts have helped communities cope with water scarcity and ensure a sustainable water future.

Incorporating gender perspectives into environmental policies and programs is essential for addressing environmental challenges effectively and ensuring equitable access to resources (UNEP, 2020). In addition to these on-the-ground activities, Indian women have been influential in policy advocacy and environmental governance. They have actively participated in ecological movements, mobilised communities, and influenced policy decisions. Women-led organisations have played a crucial role in shaping environmental policies, highlighting the importance of gender equality and social justice in environmental decision-making processes.

Indian women have demonstrated remarkable leadership and dedication to environmental protection. They have raised awareness, transformed practices, and inspired positive change in their communities and beyond through their initiatives and actions. Their contributions continue to drive sustainable development and conservation efforts, creating a more environmentally conscious and resilient India.

Sunita Narain: An environmentalist and director-general of the Centre for Science and Environment (CSE). She has been actively involved in environmental activism in India and also played a pivotal role in promoting sustainable development, advocating for clean air and water, and raising awareness about environmental issues.

Vandana Shiva: Vandana Shiva is an environmental activist, eco-feminist, and founder of Navdanya, a movement promoting seed sovereignty and organic farming. She has been a vocal advocate for sustainable agriculture, biodiversity conservation, and the rights of farmers in India.

Saalumarada Thimmakka: Known as the "Tree Woman of India," Saalumarada Thimmakka has dedicated her life to planting and caring for trees. She has planted thousands of trees in Karnataka, India, and has been recognised globally for her efforts in environmental conservation and creating awareness about the importance of tree planting.

Sheeba Sen: Sheeba Sen is an environmentalist and the founder of the 'Green Hope Foundation,' an organisation working towards environmental conservation and sustainability. She is actively involved in various environmental initiatives, including waste management, biodiversity conservation, and raising awareness about climate change.

Latika Thukral: Latika Thukral is the co-founder of the organisation 'I Am Gurgaon,' which focuses on improving the quality of life in Gurgaon, India, through environmental conservation and sustainability. She has been working on issues such as urban greening, waste management, and promoting sustainable transportation options.

Archana Soreng: Archana Soreng is an indigenous rights activist and has actively engaged in environmental advocacy. She has represented indigenous communities at the United Nations and is recognised for promoting climate justice and indigenous knowledge systems.

Shubhendu Sharma: Although not a woman leader, Shubhendu Sharma's work has been strongly supported by women-led organisations. He founded 'Afforestt,' a company that creates native, self-sustaining forests in urban spaces. The sustainability principles drive the organisation and are supported by women-led self-help groups.

These women leaders and their environmental activities have had a significant impact on promoting sustainability, conservation, and awareness about environmental issues in India. Their efforts have inspired positive change and empowered communities to act for a greener, more sustainable future.

Environmental Management Activities in India

Environmental management in India is a critical field that addresses ecological challenges, promotes sustainable practices, and conserves the country's rich biodiversity. With a diverse range of ecosystems and a large population, India faces various environmental issues that require effective management and conservation efforts. Here are some critical aspects of environmental management in India:

Policy and Governance: India has developed a robust legal and policy framework for environmental management. The Indian MoEFCC (Ministry of Environment, Forest and Climate Change) is responsible for formulating and implementing environmental policies and regulations at the national level. State and local governments also play a vital role in enforcing environmental laws and ensuring compliance.

Environmental Impact Assessment (EIA): Environmental Impact Assessment is a crucial tool used in India to evaluate proposed development projects' potential ecological and social impacts. It helps in decision-making by identifying and mitigating the adverse effects of projects on the environment and local communities.

Biodiversity Conservation: India is known for its rich biodiversity, including diverse flora and fauna. The country has established a network of national parks, wildlife sanctuaries, and protected areas to safeguard its biodiversity. Efforts are taken to conserve endangered species, restore ecosystems, and promote sustainable land and resource management practices.

Pollution Control: India faces significant challenges in managing pollution, including air pollution, water pollution, and solid waste management. Initiatives such as the National Clean Air Program (NCAP) and the Swachh Bharat Abhiyan (Clean et al.) have been launched to address these issues. Efforts are also being made to promote sustainable waste management practices and reduce plastic pollution.

Climate Change Mitigation and Adaptation: India is powerless against the consequences of climate change, including harsh weather events, rising temperatures, and changing rainfall patterns. The country is actively involved in global climate change negotiations and has set targets for reducing greenhouse gas emissions. Initiatives like the National Action Plan on Climate Change (NAPCC) focus on mitigation and adaptation strategies.

Sustainable Development: The concept of sustainable development is central to environmental management in India. Efforts are made to balance economic growth with environmental sustainability and social well-being. Sustainable practices are promoted in agriculture, energy, transport, and urban planning sectors to ensure long-term environmental and social benefits.

Public Participation and Awareness: Public participation and awareness are crucial in environmental management. Efforts are made to engage local communities, NGOs, and stakeholders in decision-making processes. Awareness campaigns, educational programs, and environmental initiatives at the grassroots level aim to create a sense of environmental responsibility among citizens.

International Cooperation: India actively participates in international environmental agreements and collaborations. The country collaborates with other nations and international organisations to address transboundary environmental issues, share best practices, and access financial and technical support for environmental initiatives.

Environmental management in India is an ongoing process that requires continuous efforts, innovation, and collaboration among various stakeholders. By addressing environmental challenges, promoting sustainable practices, and conserving natural resources, India aims to achieve a greener and more sustainable future for its citizens.

With the presence of a healthy ecosystem, the existence of human beings is possible. All living and non-living components comprise our environment. Environmental preservation has become crucial to battle climate change and global warming. Sustainable development is the need of the hour to save Mother Earth from the repercussions of industrialisation (Marshall, 2012). Ecological sustainability is a practice that paves the course for guarding the environment and natural resources in the personal, organisational, and governmental classes. Various core ecological issues are taking a heavy toll on human lives. From overpopulation, hydrological issues, ozone depletion, and global warming to deforestation, desertification, and pollution pose a severe threat to humankind. Suppose environmental conservation is becoming an effective mass movement (Mitchell, 2005).

The Interdependence of women with nature can be evolved from the following observation.

- Women have a base and skills for natural resource management
- They are better managers of natural resources
- They have a sensitivity towards Environmental
- They have high ecological consciousness

Natural resource depletion and environmental degradation directly affect women's time, income, health and social support system.

The knowledge of women regarding forest products, which have high herbal and medicinal value gained through long experience, is disappearing with deforestation. Unconscious and ruthless exploitation of natural resources, which resulted in environmental degradation, directly impacts women.

ROLE OF WOMEN IN THE CONSERVATION OF ENVIRONMENT IN INDIA

Environmentalists are those who work towards the betterment of our environment. These people have lobbied for environmental protection when the commoner would rather sleep peacefully and care for his benefit. These people have worked for a cause greater than them. By saving the environment, they have committed and impacted us, too. The origin of the environmental protection movement in India dates back to the Kehjrali movement and gained momentum through the Chipko movement, Appiko movement, Save Silent Valley movement and Narmada Bachao Andolan. The major trend in the environmental protection movements in India emphasises that environment movements in India reflect the trend that most participants are women, Adivasis, and poor people.

"Women play a central role in sustainable agricultural practices, including seed saving, organic farming, and soil conservation, contributing to food security and environmental sustainability", [2]FAO, 2011. Women and the environment are closely bounded, and the intimate relation between women and nature led to the theory of ecofeminism, a field bridging ecological ethics and feminism that seeks to explore the conceptual connections between environmental degradation and sexist oppression. Through their role as farmers and collectors of water and firewood, women have a close relationship with their local environment. Women, children and marginalised sections are the prime victims of environmental degradation, especially during natural disasters. Thus, women more actively participate in environmental protection than men, as women are more directly affected and influenced by nature than men. Women have been involved in several governmental and nongovernmental forestry and environment programs.

Chipko movement / Vriksha Andolan / Vriksha Mitra
Community forestry programs
Social forestry programs
Individual conservation programs
SHG conservation programs
Green-Belt movement
Keep the city clean programs
Green India Clean India programs

Women, especially in rural regions, are concerned with household activities like collecting food, water, grass and fuel, which enrich their wisdom of the environment, thus promoting them to execute the appropriate conservation procedures and technologies.

ENVIRONMENTAL MOVEMENTS LED BY INDIAN WOMEN

Bishnoi's First Environmentalists of India

The Bishnoi community was one of the first environmentalist communities in India. The Bishnois follow ecological principles and have a long history of practising environmental conservation. Here is some information about the Bishnois and their contributions:

The Bishnoi community was founded by Guru Jambheshwar, also known as Jambhoji, in the 15th century in the western state of Rajasthan, India. The word "Bishnoi" is derived from "bis" (twenty) and "noi" (nine), which refer to the 29 principles or rules laid down by Guru Jambheshwar for the community's followers. The core tenet of the Bishnoi faith is to protect and conserve the environment. The community practices strict environmental and wildlife conservation measures, even predating modern ecological movements. Their principles include reverence for all forms of life, prohibition of cutting green trees, and the protection of animals and birds.

The Bishnois are well-known for their efforts to preserve the blackbuck antelope, a species found in the region. In 1730, a group of Bishnois sacrificed their lives while protecting a group of blackbucks from being hunted by the then Maharaja of Jodhpur. This event, known as the Khejarli massacre, is a significant milestone in India's environmental conservation history. Besides their conservation efforts, the Bishnoi community promotes sustainable agriculture and practices organic farming methods. They emphasise the importance of water conservation, afforestation, and protection of natural resources.

The Bishnois' commitment to environmental preservation has inspired many individuals and communities in India and beyond. Their values and principles align closely with the modern-day environmental movements, highlighting the significance of their early contributions as environmentalists in India. It is important to note that while the Bishnoi community has played a crucial role in environmental conservation, they are not the only environmentalists in India. Various individuals, organisations, and communities have historically contributed to ecological preservation.

Chipko Movement

The Chipko movement, originating in the early 1970s in India, is a renowned environmental movement that gained international recognition for its non-violent protests and dedication to forest conservation. The term "Chipko" translates to "hug" or "cling," symbolising the activists' practice of embracing trees to prevent them from being felled by loggers. This movement emerged in response to rampant deforestation and the exploitation of natural resources in the Himalayan region of Uttarakhand, then known as Uttar Pradesh.

The Chipko movement was born out of the efforts of local communities, particularly women, who were directly affected by the degradation of their natural surroundings. They witnessed the devastating consequences of deforestation, including soil erosion, water scarcity, and a decline in biodiversity, severely impacting their livelihoods. Led by visionary environmentalists such as Sunderlal Bahuguna, the movement quickly gained momentum, capturing people's attention across India and the world.

What set the Chipko movement apart was its peaceful and inclusive approach. Women at the forefront of the protests formed human chains around the trees, refusing to allow them to be cut down. Their hugging the trees showcased their deep reverence for nature and their determination to protect it. The movement was not limited to physical demonstrations; it also involved advocacy, raising awareness, and influencing policy decisions.

The Chipko movement successfully brought the environmental issues faced by the Himalayan communities into the national spotlight. It compelled the government to acknowledge the importance of ecological conservation and prompted the formulation of more sustainable forest policies. One significant outcome of the movement was the introduction of a 15-year ban on tree felling in the region, which aimed to preserve the delicate ecosystem and ensure the sustainable use of forest resources.

Moreover, the Chipko movement inspired similar grassroots movements in other parts of India and worldwide. Its legacy can be seen in subsequent environmental campaigns focused on dam construction, mining activities, and land encroachment. The movement highlighted the power of collective action and demonstrated that ordinary citizens can bring about significant change through peaceful means.

Appiko Movement (1983)

The Appiko movement, the Uttara Kannada movement, emerged in India in 1983 as a significant environmental movement focused on forest conservation and sustainable development. The movement occurred in the Uttara Kannada district of Karnataka, a region renowned for its rich biodiversity and lush forests. The Appiko movement responded to the indiscriminate logging and deforestation by timber companies, causing irreparable damage to the ecosystem and threatening the livelihoods of local communities dependent on the forests. The term "Appiko" translates to "embrace" in the Kannada language, symbolising the activists' commitment to hugging the trees to prevent their felling, similar to the Chipko movement.

The movement was led mainly by local villagers, youth, and activists, with Pandurang Hegde playing a pivotal role as one of its key leaders. The participants of the Appiko movement engaged in peaceful protests, including tree-hugging, sit-ins, and demonstrations, to draw attention to environmental degradation and demand immediate action to protect the forests. The Appiko movement adopted a multifaceted approach, combining grassroots mobilisation, legal activism, scientific research, and public awareness campaigns. The activists aimed to address the ecological crisis while also advocating for the rights of local communities and promoting sustainable forest management practices.

The movement raised widespread awareness about the importance of forest conservation and the detrimental effects of unchecked logging. It brought the issue to the forefront of public discourse, engaging various stakeholders, including government authorities, environmental organisations, and the general public. The movement's efforts were instrumental in pressuring the government to take action and formulate more stringent regulations for forest conservation and sustainable development.

As a result of the Appiko movement, the government of Karnataka implemented significant policy changes. It imposed stricter controls on logging activities, banned the export of raw timber, and initiated measures to involve local communities in decision-making processes concerning the management of forests. The movement's impact extended beyond the Uttara Kannada district, inspiring similar environmental campaigns nationwide and highlighting the importance of community-led initiatives for sustainable development. The Appiko movement stands as a testament to the power of grassroots activities and the ability of individuals and communities to effect change. It showcased the value of non-violent protests and the potential for local activism to influence policy decisions and drive environmental conservation efforts.

Today, the legacy of the Appiko movement continues to inspire environmental activists and conservationists globally. Its emphasis on sustainable development, community participation, and the protection of natural resources resonates in the face of ongoing environmental challenges. The movement's impact on policy formulation and public consciousness serves as a reminder of the significance of preserving our forests and the urgent need for sustainable practices to ensure the well-being of both people and the planet.

Silent Valley Movement (1976)

The Silent Valley movement, which took place in 1976, was a significant environmental movement in India aimed at preserving the Silent Valley region in the southwestern state of Kerala. The campaign sought to protect the rich biodiversity and unique ecosystem of the Silent Valley from a proposed hydroelectric project that would have led to massive deforestation and ecological destruction. The Silent Valley, located in the Palakkad district of Kerala, is a pristine tropical rainforest known for its lush vegetation, rare flora and fauna, and the Silent Valley National Park. The proposed hydroelectric project, called the Silent Valley Hydroelectric Project, intended to construct a dam across the Kuntipuzha River within the Silent Valley, which would have resulted in flooding a significant portion of the forest.

The movement was spearheaded by a coalition of environmentalists, scientists, activists, and concerned citizens who recognised the ecological importance of the Silent Valley and its irreplaceable value to the region. Their primary objective was to halt dam construction and protect the forest from irreversible damage. The Silent Valley movement gained momentum through public awareness campaigns, peaceful protests, and advocacy efforts. Activists highlighted the ecological significance of Silent Valley, emphasising the need to preserve its unique biodiversity, which included several endangered species of plants, animals, and birds. They raised concerns about the potential adverse impacts of the dam on the ecosystem, water resources, and the livelihoods of local communities.

The movement garnered substantial support from various quarters, including artists, writers, scientists, and political leaders. Notable figures such as poet and environmentalist Sugathakumari played a prominent role in mobilising public sentiment and rallying support for the cause. The Silent Valley movement's efforts successfully drew national and international attention to the issue. It generated widespread public outcry against the proposed dam and pressured the government to reconsider the project. Scientists and experts provided evidence-based arguments, highlighting the ecological fragility of the region and the importance of preserving it.

Ultimately, the Silent Valley movement achieved a significant victory when the government of Kerala abandoned the dam's construction in 1983. The Silent Valley National Park was subsequently established in 1984, ensuring the long-term protection of the forest and its invaluable biodiversity. The Silent Valley movement is a testament to the power of public mobilisation and environmental activism in safeguarding natural habitats. It showcased the importance of grassroots actions, scientific evidence, and public awareness in influencing policy decisions related to environmental conservation.

Today, the Silent Valley National Park is recognised as one of India's most ecologically significant regions. It is a crucial habitat for rare and endangered species, attracting nature enthusiasts, researchers, and tourists worldwide. The Silent Valley movement's impact reverberates beyond its immediate success, inspiring subsequent environmental activities in India and highlighting the importance of preserving ecologically sensitive areas. It serves as a reminder of the ongoing need for collective efforts to protect and conserve our natural heritage for future generations.

Narmada Bachao Andolan (1985)

The Narmada Bachao Andolan (NBA), which translates to the "Save Narmada Movement," is a prominent social and environmental movement that began in 1985 in India. The campaign aimed to address the issues arising from the construction of large dams in Gujarat and Madhya Pradesh named the Sardar Sarovar Dam. The Narmada River, one of the longest rivers in India, is considered sacred and is home to numerous tribal communities, forests, and diverse ecosystems. The proposed dam projects, including the Sardar Sarovar Dam, sought to harness the river's water resources for irrigation, hydropower generation, and water supply. Led by activist Medha Patkar and supported by various organisations and affected communities, the Narmada Bachao Andolan aimed to highlight the negative impacts of the dams on the environment, ecology, and displaced communities. The movement focused on advocating for the rights of the local tribal communities, farmers, and fisherfolk who would be adversely affected by the dam construction.

The Narmada Bachao Andolan raised concerns about forced displacement, loss of fertile agricultural land, disruption of livelihoods, and ecological damage caused by the dam projects. The activists argued that the proposed dams would guide the expulsion of thousands of families, primarily from marginalised and tribal communities, without adequate rehabilitation and compensation. The movement employed various non-violent protest strategies, including rallies, hunger strikes, sit-ins, and public awareness campaigns, to draw attention to the issues and push for a reassessment of the dam projects. NBA activists organised mass mobilisations and engaged in dialogue with government authorities, demanding a comprehensive rehabilitation policy for the affected communities.

The Narmada Bachao Andolan faced significant challenges and witnessed several legal battles. The movement's activists and affected communities filed numerous court petitions to halt or reassess the dam projects. The legal proceedings led to several landmark judgments and court orders, acknowledging the need for comprehensive rehabilitation measures and environmental assessments. The Narmada Bachao Andolan is recognised for its significant role in advocating for the rights of displaced communities and highlighting the need for sustainable development and comprehensive rehabilitation policies. It inspired similar movements in India and globally, raising awareness about large infrastructure projects' social and environmental costs.

The legacy of the Narmada Bachao Andolan continues to shape debates and policies concerning dam construction, indigenous rights, and environmental justice in India. It emphasises balancing development with social and ecological considerations and the need for inclusive decision-making processes considering marginalised communities' concerns and voices.

LITERATURE REVIEW

Hannam (1999), This paper examines the impact of various social, economic, and political changes on the work of the Indian Forest Service, the elite organisation concerned with environmental management in India. This paper pursues to correct this by concentrating on three key challenges:

Due to economic liberalisation, the work the forest officer is expected to undertake has changed beyond recognition.
Political manipulation and corresponding accusations of corruption have increased.
Social changes and their impact on recruitment into the Indian Forest Service.

It is concluded that, despite the many challenges the Indian Forest Service has faced over the years, it is remarkably resilient and remains the most potent agent in rural India.

Khatibi and Indira (2011). The present paper intends to study the experiences of some NGOs in the Mysore district of Karnataka State. This study shows an affinity between environmental management and women's empowerment in ecological activities and how women can become voices for advancing environmental sustainability.

Asteria et al. (2016), This study is about the experience of women's role in environmental management to raise environmental security and form of women's emancipation movement. Ecological concerns are conducted by residents of urban women who become environmental activists based on environmental literacy. Pro-environment activities undertaken as a form of responsibility for environmental concern, such as eco-management, educational, and economic action, by persuading residents to proactively and consistently continue environmental management and develop a sense of community in shaping the networks of ecological concern in the local context for global effect (Lohani & Aburaida, 2017).

Research Objectives

To analyse the role of women leaders in shaping policies in Indian environmental management.
To evaluate the significant challenges and contributions of women leaders in the environmental management of India

RESEARCH METHODOLOGY

Research methodology is the strategies or techniques to specify, select, procedure, and research information about a topic. Also, it is the science of systematically studying how research is done (Pandey & Pandey, 2015). This study is entirely based on qualitative data. Data have been collected from the secondary source. The present study mainly considers secondary data for analysing the influence of Indian women leaders in environmental activities. Existing data is summarised and collected with accurate data. The secondary data were collected from various publications of economic reviews, management journals, Indian management studies, websites, etc.

RESULTS AND ANALYSIS

Women's empowerment in environmental activities in India has been a significant focus, aiming to promote gender equality, enhance women's participation, and amplify their impact on sustainability and conservation efforts. Several initiatives and strategies are implemented to empower women in environmental activities, fostering their leadership and ensuring their active involvement. Here are some critical aspects of women's empowerment in this context:

Skill Development and Capacity Building: Programs and workshops were organised to enhance the technical skills and knowledge of women in areas such as sustainable agriculture, waste management, renewable energy, and biodiversity conservation. These initiatives equip women with the necessary expertise to engage in environmental activities actively.

Entrepreneurship and Livelihood Opportunities: Efforts are made to create entrepreneurship and livelihood opportunities for women in the environmental sector. Women are encouraged to start sustainable businesses, such as organic farming, eco-tourism, handicrafts from recycled materials, and sustainable fashion, enabling them to become financially independent and contribute to environmental conservation.

Leadership and Decision-making Roles: Women are encouraged to take up leadership roles and participate in decision-making processes related to environmental management. Their representation in government bodies, community organisations, and ecological committees is promoted to ensure their voices are heard, perspectives are considered, and their expertise is utilised in shaping environmental policies and practices.

Awareness and Education: Environmental education and awareness programs are specifically designed for women, focusing on their rights, roles, and responsibilities in environmental conservation. These programs aim to raise awareness about environmental issues, sustainable practices, and the interlinkages between gender equality and ecological sustainability.

Access to Resources and Funding: Efforts are made to ensure women have equal access to resources, including land, credit facilities, and technology, to engage in environmental activities. Special funds and grants are provided to support women-led ecological initiatives and projects, enabling them to implement sustainable practices and conservation efforts effectively.

Networking and Collaboration: Platforms for networking, collaboration, and knowledge sharing are created to connect women involved in environmental activities. Women-led organisations and networks are established to foster mutual support, exchange experiences, and amplify the collective impact of their efforts.

Recognition and Celebrating Women's Contributions: Women's contributions to environmental activities are acknowledged, celebrated, and showcased to inspire others and create role models. Awards and recognition programs are instituted to honour the achievements of women leaders in environmental sustainability and conservation.

Through these empowerment strategies, women in India increasingly become agents of change in environmental activities. Their active participation, leadership, and contributions are crucial in promoting sustainable development, addressing environmental challenges, and fostering a more inclusive and equitable approach to environmental management.

CONCLUSION

Indian women have demonstrated a solid commitment to environmental management and sustainability. Their contributions to ecological movements, conservation efforts, and advocacy have raised awareness, protected natural resources, and promoted sustainable practices. "Women's traditional knowledge and practices have been vital in biodiversity conservation, particularly in preserving indigenous plant varieties and herbal medicines" (Howard, 2003). Women have been at the forefront of various environmental activities, such as the Chipko movement, the Narmada Bachao Andolan, and initiatives led by organisations like Navdanya. They have actively participated in protests, awareness campaigns, and legal battles to protect forests, rivers, and communities from the adverse impacts of development projects.

Furthermore, women have played a vital role in sustainable agriculture and food systems, emphasising organic farming practices, seed sovereignty, and biodiversity conservation. Their engagement in self-help groups and community-based initiatives has promoted sustainable livelihoods, waste management, and water conservation in rural areas. Women have also been critical advocates for climate change action, raising awareness about its impacts and advocating for resilience-building measures. Their involvement in grassroots movements, organisations, and conferences has fostered a greater understanding of climate change issues and the need for sustainable solutions.

The environmental management of Indian women is not limited to specific movements or initiatives. It permeates various aspects of their lives, from everyday practices in households to their roles as educators, community leaders, and policymakers. Their concerns for nature encompass conservation, sustainable development, community resilience, and social justice .

Recognising and including women's perspectives and expertise in environmental decision-making processes is crucial for effective and holistic environmental management in India. Their deep connections with nature, traditional knowledge, and caregiving roles make them valuable stakeholders in shaping sustainable policies and practices.

Moving forward, it is imperative to continue supporting and empowering women in environmental management. By promoting gender equality, ensuring representation in decision-making processes, and providing access to education and resources, Indian women can continue to drive positive change and contribute to a more sustainable and resilient future for the country's environment and communities.

RECOMMENDATIONS

Indeed, here are some recommendations for improving Indian environmental management with a focus on the contributions of women leaders:

Encourage government agencies, NGOs, and environmental organisations to adopt gender-inclusive policies that ensure the active participation of women in environmental decision-making processes. This can be achieved by establishing gender-sensitive committees, task forces, and forums to address environmental issues.

Invest in programs that empower women to take on leadership roles in environmental management. This includes training, mentoring, and networking opportunities for women interested in environmental science, policy, and advocacy careers.

Provide financial and logistical support to women-led conservation and sustainability projects. These initiatives can range from community-based resource management to sustainable agriculture and biodiversity conservation efforts.

Promote awareness campaigns that highlight the importance of gender equality in environmental management. Educate the public, policymakers, and the media on how women's involvement can lead to more effective and sustainable environmental solutions.

Invest in research and data collection efforts to better understand the role of women in environmental management. This includes studying the impact of women's leadership on conservation outcomes and identifying barriers they face.

Promote science, technology, engineering, and mathematics (STEM) education among girls and women. Support scholarships and mentorship programs to increase the representation of women in environmental science and research.

Recognise the critical role of women in climate change mitigation and adaptation efforts. Ensure that climate policies and initiatives include gender-responsive strategies and solutions.

These recommendations aim to harness the potential of women leaders in Indian environmental management, ultimately leading to more effective, equitable, and sustainable approaches to addressing environmental challenges.

REFERENCES

Agarwal, B. (2001). Participatory Exclusions, Community Forestry, and Gender: An Analysis for South Asia and a Conceptual Framework. *World Development*, *29*(10), 1623–1648. doi:10.1016/S0305-750X(01)00066-3

Agrawal, A., & Gibson, C. C. (1999). Enchantment and Disenchantment: The Role of Community in Natural Resource Conservation. *World Development*, *27*(4), 629–649. doi:10.1016/S0305-750X(98)00161-2

Asteria, D., Herdiansyah, H., & Apriana, I. W. A. (2016). Women's Environmental Literacy as Social Capital in Environmental Management for Environmental Security of Urban Area. In *IOP Conference Series: Earth and Environmental Science*. IOP Publishing. 10.1088/1755-1315/30/1/012014

Fraser, E. D., Dougill, A. J., Mabee, W. E., Reed, M., & McAlpine, P. (2006). Bottom-up and top-down: Analysis of participatory processes for sustainability indicator identification as a pathway to community empowerment and sustainable environmental management. *Journal of Environmental Management*, *78*(2), 114–127. doi:10.1016/j.jenvman.2005.04.009 PMID:16095806

Hannam, K. (1999). Environmental management in India: Recent challenges to the Indian Forest Service. *Journal of Environmental Planning and Management*, *42*(2), 221–233. doi:10.1080/09640569911226

Howard, P. L. (2003). *Women and Plants: Gender Relations in Biodiversity Management and Conservation*. Zed Books.

Khatibi, F. S., & Indira, M. (2011). Empowerment of women through self-help groups and environmental management: Experiences of NGOs in Karnataka State, India. *Journal of Human Ecology (Delhi, India)*, *34*(1), 29–40. doi:10.1080/09709274.2011.11906366

Lohani, M., & Aburaida, L. (2017). Women empowerment: A key to sustainable development. *The Social ION*, *6*(2), 26–29. doi:10.5958/2456-7523.2017.00006.4

Marshall, G. (2012). *Economics for collaborative environmental management: renegotiating the commons*. Routledge. doi:10.4324/9781849771047

Mitchell, B. (2005). Participatory partnerships: Engaging and empowering to enhance environmental management and quality of life? *Social Indicators Research*, *71*(1-3), 123–144. doi:10.100711205-004-8016-0

Pandey, P., & Pandey, D. M. (2015). *Research Methodology: Tools and Techniques*. Bridge Center.

UNEP. (2020). *Gender Equality and the Environment: Policy and Strategy*. UNEP. https://wedocs.unep.org/bitstream/handle/20.500.11822/7655/-Gender_equality_and_the_environment_Policy_and_strategy-2015Gender_equality_and_the_environment_policy_and_strategy.pdf

Chapter 5
Breaking Barriers and Fostering Leadership:
Empowering Women in Environmental Management

Ann Baby
Rajagiri College of Social Sciences (Autonomous), Kalamassery, India

Binoy Joseph
Rajagiri College of Social Sciences (Autonomous), Kalamassery, India

ABSTRACT

Environmental management has emerged as a critical field of study and practice, aiming to address the challenges posed by environmental degradation and climate change. Women have played an increasingly prominent role in this domain, contributing significantly to sustainable development and environmental conservation. This research explores the opportunities and challenges faced by women leaders in environmental management, aiming to highlight their contributions, identify barriers, and proposes strategies to promote gender equality within the field. The study draws upon academic literature, case studies, and expert opinions to present a comprehensive analysis of the topic.

INTRODUCION

Environmental management has become a pivotal aspect of our global efforts to combat environmental degradation and address the challenges posed by climate change. Women's involvement and leadership in this field have increased significantly over the years, highlighting their dedication and expertise in sustainable development and environmental conservation (Abdulrahman et al., 2020). However, despite their growing presence, women leaders in environmental management still face unique opportunities and challenges that warrant thorough examination. This research seeks to explore the opportunities and challenges encountered by women in leadership roles within the environmental management domain. By

DOI: 10.4018/978-1-6684-5986-7.ch005

understanding these dynamics, the aim is to promote gender equality, enhance women's contributions, and foster a more inclusive and effective approach to environmental stewardship (Warren et al., 2019).

This research will explore the evolving role of women in environmental management and their significant contributions to sustainable development. It will shed light on the opportunities that have opened up for women in this field, such as education and capacity building programs, networking opportunities and advocacy for gender recognition. Additionally, the study will address the various challenges faced by women leaders, including gender bias, limited access to resources, and difficulties in achieving work-life balance. By examining real-world case studies, the study will provide practical insights into overcoming these barriers.

Furthermore, the study will present strategies aimed at promoting gender equality in environmental management. These strategies will encompass creating inclusive work environments, implementing gender quotas, and encouraging male allies in the pursuit of gender equality. The role of governments and non-governmental organizations (NGOs) in promoting gender equality in the field are explored, along with the potential impacts of greater gender diversity in environmental leadership. Overall, this research will contribute to the ongoing discourse on gender equality in environmental management and provide valuable insights to empower women leaders in the field. It is hoped that by understanding the challenges and leveraging the opportunities, we can work towards a more inclusive and sustainable future for all.

Research Objective

The primary objective of this research is to conduct a comprehensive analysis of the opportunities and challenges faced by women leaders in environmental management. The specific goals are as follows:

To assess the historical involvement of women in environmental management and highlight their significant contributions to sustainable development and conservation efforts.

To identify the existing opportunities available to women leaders in the environmental management field, such as educational programs, skill development initiatives, networking platforms, and recognition programs.

To explore the various challenges and barriers that hinder women's progress and representation in leadership positions within environmental organizations, including gender bias, lack of access to resources, work-life balance issues, and underrepresentation in decision-making roles.

To examine relevant case studies of successful women-led environmental initiatives and instances where gender barriers are effectively overcome, drawing valuable insights from real-world experiences.

To propose strategies and best practices that could be implemented to promote gender equality in environmental management, with a focus on creating inclusive work environments, introducing gender quotas, encouraging female participation in STEM fields, and fostering male allies in the pursuit of gender equality.

To evaluate the role of governments and non-governmental organizations (NGOs) in supporting gender equality initiatives in environmental management, analyzing existing policies, and exploring best practices.

Methodology

The research methodology for this study will encompass a multi-dimensional approach to provide a comprehensive analysis of the topic. The following steps are undertaken:

Literature Review: Extensive literature research are conducted to gather existing scholarly works, case studies, reports, and data relevant to women's participation in environmental management. This will form the foundation of the research, allowing for a thorough understanding of the historical context and the current state of affairs.

Case Studies: A selection of case studies are analyzed, focusing on successful women-led environmental initiatives and examples of overcoming gender barriers. These case studies will illustrate real-world scenarios and offer practical strategies for promoting gender equality.

Data Analysis: The collected data is analysed to derive meaningful conclusions and identify key themes related to opportunities and challenges faced by women leaders in environmental management.

Recommendations: Based on the findings, the study will propose actionable recommendations and strategies to foster gender equality in environmental leadership roles, considering the perspectives of women leaders, organizations, and policymakers.

The research's methodology will ensure a comprehensive and evidence-based approach, shedding light on the experiences of women leaders in environmental management and offering actionable steps towards a more inclusive and sustainable future.

Women in Environmental Management

Women in environmental management have played an increasingly prominent and influential role in addressing environmental challenges and promoting sustainability. Over the years, their contributions have significantly evolved, breaking barriers and shattering stereotypes to advance the cause of environmental conservation and management (Birindelli et al., 2019).

Historical Overview of Women's Involvement

Throughout history, women have been integral to environmental management, even if their roles were often overlooked or marginalized. In traditional societies, women were responsible for gathering food, water, and other natural resources, making them keen observers of their surroundings. They possessed invaluable knowledge about ecosystems, plant identification, and sustainable agricultural practices. However, as modern environmental management structures emerged, women's involvement was often limited due to societal norms and discriminatory practices(Shrestha et al., 2023).

In the 20th century, the women's rights movement facilitated a shift towards greater gender equality, enabling women to play more active roles in environmental management. Female pioneers emerged, advocating for conservation and sustainability, such as Rachel Carson, whose book "Silent Spring" ignited the modern environmental movement. Despite facing barriers, women's involvement in environmental management steadily increased, leading to their prominent positions in various environmental organizations and initiatives today(Ramirez-Santos et al., 2023).

Women's Contributions to Environmental Management

Women leaders in environmental management have made significant contributions to tackling complex environmental challenges. Their multidimensional approach often combines scientific expertise with a holistic understanding of social and community dynamics. Women have shown remarkable resilience, creativity, and empathy in addressing environmental issues that affect vulnerable populations, including women and children.

Many women-led initiatives have focused on sustainable resource management, afforestation, climate change adaptation, and renewable energy adoption. Their emphasis on inclusive and participatory decision-making has strengthened community engagement in environmental protection and sustainable development projects. Women have also played essential roles in conservation efforts, advocating for the protection of biodiversity and endangered species, as well as promoting responsible waste management and pollution control (Bullough et al., 2022).

The Importance of Gender Diversity in Leadership Roles:

Gender diversity in leadership roles within environmental management is of paramount importance for several reasons:

Diverse Perspectives: Women leaders bring unique perspectives and approaches to environmental problem solving, stemming from their distinct experiences, knowledge, and values. Incorporating these diverse viewpoints leads to more comprehensive and innovative solutions. Frameworks have been rooted in recognition theory, highlighting principles like reciprocal understanding, diverse perspectives, trust, and integrity. In addressing barriers to female gender inclusion, four transformative stages namely Awareness-building and reflection, Vision development, Reevaluation of management concepts, and Action-driven implementation through an integrated Human Relations Management system, translating principles into observable behaviors while nurturing and recognizing inclusive actions (Pless & Maak, 2004).

Inclusive Decision-Making: Gender diversity in leadership ensures that environmental policies and initiatives consider the needs and concerns of all stakeholders. Inclusive decision-making processes lead to more equitable and sustainable outcomes. Women's pivotal role in driving positive change for the environment, women's leadership in fostering sustainability and her role in emphasizing the link between gender equality and eco-friendly practices has been studied through case studies and research (Balabantaray, 2023). Such research emphasizes the unique perspectives women offer and advocates for their recognition in environmental decision-making, promoting harmony with nature.

Role Models and Inspiration: Women leaders in environmental management serve as role models for the next generation, inspiring more young women to pursue careers in the field. Increased representation encourages girls to envision themselves as future environmental leaders. Research explores a 3-day design-thinking workshop in Japan, aiming to shift female youths' STEM (Science, Technology, Engineering, and Math) perceptions. Results indicate increased interest in engineering, enhanced creative confidence, and more positive STEM views(Kijima et al., 2021). Short interventions can empower female youths in STEM and revitalize education for greater gender inclusivity, fostering confidence and creativity.

Addressing Environmental Challenges Effectively: Environmental problems affect men and women differently due to social and economic disparities. Gender diversity in leadership enables more targeted and effective responses to these challenges. COVID-19 has amplified gender disparities globally. Women faced higher employment loss, shouldered caregiving responsibilities, and saw increased gender-based violence. Girls dropped out of school more than boys. However, vaccine hesitancy became similar by September 2021. Addressing these disparities demands policies promoting women's participation in the workforce and education(Flor et al., 2022).

Organizational Performance: Studies have shown that gender-diverse teams enhance organizational performance and foster creativity and collaboration. In environmental management, where complex issues demand innovative solutions, diverse leadership teams can lead to improved outcomes. Exploring 10,334 U.S. listed firms, a positive link between board gender diversity (BGD) and environmental innovation has been discovered. This connection is stronger in less profitable companies and eco-sensitive industries. The findings, considering gender theories, emphasize the significance of BGD in fostering environmental progress, offering insights for policymakers(Nadeem et al., 2020).

In conclusion, women have been intrinsic to environmental management throughout history, and their contributions continue to shape the field positively. Emphasizing gender diversity in leadership roles is not only a matter of equity but also crucial for enhancing environmental efforts and achieving sustainable outcomes. Empowering women leaders in environmental management is a collective responsibility that benefits society, the environment, and future generations (Botella et al., 2019; Brahma et al., 2021; Yarram & Adapa, 2021).

Opportunities for Women Leaders in Environmental Management

Opportunities for women leaders in environmental management are continually growing as society recognizes the importance of gender diversity in addressing global environmental challenges.

Education and Skill Development: Access to quality education and skill development programs plays a critical role in empowering women to become leaders in environmental management. Educational institutions and organizations offer various academic programs and workshops that focus on environmental sciences, conservation, sustainable development, and policy-making. By participating in these programs, women can gain the knowledge and expertise necessary to excel in leadership roles within the field.

Networking and Mentorship: Networking opportunities and mentorship programs are vital for fostering the growth and visibility of women leaders in environmental management. Professional conferences, workshops, and seminars provide platforms for women to connect with like-minded individuals, share experiences, and learn from established leaders. Mentorship programs pair experienced environmental professionals with emerging women leaders, providing guidance, support, and career advice to navigate the challenges they may face.

Capacity Building Programs: Capacity-building programs aim to enhance women's leadership skills, decision-making abilities, and project management capabilities. These programs often focus on developing critical competencies, such as communication, negotiation, conflict resolution, and strategic planning. As women build their leadership capacity, they become better equipped to drive sustainable and impactful environmental initiatives.

Advocacy and Recognition: Environmental organizations and institutions can promote gender equality by actively advocating for women's representation and recognition in leadership positions. This includes publicly acknowledging the contributions of women leaders, highlighting their successes, and rewarding their achievements. Recognizing women's efforts can create positive reinforcement and motivate others to pursue leadership roles in environmental management.

Access to Funding and Resources: Securing funding and accessing resources are vital components of effective environmental management projects. Women leaders often face barriers in accessing financial resources and technical support. Providing equal opportunities for funding and access to resources ensures that women-led initiatives receive the necessary support to implement sustainable projects and programs successfully.

Breaking Gender Stereotypes: Challenging and breaking gender stereotypes within the environmental management field is essential for encouraging more women to take on leadership roles. By highlighting successful women leaders and their achievements, stereotypes that suggest men are better-suited roles can be dismantled, fostering a more inclusive environment.

Inclusive Recruitment Practices: Environmental organizations should adopt inclusive recruitment practices that actively seek out and encourage women to apply for leadership positions. This includes removing unconscious biases from hiring processes and creating diverse selection committees that value diverse perspectives and experiences.

International Opportunities: Global environmental challenges require international cooperation and collaboration. Providing women with opportunities to engage in international environmental projects, conferences, and research collaborations fosters cross-cultural exchange and enhances their leadership skills on a global scale.

Policy and Advocacy: Women leaders can actively engage in policy development and advocacy efforts related to environmental management. By participating in decision-making processes and advocating for gender-responsive policies, women can influence environmental agendas at regional, national, and international levels.

Work-Life Balance Support: Supportive work environments that consider the work-life balance of women leaders are crucial for their career advancement and well-being. Flexible working arrangements, family-friendly policies, and adequate parental leave can help retain and support talented women in leadership roles.

By embracing these opportunities and implementing supportive measures, environmental organizations can create an inclusive and gender-diverse leadership landscape that harnesses the full potential of women leaders in environmental management. This will result in more effective and sustainable solutions to the pressing environmental challenges of our time (Akinsemolu & Olukoya, 2020; Botella et al., 2019; Mumba, 2009; Pierli et al., 2022; Robert & Brown, 2004).

Challenges Faced by Women Leaders in Environmental Management

Women leaders in environmental management encounter various challenges that hinder their full participation and impact within the field. Some of the key challenges faced by women in leadership roles include:

Gender Bias and Stereotypes

Deep-rooted gender biases and stereotypes can undermine the credibility and capabilities of women leaders in environmental management. Women may face skepticism about their scientific knowledge, decision-making abilities, and suitability for leadership positions. These biases can lead to unfair evaluations, limited opportunities for advancement, and reduced access to resources.

Lack of Access to Resources

Women leaders often encounter obstacles in accessing financial resources, technology, and research facilities. Inadequate funding and limited access to necessary resources can hinder the implementation of impactful environmental projects and initiatives, thereby affecting their ability to make significant contributions to the field.

Work-Life Balance

Balancing professional responsibilities with personal and family commitments remains a significant challenge for women leaders in environmental management, as it does in many other professions. The demanding nature of environmental work, which often involves fieldwork and extended hours, can create difficulties in maintaining a healthy work-life balance.

Underrepresentation in Decision-Making Positions

Women continue to be underrepresented in decision-making positions within environmental organizations and government agencies. The lack of representation diminishes their influence on policy formulation and resource allocation, perpetuating a gender imbalance in the sector.

Networking and Mentorship Gaps

Limited access to networks and mentorship opportunities can hinder women's professional growth and career progression. Networking and mentorship play pivotal roles in providing guidance, support, and exposure to new opportunities, all of which are essential for advancing in leadership positions.

Implicit Bias and Discrimination

Implicit bias and discrimination, often unintentional and unconscious, can affect how women leaders are perceived and treated in the workplace. This can manifest in the form of unequal pay, exclusion from decision-making discussions, and less access to training and development opportunities.

Stereotyping of "Women's Issues"

Environmental issues that disproportionately affect women, such as access to water, energy, and healthcare, may be perceived as secondary or less critical compared to other environmental challenges. As a result, women leaders advocating for such issues may face difficulty gaining support and recognition for their efforts.

Lack of Role Models

A scarcity of visible female role models in environmental leadership can discourage aspiring women from pursuing careers in the field. The absence of relatable and successful women leaders may lead to self-doubt and reduced confidence among women seeking to advance in their environmental management careers.

Family and Societal Pressures

Cultural norms and societal expectations can exert pressure on women leaders to prioritize family responsibilities over their professional pursuits. This can create internal conflicts and external challenges that hinder their career advancement and commitment to environmental leadership roles.

Glass Ceiling Effect

The existence of an invisible barrier, known as the glass ceiling, can impede women's upward mobility into top leadership positions within environmental organizations. This phenomenon often prevents women from reaching the highest echelons of decision-making, limiting their influence on shaping environmental policies and strategies.

Addressing these challenges requires collective efforts from environmental organizations, governments, and societies as a whole.

Implementing gender-inclusive policies, fostering supportive work environments, promoting mentorship programs, and challenging gender stereotypes are crucial steps toward overcoming these obstacles and empowering women leaders in environmental management. By promoting gender equality, we can unlock the full potential of diverse leadership and enhance the effectiveness of environmental conservation and sustainability efforts (Akinsemolu & Olukoya, 2020; Bhakta et al., 2020; Burkhardt et al., 2020; Robert & Brown, 2004).

Gender equality's potential to promote state environmentalism has been overlooked in prior environmental politics research. To bridge this gap, the study examines the link between female parliamentary representation and environmental treaty ratification across many countries. Results reveal that nations with more women in Parliament tend to ratify environmental treaties more often. This underscores the need to integrate gender analysis into environmental politics, aligning with feminist theories that connect nature and women's exploitation (Norgaard & York, 2005).

Case Studies

This section discusses about a few case studies, its opportunities and challenges faced by women leaders who are active in environmental management. Chad, Kenya, Brazil and India are the countries taken into consideration for the case studies as they differ significantly in terms of geography, culture, economy, and many other aspects. However, one common factor among them is that two of them are located on the African continent (Chad and Kenya are in Africa), the South American continent (Brazil is in South America) and the Asian continent (India is in Asia). They each have unique historical, social, and economic contexts that shape their individual characteristics and challenges.

Case Study One: Dr. Wangari Maathai - Green Belt Movement (Kenya)

Dr. Wangari Maathai was a Kenyan environmental activist and founder of the Green Belt Movement. In the late 1970s, she observed the increasing degradation of Kenya's environment due to deforestation and soil erosion, which also had severe impacts on rural communities, particularly women. In response, she initiated the Green Belt Movement in 1977, which focused on tree planting, environmental conservation, and community empowerment (Shattuck, 2022).

Opportunity: Dr. Maathai recognized the potential of women to drive positive change and contribute to environmental conservation. She saw tree planting as a simple yet effective way for women to address deforestation while also gaining economic benefits from the sale of tree products.

Achievements: Under Dr. Maathai's leadership, the Green Belt Movement empowered thousands of women to plant over 30 million trees across Kenya. Through this initiative, women gained economic independence, improved access to resources, and participated actively in decision-making processes related to environmental management. The movement's success also contributed to Dr. Maathai becoming the first African woman to receive the Nobel Peace Prize in 2004.

Case Study Two: Marina Silva - Amazon Rainforest Conservation (Brazil)

Marina Silva is a Brazilian environmentalist and politician known for her advocacy for the protection of the Amazon rainforest. She served as the Minister of Environment in Brazil from 2003 to 2008 and played a pivotal role in implementing policies to combat deforestation and promote sustainable development (Silva, 2019).

Opportunity: As the Minister of Environment, Marina Silva had a unique opportunity to influence Brazil's environmental policies and address the challenges posed by deforestation in the Amazon rainforest.

Achievements: During her tenure as Minister, Marina Silva strengthened enforcement against illegal logging, encouraged sustainable practices in the forest sector, and increased protected areas in the Amazon rainforest. Her efforts contributed to a significant reduction in deforestation rates and enhanced international recognition of the importance of Amazon conservation.

Case Study 3: Hindou Oumarou Ibrahim - Indigenous Environmental Activism (Chad)

Hindou Oumarou Ibrahim is an environmental activist and indigenous rights advocate from Chad. As a member of the Mbororo pastoralist community, she has been actively involved in addressing environmental challenges in the Sahel region, including desertification and climate change (Gore et al., 2020).

Opportunity: Hindou Ibrahim recognized the critical role of indigenous knowledge and practices in addressing environmental challenges and advocating for climate action.

Achievements: She co-founded the Association for Indigenous Women and Peoples of Chad (AFPAT), which focuses on integrating indigenous knowledge with scientific data for sustainable land management. Hindou Ibrahim has also been a prominent advocate for the inclusion of indigenous perspectives in international climate negotiations, emphasizing the importance of recognizing indigenous land rights in climate solutions.

Case Study Four: Sunita Narain - Centre for Science and Environment (India)

Sunita Narain is an Indian environmentalist and activist known for her work in sustainable development, environmental policy advocacy, and promoting eco-friendly practices. She is the Director-General of the Centre for Science and Environment (CSE), a prominent environmental research and advocacy organization in India.

Opportunity: Narain's efforts in promoting sustainable farming practices, Organic Farming Advocacy, Campaign against Pesticides, Rainwater Harvesting, Advocacy for Sustainable Food Systems and Promotion of Indigenous Farming Practices has been substantial.

Achievements: Sunita Narain's contributions are marked by her commitment to bridging the gap between science, policy, and grassroots action. Through her research, writing, and advocacy, she has raised awareness about the importance of sustainable farming practices for ecological resilience, food security, and the overall well-being of communities. Her work also underscores the vital role of women in agriculture and the environment, as they are often central to implementing and adapting sustainable practices at the community level.

These case studies highlight the impactful contributions of women in environmental management. Through their leadership, advocacy, and community engagement, they have driven positive change, promoted sustainable practices, and advocated for the protection of the environment and the well-being of communities. Their achievements demonstrate the potential of women in shaping a more sustainable and equitable future for the planet.

Much like global regions, Kurdistan faces environmental hurdles, spanning climate change to local issues. Kurdish environmental activists, united in NGOs, safeguard the region. This study spotlights emerging environmental activism in Rojhelat/Eastern Kurdistan, delving into its objectives, growth, and socio-political context. Inspired by "subaltern environmentalism," it examines Chya Green Association's evolution, activity, and discourse (Hassaniyan, 2021).

Strategies for Promoting Gender Equality in Environmental Management

Promoting gender equality in environmental management is essential for creating a more inclusive and effective approach to addressing environmental challenges. Here are some strategies to foster gender equality in this field.

Creating Inclusive Work Environments

Environmental organizations and agencies should actively work towards fostering inclusive work environments that value and promote gender equality. This can be achieved by implementing policies that prevent discrimination and bias, ensuring equitable pay, and providing equal opportunities for professional development and career advancement regardless of gender. Leadership should lead by example, championing gender diversity and ensuring diverse representation in decision-making positions.

Establishing Gender Quotas and Affirmative Action

Consider implementing gender quotas or affirmative action policies to increase the representation of women in leadership roles within environmental organizations and governmental bodies. While some may debate the effectiveness of quotas, they can serve as a temporary measure to address gender imbalances and create more opportunities for qualified women leaders. As gender representation improves, the focus can shift towards fostering merit-based selection processes.

Encouraging Female Participation in STEM Fields

Women's representation in Science, Technology, Engineering, and Mathematics (STEM) fields directly impacts their involvement in environmental management. Encouraging and supporting girls and women to pursue STEM education and careers will expand the talent pool of women with relevant expertise in environmental sciences, technology, and policy-making.

Fostering Male Allies in Gender Equality

Promoting gender equality is not solely the responsibility of women. Men in leadership positions can play a crucial role as allies by actively supporting and advocating for women's leadership and gender diversity initiatives. Male leaders can serve as mentors, sponsors, and advocates, contributing to a more inclusive and supportive work culture.

Providing Mentorship and Networking Opportunities

Mentorship programs that pair experienced environmental professionals with aspiring women leaders can be instrumental in nurturing talent and providing guidance on career development. Networking opportunities, such as conferences and workshops, should be designed to facilitate connections and collaborations among women leaders and potential mentors.

Implementing Gender-Responsive Policies and Projects

Environmental projects and policies should be designed with gender responsiveness in mind, considering the differing impacts and needs of men and women. Integrating a gender perspective can enhance the effectiveness and inclusivity of environmental initiatives, leading to more sustainable outcomes.

Advocating for Gender-Responsive Funding

Funding agencies and donors should prioritize gender-responsive projects and initiatives. Providing funding specifically dedicated to women-led environmental projects can increase their visibility and impact. Moreover, funding criteria should consider the gender mainstreaming approach to ensure that projects address gender disparities and contribute to women's empowerment.

Promoting Gender Equality in Environmental Education

Environmental education programs should integrate gender equality principles, highlighting the contributions of women to environmental conservation and emphasizing the importance of diverse perspectives in tackling environmental challenges. Encouraging gender-balanced participation in environmental education will contribute to a more inclusive and sustainable environmental workforce.

Conducting Gender Analysis and Research

Integrating gender analysis in environmental research can lead to better-informed policies and strategies. Research should explore gender-related barriers and opportunities in environmental management, providing evidence to support gender equality initiatives and advocate for women's representation in decision-making.

Encouraging Women's Participation in Environmental Leadership Platforms

Encouraging and supporting women's participation in international environmental leadership platforms, conferences, and negotiations can amplify their voices on global environmental issues. Increased representation of women in international forums contributes to a more inclusive and diverse approach to global environmental governance.

By implementing a combination of these strategies, the environmental management field can work towards creating a more equitable and inclusive environment where women's leadership is recognized, valued, and empowered. Achieving gender equality in environmental management will not only lead to better environmental outcomes but also foster a more sustainable and just society(Birindelli et al., 2019; Botella et al., 2019; Ikram et al., 2019; Pierli et al., 2022; Yarram & Adapa, 2021).

Policy Formulation and Implementation by Governments

Governments play a vital role in promoting gender equality in environmental management through the formulation and implementation of policies. They should adopt and enforce policies that support women's representation and leadership in environmental organizations, public agencies, and decision-making bodies. These policies may include gender quotas, gender mainstreaming guidelines, and initiatives to address workplace discrimination and biases(Abdulrahman et al., 2020; Galizzi et al., 2021; Mensah, 2020; Shrestha et al., 2023).

Gender-Responsive Budgeting

Governments can allocate resources specifically for gender-responsive environmental projects and initiatives. Gender-responsive budgeting ensures that environmental funding considers the differing needs and impacts on men and women, promoting inclusive and equitable resource allocation.

Support for Women's Capacity Building

Governments should invest in training programs, workshops, and scholarships aimed at enhancing women's skills and capacities in environmental sciences, policy development, and leadership. By supporting women's capacity building, governments can increase the pool of qualified women leaders in environmental management.

Awareness Campaigns

Governments can launch public awareness campaigns to challenge gender stereotypes and promote women's contributions to environmental management. These campaigns can highlight the achievements of women leaders in the field and showcase the importance of gender diversity for effective environmental solutions.

Reporting and Monitoring

Governments can establish mechanisms for monitoring and reporting on gender representation and equality in environmental management. Regular gender-disaggregated data collection and reporting will allow for progress assessment and identify areas that require further attention.

Can NGOs Help?

NGOs play a critical role in environmental management by complementing government efforts, filling gaps in resources and expertise, and providing a platform for citizen engagement. Their agility, expertise, and focus on environmental issues make them essential partners in building a more sustainable and resilient planet (Faroque et al., 2022; Gupta et al., 2020; Mccormick, 2010).

Advocacy and Awareness

NGOs can be powerful advocates for gender equality in environmental management. They can raise awareness about gender disparities within the field, promote gender-inclusive policies, and advocate for the equal representation of women in leadership roles.

Providing Mentorship and Support

NGOs can establish mentorship programs to support women leaders and provide guidance in navigating their careers in environmental management. Additionally, NGOs can offer networking opportunities and professional development resources for women to enhance their skills and connections.

Capacity Building and Training

NGOs can organize capacity-building workshops and training programs that focus on leadership development, project management, and environmental sciences for women. By building the capacity of women leaders, NGOs can enhance their contributions to environmental initiatives.

Highlighting Best Practices

NGOs can highlight successful case studies of women-led environmental projects to highlight the impact of gender diversity in leadership roles. By highlighting best practices, NGOs can inspire other organizations to adopt gender-inclusive approaches.

Collaboration and Partnerships

NGOs can collaborate with governmental agencies, international organizations, and other stakeholders to promote gender equality in environmental management. By working together, these entities can pool resources and expertise to drive meaningful change.

Conducting Research and Gender Analysis

NGOs can conduct research and gender analysis on environmental issues to uncover gender-specific challenges and opportunities. Research findings can inform policies and initiatives aimed at fostering gender equality in environmental management.

Gender-Responsive Project Design

NGOs should ensure that their environmental projects and interventions are gender-responsive; taking into account the differing needs and influences men and women. This approach will lead to more effective and inclusive project outcomes.

Overall, both governments and NGOs have critical roles to play in promoting gender equality in environmental management. By working together, they can create an enabling environment that empowers women leaders, fosters diversity, and enhances the sustainability of environmental initiatives.

Future Opportunities for Women Leaders in Environmental Management

Opportunities for women leaders in environmental management appears promising yet complex. As gender equality becomes a more prominent global agenda, there is growing recognition of women's contributions to environmental management. More organizations and governments are committed to promoting gender diversity in leadership positions, providing greater opportunities for women to excel in the field. Advancements in environmental technology and data analytics are creating new avenues for women to make significant contributions. Women leaders with expertise in technology and data-driven decision-making are well positioned to address complex environmental challenges effectively. The growing emphasis on global collaboration and environmental governance offers opportunities for women leaders to participate in international platforms and negotiations. Increased representation of women in global forums can bring diverse perspectives to the table and influence the development of more inclusive and equitable environmental policies. The rise of social entrepreneurship presents opportunities for women leaders to initiate and lead innovative environmental projects. Women entrepreneurs who leverage their skills and passion for environmental conservation can drive impactful initiatives with both social and ecological benefits.

Future Challenges for Women Leaders in Environmental Management

Several trends and developments are shaping the landscape, presenting new possibilities for women's leadership while also requiring proactive efforts to overcome persistent challenges. Although progress has been made, gender bias and stereotypes persist in environmental management. Overcoming these deeply ingrained attitudes will require ongoing efforts to challenge societal norms, raise awareness, and foster inclusive work environments. While technological advancements offer opportunities, they may also present challenges, especially for women who may face barriers in accessing and mastering new technologies. Ensuring equal access to technology and providing training and support are essential. Despite advancements, the gender gap in Science, Technology, Engineering, and Mathematics (STEM) fields remains an obstacle to women's leadership in environmental management. Encouraging more girls and women to pursue STEM education and careers are crucial to bridge this gap. Women leaders in environmental management who face multiple forms of discrimination due to factors such as race, ethnicity, and socio-economic background may experience additional barriers. Ensuring that gender equality initiatives are inclusive and consider intersectional challenges are vital. Sustainable funding for women-led environmental initiatives remains a challenge. Providing consistent and long-term financial support is crucial to ensure the success and continuity of women's leadership projects.

CONCLUSION

The future for women leaders in environmental management is a mix of opportunities and challenges. Advancements in gender equality, technology, and international collaboration offer unprecedented prospects for women to contribute to sustainable environmental solutions on a global scale. However, addressing gender bias, bridging the gender gap in STEM fields, and ensuring sustainable funding for women-led initiatives will require continued dedication and collaboration from governments, NGOs, environmental organizations, and society as a whole. Creating a more inclusive and equitable environmental management sector will not only empower women leaders but also drive more effective and sustainable environmental solutions. By overcoming challenges and leveraging opportunities, women leaders can play a transformative role in shaping a more sustainable and resilient future for our planet.

REFERENCES

Abdulrahman, M., Hussein, A., & Amoush, M. (2020). Female Characteristics and Their New Roles in Leadership. *Journal of Business and Management Sciences*, *8*(2), 38–47. doi:10.12691/jbms-8-2-1

Akinsemolu, A. A., & Olukoya, O. A. P. (2020). The vulnerability of women to climate change in coastal regions of Nigeria: A case of the Ilaje community in Ondo State. *Journal of Cleaner Production, 246.*

Balabantaray, S. R. (2023). *Women's Leadership and Sustainable Environmental Initiatives: A macroscopic investigation from Ecofeminism framework Women's Leadership and Sustainable Environmental Initiatives: A macroscopic investigation from Ecofeminism framework. August.* Research Gate.

Bhakta, H., Raja, K., Shankar, V. R., & Prakash, V. (2020). *Since January 2020 Elsevier has created a COVID-19 resource centre with free information in English and Mandarin on the novel coronavirus COVID- 19. The COVID-19 resource centre is hosted on Elsevier Connect, the company's public news and information* . Research Gate.

Birindelli, G., Iannuzzi, A. P., & Savioli, M. (2019). The impact of women leaders on environmental performance: Evidence on gender diversity in banks. *Corporate Social Responsibility and Environmental Management*, *26*(6), 1485–1499. doi:10.1002/csr.1762

Botella, C., Rueda, S., López-Iñesta, E., & Marzal, P. (2019). Gender diversity in STEM disciplines: A multiple factor problem. *Entropy (Basel, Switzerland), 21*(1), 1–17. doi:10.3390/e21010030 PMID:33266746

Brahma, S., Nwafor, C., & Boateng, A. (2021). Board gender diversity and firm performance: The UK evidence. *International Journal of Finance & Economics, 26*(4), 5704–5719. doi:10.1002/ijfe.2089

Bullough, A., Guelich, U., Manolova, T. S., & Schjoedt, L. (2022). Women's entrepreneurship and culture: Gender role expectations and identities, societal culture, and the entrepreneurial environment. *Small Business Economics, 58*(2), 985–996. doi:10.100711187-020-00429-6

Burkhardt, K., Nguyen, P., & Poincelot, E. (2020). Agents of change: Women in top management and corporate environmental performance. *Corporate Social Responsibility and Environmental Management, 27*(4), 1591–1604. doi:10.1002/csr.1907

Faroque, A. R., Quader, E., Gani, M. O., & Mortazavi, S. (2022). The Role of NGOs in CSR. *Comparative CSR and Sustainability, November*, 109–125.

Flor, L. S., Friedman, J., Spencer, C. N., Cagney, J., Arrieta, A., Herbert, M. E., Stein, C., Mullany, E. C., Hon, J., Patwardhan, V., Barber, R. M., Collins, J. K., Hay, S. I., Lim, S. S., Lozano, R., Mokdad, A. H., Murray, C. J. L., Reiner, R. C. Jr, Sorensen, R. J. D., & Gakidou, E. (2022). Quantifying the effects of the COVID-19 pandemic on gender equality on health, social, and economic indicators: A comprehensive review of data from March, 2020, to September, 2021. *Lancet, 399*(10344), 2381–2397. doi:10.1016/S0140-6736(22)00008-3 PMID:35247311

Galizzi, G., Meliou, E., & Steccolini, I. (2021). Theme: Experiences and challenges with gender budgeting and accounting. Moving towards gender-responsive forms of accountability?: Editorial: Institutionalizing gender budgeting—contemporary experiences and future challenges. *Public Money & Management, 41*(7), 499–501. doi:10.1080/09540962.2021.1971862

Gore, T., Alestig, M., & Ratcliff, A. (2020). Confronting Carbon Inequality. *Oxfam Media Briefing, September*. https://oxfamilibrary.openrepository.com/bitstream/handle/10546/621052/mb-confronting-carbon-inequality-210920-en.pdf

Gupta, D., Lele, S., & Sahu, G. (2020). Promoting a responsive state: The role of NGOs in decentralized forest governance in India. *Forest Policy and Economics, 111*(January), 102066. doi:10.1016/j.forpol.2019.102066

Hassaniyan, A. (2021). The environmentalism of the subalterns: A case study of environmental activism in Eastern Kurdistan/Rojhelat. *Local Environment, 26*(8), 930–947. doi:10.1080/13549839.2021.1933927

Ikram, M., Zhou, P., Shah, S. A. A., & Liu, G. Q. (2019). Do environmental management systems help improve corporate sustainable development? Evidence from manufacturing companies in Pakistan. *Journal of Cleaner Production, 226*, 628–641. doi:10.1016/j.jclepro.2019.03.265

Kijima, R., Yang-yoshihara, M., & Maekawa, M. S. (2021). Using-design-thinking-to-cultivate-the-next-generation-of-female-STEAM-thinkersInternational-Journal-of-STEM-Education.pdf. *International Journal of STEM Education, 6*. https://stemeducationjournal.springeropen.com/articles/10.1186/s40594-021-00271-6

Mccormick, J. (2010). *International Regimes*. Research Gate.

Mensah, I. K. (2020). Impact of Government Capacity and E-Government Performance on the Adoption of E-Government Services. *International Journal of Public Administration*, *43*(4), 303–311. doi:10.1080/01900692.2019.1628059

Nadeem, M., Bahadar, S., Gull, A. A., & Iqbal, U. (2020). Are women eco-friendly? Board gender diversity and environmental innovation. *Business Strategy and the Environment*, *29*(8), 3146–3161. doi:10.1002/bse.2563

Norgaard, K., & York, R. (2005). Gender equality and state environmentalists. *Gender & Society*, *19*(4), 506–522. doi:10.1177/0891243204273612

Pierli, G., Murmura, F., & Palazzi, F. (2022). Women and Leadership: How Do Women Leaders Contribute to Companies' Sustainable Choices? *Frontiers in Sustainability*, *3*(July), 1–10. doi:10.3389/frsus.2022.930116

Pless, N. M., & Maak, T. (2004). Building an inclusive diversity culture: Principles, processes and practice. *Journal of Business Ethics*, *54*(2), 129–147. doi:10.100710551-004-9465-8

Ramirez-Santos, A. G., Ravera, F., Rivera-Ferre, M. G., & Calvet-Nogués, M. (2023). Gendered traditional agroecological knowledge in agri-food systems: A systematic review. *Journal of Ethnobiology and Ethnomedicine*, *19*(1), 1–19. doi:10.118613002-023-00576-6 PMID:37024903

Robert, B., & Brown, E. B. (2004).Covariance Structure Analysis of Health-Related Indicators in Elderly People Living at Home with a Focus on Subjective Health Perception, (1).

Shattuck, A. (2022). One Person, One Tree: Wangari Maathai and Ecofeminism in the Greenbelt Movement. *Women Leading Change: Case Studies in Women, Gender, and Feminism*, *6*(2), 21–35.

Shrestha, R. K., L'Espoir Decosta, J. N. P., & Shrestha, R. (2023). Creating value networks through a learning society for Indigenous women tourism entrepreneurs. *International Journal of Contemporary Hospitality Management*. doi:10.1108/IJCHM-03-2022-0317

Silva, M. (2019). The Hollow Consensus on the Amazon. *Jordan Times*.

Warren, M. A., Donaldson, S. I., Lee, J. Y., & Donaldson, S. I. (2019). Reinvigorating Research on Gender in the Workplace Using a Positive Work and Organizations Perspective. *International Journal of Management Reviews*, *21*(4), 498–518. doi:10.1111/ijmr.12206

Yarram, S. R., & Adapa, S. (2021). Board gender diversity and corporate social responsibility: Is there a case for critical mass? *Journal of Cleaner Production*, *278*, 123319. doi:10.1016/j.jclepro.2020.123319

Chapter 6
Opportunities and Challenges for Women Leaders in Environmental Management

Swarnadeep Maity
https://orcid.org/0000-0001-5352-4447
NSB Academy, India

ABSTRACT

Women leaders in environmental management face unique opportunities and challenges in both developed and developing countries. In developed countries, opportunities include progressive policies, well-established institutions, and greater access to resources. Women can access quality education and training programs, legal frameworks promoting gender equality, and professional networks. However, in developed countries, women face challenges such as deep-rooted gender biases, stereotypes, and the 'glass ceiling' phenomenon, which limits women's representation in top decision-making positions. In developing countries, women leaders can engage with local communities and integrate traditional knowledge and sustainable practices into environmental management strategies. Limited resources and infrastructure in developing countries foster innovation and creativity, allowing them to develop contextualized solutions that address environmental challenges efficiently. However, challenges in developing countries include traditional gender roles, cultural norms, and resource constraints.

INTRODUCTION

Women's vital role in environmental management has earned widespread acknowledgment in recent decades. Women have emerged as essential stakeholders, leaders, and change agents in the pursuit of sustainable development and environmental preservation as the global community grapples with the rising difficulties posed by climate change, resource depletion, and ecological degradation. Despite their growing involvement and achievements, women's presence in leadership roles in environmental management remains disproportionately low, impeding their full potential and achieving transformative effects. The purpose of this chapter is to look at the opportunities and challenges that women confront

DOI: 10.4018/978-1-6684-5986-7.ch006

as environmental leaders. This chapter dives into the many factors that impact their experiences, covering concerns such as gender bias, cultural barriers, institutional constraints, and societal expectations, which frequently restrict women's professional advancement and recognition as prominent environmental leaders. Furthermore, this chapter hopes to shed light on the beneficial effects that gender diversity and inclusivity may have on environmental decision-making processes and generate innovative solutions to complicated ecological problems and the future of women leaders in environmental management. To provide a complete knowledge of the current condition of women's leadership in environmental management, the chapter focuses on a comprehensive analysis of existing literature, case studies, and expert insights. We hope to highlight the untapped potential of women leaders in encouraging sustainable development, conservation, and climate resilience by exploring success stories and best practises from many locations and industries.

LITERATURE REVIEW

In recent years, there has been a growing interest in the role of women in leadership roles in the field of environmental management. As public awareness of climate change, sustainable development, and environmental preservation has grown, so has the demand for strong leadership in these areas. However, women continue to be underrepresented in leadership roles in environmental management when compared to their male colleagues. The purpose of this literature review is to investigate the opportunities and challenges that women leaders confront in the context of environmental management.

Jackson (1993) contends that women's environmental interactions cannot be understood apart from men's, and that community service and natural resource defence by women have political, ideological, social, and economic causes. The essay also emphasises that women's environmental knowledge is a social product, and that gendered analysis of property relations is required to understand men and women's distinct incentives for conservation. Finally, it argues that gendered environmental decision making should be viewed through the lens of both cooperative and conflicting intrahousehold connections.

Kimball et al. (2012) contributed to the theory on the impact of leadership characteristics on corporate behaviour by investigating the effects of female leadership on corporate environmental performance in general. The authors discovered that organisations that include women in their top management team and board of directors have stronger environmental performance, with the board having a greater impact. Furthermore, adding women to a company's senior management has an impact on its environmental performance only if the company also has women on its board of directors. In particular, the authors explore how these findings contribute to studies of upper-level management and female leadership in major corporations. According to the article, gender diversity has positive effects on company environmental performance when it is present in leadership positions.

Mensah et al. (2014) found female traditional leaders in community planning and development has been a topic of interest in recent years. Several studies had highlighted the significant contributions of queen mothers in promoting sustainable development in their communities. A study conducted in Ghana found that queen mothers played a crucial role in mobilizing resources, promoting education, and advocating for the rights of women and children. Another study in Nigeria found that female traditional leaders were effective in promoting community participation in development projects and enhancing the well-being of their communities. However, despite their contributions, female traditional leaders often face challenges such as limited access to resources, cultural barriers, and gender-based discrimination.

Overall, the literature suggested that female traditional leaders had the potential to play a significant role in community planning and development.

Women leadership and corporate environmental strategies were investigated by Glass et al. (2016). The study examines the gender composition of leaders and its effects on corporate environmental practice, contributing to the literature on corporate governance and environmental performance. This study examines several hypotheses concerning the effect of gender diversity among top leaders on organizational environmental outcomes using a dataset of Fortune 500 CEOs and boards of directors over a ten-year period. Research indicates that firms with gender diverse management teams are more likely to pursue environmentally friendly strategies than other firms. The study provides valuable insight on how gender diversity in leadership can positively impact corporate environmental performance.

From 2011 to 2016, Birindelli et al. (2019) examined the impact of female CEOs on environmental performance in a sample of 96 listed banks in the EMEA (Europe, Middle East, and Africa) area. Women on the board of directors, chief executive officer gender, and the connection between these two elements are used to investigate gender diversity in leadership positions. By examining three theoretical perspectives: gender difference, critical mass, and homophily, this study shaded light on conflicting results in previous studies. The major findings implied that there was a nonlinear relationship between female directors and bank environmental performance, and that female chief executive officers have a strategic role in influencing this relationship, corroborating the banking sector's homophily perspective.

Shinbrot et al. (2019) found that the treatment and perceptions of women were the most frequently cited of the four key themes mentioned: paternal structure, treatment and perceptions of women, lack of access, and internal constraints. Nonetheless, these women, and men as well, believe that patriarchal institutions, bad attitudes, and adversarial treatment remain major hurdles for women in leadership roles, particularly in international places where they attempt to make a difference. However, the interchangeable usage of the terms gender and women, which is not covered in their study.

According to Atchison and Down (2019), men and boys regularly show lower levels of environmental concern than women and girls. Separately, research shows that women in elected office give greater attention to the problems that are important to their female constituency. Interestingly, the research has paid little consideration to the function of women officeholders on environmental policy, despite the stability of the gender disparity in views towards environmental issues. This paper's objective was to begin filling in that gap. Our preliminary findings suggest that women officeholders are related to adopting higher environmental standards using aggregated cross-section time series analyses of environmental requirements in 18 Western parliamentary democracies (1990-2012).

Lu and Herremans (2019) presented a study on the association between board gender diversity and environmental performance in US corporations. The study employs a rigorous research approach and statistical methodologies to analyse data from the S&P 1500 composite index. The data imply that gender diversity on boards is favourably associated with environmental performance, particularly in companies with a higher environmental effect. The study also emphasises the need of evaluating the context in which enterprises operate, as well as the need to investigate the social and environmental components of sustainability independently. To validate its findings, the study conducts a content analysis of board members' biographies.

Justin et al. (2020) examines the roles and strategies of 25 female environmentalists from throughout the world in the twentieth and twenty-first centuries. They were picked based on the numerous honours and accolades they have garnered in previous years. The roles of female environmentalists prior to becoming environmentalists were discovered to be diverse, ranging from politicians to researchers to high school students. Their duties after committing to environmental conservation and nurturing varied from advocacy to activism, policy efforts, environmental research, and so forth.

The purpose of Barrios et al. (2020) research is to examine women's leadership in marginalised Colombian Caribbean fishing communities and its impact in attaining sustainable livelihoods. Women leaders emphasized the importance of empowerment, sustainable development, and their roles in helping the community. They were unable to identify tangible benefits for their communities, emphasizing the necessity for long-term solutions. The study emphasizes the need of empowering women leaders to think beyond short-term rewards. Male leaders recognized women's leadership potential and the critical role they play in community development. Obtaining the support of male leaders is critical for the success of women leaders as a coherent group. More research and community participation are needed to discover routes for future community well-being and sustainable development.

Buss et al. (2021) examined the relationship between board gender diversity and environmental performance in US firms. The study analyses data from the S&P 1500 composite index using a rigorous research strategy and statistical methodologies. According to the statistics, gender diversity on boards is positively associated with environmental performance, especially in companies with a higher environmental impact. The study also underlines the importance of examining the context in which businesses operate, as well as independently investigating the social and environmental components of sustainability.

Because prior research has been ambiguous, the relationship between women on boards and corporate environmental performance is perplexing. De Masi et al. (2022) investigated this association using the homophily and self-construct theories. They study how the importance of women directors in improving environmental performance varies with the type of their nomination background using data from Italian FTSE-MIB companies from 2008 to 2017. To better understand the mechanisms underlying the importance of women on boards in environmental performance, the study categorised women based on their nomination history. Our panel regression results showed that after a threshold of three women on boards was reached, family women directors have a detrimental impact on company environmental performance.

Liu et al. (2022) examined the relationship between gender diversity among directors and corporate environmental performance in Chinese listed companies. The study finds that an increase in the proportion of female directors is positively associated with improved environmental performance, and that this relationship is moderated by ownership structure. The study focused on agency theory, resource dependence theory and ownership theory. agency theory suggests that gender diversity among directors can help mitigate agency problems and improve firm performance, while resource dependence theory suggests that gender diversity can help firms access diverse resources and improve their ability to adapt to changing environments. Ownership theory, on the other hand, suggests that state-owned enterprises may be more likely to benefit from gender diversity due to their unique ownership structure and goals.

The energy sector is a major source of carbon emissions and a major polluter on a worldwide scale. As a result, the key concerns of Abd Majid and Jaaffar, (2023) study were the energy industry and women and climate change. From 2018 to 2020, this study looked at the impact of women's leadership on carbon disclosure among the top 100 global energy leaders. They unambiguously acknowledged the impact of female leadership, as measured by the number of female board members who were industry specialists, serve as consultants, and act as proxy community leaders for global energy leaders. The empirical data revealed that among the world's top energy leaders, women leaders boosted the level of carbon disclosure.

Apart from Asian Women there were some African environmentalists who dedicated their lives for the sustainable development. Among them most recognized environmentalist was Wangari Maathai.

Wangari Maathai (Kenya): Wangari Maathai was a Kenyan environmentalist, activist, and Nobel Peace Prize winner. Maathai, who was born on April 1, 1940, in Nyeri, Kenya, was the first African woman to be awarded the Nobel Peace Prize in 2004 for her exceptional contributions to sustainable development, democracy, and peace. She is best known for establishing the Green Belt Movement, a grassroots organisation dedicated to environmental preservation, community empowerment, and women's rights. Maathai mobilised local communities to plant trees, battle deforestation, and promote sustainable land management practises through the Green Belt Movement. Her actions not only helped to restore degraded habitats, but they also empowered women by providing them with income-generating opportunities. Maathai's work highlighted the interconnectedness of environmental deterioration, poverty, and social injustice.

OPPORTUNITIES & CHALLENGES FOR WOMEN ENVIRONMENTALISTS IN DEVELOPING COUNTRIES

In their quest of environmental protection and action, Asian women environmentalists confront both possibilities and problems. Consider the following crucial points:

Opportunities

Representation and Empowerment: Asian women can contribute varied viewpoints and experiences to the environmental movement, resulting in a more inclusive and comprehensive approach to environmental problem-solving.

Community Engagement: Asian women frequently have deep relationships to their communities and may effectively engage and mobilise people at the grassroots level, encouraging local environmental projects and sustainable practises.

Education and knowledge: Through education, outreach programmes, and media, Asian women may play an important role in raising knowledge about environmental issues and encouraging eco-consciousness in their communities.

Leadership and Decision-Making Roles: Opportunities for Asian women to take up leadership roles in environmental organisations, government agencies, and corporate sectors are increasing, influencing policy decisions, and advancing sustainable practises according to Nasrin (2012).

Women Environmentalists in Developing Countries

Asian women have made important contributions to the fields of ecology and sustainability on a global and national scale. Here are a few well-known Asian female environmentalists:

Vandana Shiva (India): Vandana Shiva is an eco-feminist and environmental activist from India. She has advocated for sustainable agriculture and biodiversity conservation while campaigning against genetically engineered crops. She is well-known for her work to promote seed sovereignty and preserve traditional farming practises. Vandana Shiva is a well-known environmentalist, researcher, and activist who has spent her life advocating for sustainable agriculture, biodiversity protection, and social justice. Shiva, who was born on November 5, 1952, in Dehradun, India, has dedicated her life to addressing serious environmental and social challenges. She founded **Navdanya**, an organisation that promotes seed saving and biodiversity preservation while advocating for farmers' and indigenous communities' rights. Shiva's work focuses on opposing agricultural industrialization and the domination of genetically modified crops, emphasising the value of seed sovereignty and the preservation of traditional farming practises. Her publications and presentations have drawn light on the detrimental effects of globalisation, corporate dominance over agriculture, and the necessity for a transition to sustainable, regenerative farming systems. Vandana Shiva's environmental efforts has earned her international acclaim, including the Right Livelihood Award and the Sydney Peace Prize. Shiva's persistent devotion continues to inspire and mobilise individuals and communities around the world, emphasising the critical link between environmental sustainability, food systems, and social fairness.

Dai Qing (China): Dai Qing is a well-known Chinese environmentalist, journalist, and writer who is well-known for her activism and advocacy for environmental and social justice. Dai Qing, who was born on March 22, 1941, in Chengdu, Sichuan province, has been at the vanguard of China's environmental movements, notably in the context of large-scale dam construction. She is well known for her resistance to the Three Gorges Dam, one of the world's largest hydropower projects, as well as her efforts to encourage transparency, public engagement, and human rights protection in China's development projects. Dai Qing's activism began in the 1980s, when she began researching and writing about the Three Gorges Dam's social and environmental repercussions. She questioned the Chinese government's portrayal of the project as a symbol of progress and development, instead emphasising the relocation of millions of people, environmental devastation, and the possibility of dam failure. Despite severe control and criticism from Chinese officials, her 1989 book, **"Yangtze! Yangtze!"** gave a critical and extensive account of the dam's impacts.

Sunita Narain (India): She is an Indian environmentalist and the director-general of India's Centre for Science and Environment (CSE). She has played a critical role in campaigning for sustainable development and advancing environmental and climate change policy. Narain has been actively involved in creating national and international environmental policies. Sunita Narain is an internationally recognised environmentalist and campaigner from India. She was born on October 24, 1961, and has contributed significantly to environmental protection, sustainable development, and public policy. CSE has played a vital role in solving critical environmental challenges like as air and water pollution, climate change, and urban sustainability under her leadership. Narain has been an outspoken proponent of sustainable and equitable development, emphasising the importance of policies that prioritise environmental conservation and social justice. She has aggressively advocated for clean, inexpensive energy options, trash management measures, and environmentally friendly transportation.

Kimiko Hirata (Japan): Kimiko Hirata is a well-known environmentalist and activist from Japan. She serves as the International Director of the Kiko Network, a non-profit dedicated to battling climate change and supporting renewable energy. Hirata has played an important role in increasing awareness of the environmental and socioeconomic consequences of Japan's energy policies, notably its reliance on coal power. She has been an active participant in campaigns and advocacy attempts to move Japan towards cleaner, more sustainable energy sources. Hirata was a key figure in mobilising public opposition to new coal-fired power plants in Japan and has been an outspoken critic of the government's backing for coal industry expansion. These are just a few of the several Asian women who have made major environmental contributions. Their work exemplifies the diversity, commitment, and impact of women in addressing environmental issues and promoting sustainability.

Aleta Baun (Indonesia): Inspirational Indonesian environmentalist Aleta Baun is well-known for her valiant efforts in the fight against illegitimate mining and deforestation on the island of Timor. Baun, who was raised in a small town in eastern Timor, developed a strong bond with nature and a profound appreciation for the value of protecting the environment for future generations. When she saw the terrible effects that marble mining was having on the pristine landscapes and ecosystems of her own country in the late 1990s, she decided to become an environmental activist. The forests, rivers, and lush plains that had supported the local communities' lives for countless generations were suffering irreparable damage from illegal mining operations.

Baun bravely led a movement to safeguard her community's natural resources and cultural legacy against opposition from mining firms and even threats to her life. She encouraged the locals to support her cause, especially the women, realising the crucial function that women play in protecting and preserving the environment. Her deeply held conviction that nature and human existence are interdependent has served as the driving force for Baun's action. She has emphasised the necessity of preserving the ecosystems' fragile balance and the significance of residing in peace with the soil. For Baun, environmental protection entails not just conserving the natural environment but also her people's cultural identity and traditional knowledge.

The tireless efforts and passion of Aleta Baun have not gone unnoticed. She has been honoured for her work on a national and international level, **and in 2013 she won the Goldman Environmental Prize.**

Bharrathi Devi (India): Bharrathi Devi, also referred to as the "Solar Queen of Tamil Nadu," is a prominent Indian environmentalist and proponent of renewable energy who has done much to advance sustainable energy options in rural areas. She has gained fame and affection both domestically and abroad for her devotion to providing clean energy to the masses and to environmental preservation. She began her career in the field of renewable energy by thoroughly researching solar energy systems and other eco-friendly technology. She established "Sampark," an organisation with the goal of bringing the advantages of solar energy to isolated and neglected regions, armed with knowledge and passion.

Bharrathi Devi's work focuses on putting in solar power systems in rural regions, especially in community centres and schools. Through Sampark, she has played a key role in installing solar lighting systems and solar water pumps, providing access to electricity and clean water in formerly off-the-grid locations. Bharrathi Devi's involvement in advocacy and educational initiatives to spread the word about renewable energy and its potential to change lives extends beyond her work with Sampark. She leads workshops and training sessions to educate rural communities on the use and maintenance of solar energy, enabling them to take control of their energy requirements.

Bharrathi Devi has received various honours and awards for her dedication and creative strategies, including recognition from the United Nations and other esteemed organisations. Numerous people's lives, especially those of women and children who now have access to cleaner, more sustainable energy sources, have been significantly impacted by her work.

Laila Kasuri (Pakistan): She is an expert in sustainable development who has worked in the water and environmental sectors for more than nine years. She is in charge of a project aimed at achieving the SDGs in First Nations communities while working as a Senior Advisor for the Assembly of First Nations (AFN). Prior to joining the AFN, she had been creating and implementing projects in Bangladesh, India, Cambodia, Laos, Myanmar, and Jordan for international organisations like the World Bank and the Global Green Growth Institute. Additionally, Ms. Kasuri worked as adjunct faculty at the Lahore University of Management Sciences, where she contributed to the creation of the Centre for Water Informatics and Technology.

Water experts like Laila Kasuri are becoming more crucial as climate change has a negative impact on the developing world and a preventative approach to sustainability is needed. Kasuri is a water analyst for Global Green Growth Investment's Policy Solutions Division. She has directed research on climate-smart irrigation, flood risk reduction, and river basin planning at institutions like the World Bank, the US Army Corps of Engineers, and the UC Davis Centre for Watershed Sciences.

In the year 2013, she was awarded **"Thomas Temple Hoopes Award"** and because of her outstanding contribution towards sustainable development, in the year 2014, she received **"Medal for Excellence, US Army Corps of Engineers".**

Delima Silalahi (Indonesia): The executive director of Kelompok Studi dan Pengembangan Prakarsa Masyarakat (KSPPM), an NGO devoted to the preservation of indigenous forests in North Sumatra, is Delima Silalahi, 46. She is a Batak woman from Siborong-Borong, one of many regions in North Sumatra impacted by the removal of forests for industrial crops. Siborong-Borong is in the district of North Tapanuli. Delima joined KSPPM as a volunteer in 1999 and was an activist while still in college.

Six Indigenous groups in North Sumatra were given legal stewardship of 17,824 acres of tropical forest land because of a campaign led by Delima Silalahi. A pulp and paper firm that had partially turned this area into a monoculture, non-native, industrial eucalyptus plantation was driven out by the action of her community. To create valuable carbon sinks of biodiverse Indonesian tropical forest, the six communities have started to restore the woods.

Challenges

<u>Gender Bias and preconceptions:</u> Asian women may suffer gender biases and preconceptions in the environmental industry, limiting their possibilities and recognition. They may face difficulties in being taken seriously or perceived as capable leaders according to Piper, N. (2003).

<u>Socio-cultural Constraints:</u> As mentioned by He et al. (2012), societal expectations, cultural norms, and traditional gender roles may stymie Asian women's participation in environmental advocacy. Balancing family obligations, societal expectations, and environmental work can be difficult.

<u>Intersectionality:</u> When Asian women's gender, race, ethnicity, and other variables overlap, they may confront intersectional problems. They may confront challenges and discrimination that necessitate a diversified approach to properly address.

To address these difficulties and capitalise on possibilities, diverse parties must work together. Promoting gender equality, providing mentorship and tools, creating safe spaces, and celebrating the accomplishments of Asian women environmentalists are all critical. We can develop an inclusive and varied environmental movement that effectively confronts our time's concerns by doing so.

WOMEN ENVIRONMENTALISTS IN DEVELOPED COUNTRIES

American women have played an important part in the environmental movement, committing to safeguarding and maintaining the natural world for future generations. Katherine (2023) in her article focused on 16 women environmentalists who had made important contributions to increasing awareness about environmental issues, implementing sustainable practises, and creating positive change via their unrelenting dedication, leadership, and activism. This part focuses on the accomplishments and effects of American women environmentalists in their pursuit of a greener, more sustainable world. Barth (2020) in article "10 Woman Environmentalists You Should Know About" mentioned the details about the environmentalists and their impressive works.

Rachel Carson (1907-1964): She was an American marine biologist, writer, and environmentalist whose seminal work "Silent Spring" (1962) changed the course of environmental research. Carson uncovered the harmful effects of pesticides, particularly DDT, on wildlife and the environment through her thorough research and persuasive speech. Her work not only raised public awareness but also resulted in significant policy reforms, including the creation of the Environmental Protection Agency (EPA) and the prohibition of DDT in the United States. Carson's legacy as a passionate advocate for the natural world, as well as her ability to articulate difficult environmental issues to a wide audience, continue to inspire and drive the battle for a sustainable and healthy Earth.

Jane Goodall: She is a well-known primatologist, anthropologist, and environmental campaigner who was born on April 3, 1934, in London, England. Her remarkable chimp research in Tanzania's Gombe Stream National Park transformed our understanding of monkeys and their behaviour. Beginning in 1960, Goodall's intensive study challenged established scientific methodologies and produced astonishing insights into the social and emotional lives of chimps, demonstrating their use of tools and capacity for complex communication. Goodall has been a persistent campaigner for conservation and biodiversity as an environmentalist. In 1977, she founded the Jane Goodall Institute, which focuses on wildlife research, conservation, and community-based programmes. Goodall's Roots & Shoots programme motivates and empowers young people all around the world to take action on environmental and humanitarian challenges. She has won countless honours and accolades for her outstanding work, and her lifelong commitment to conservation continues to inspire future generations to protect and preserve the natural environment.

Winona LaDuke: She is an American environmentalist, Indigenous activist, and writer who was born on August 18, 1959. She is well-known for her unrelenting devotion to environmental sustainability and Indigenous rights. LaDuke, an Anishinaabe woman, has spent her life fighting for the preservation of Indigenous cultures, lands, and natural resources. She founded the White Earth Land Recovery Project in 1989 to promote economic self-sufficiency and environmental justice in Indigenous communities. LaDuke has been a strong opponent of extractive industries that threaten Indigenous territories and ecosystems, such as mining and pipelines.

Majora Carter: Majora Carter is an American urban regeneration strategist, environmental justice advocate, and entrepreneur who was born on October 27, 1966. Carter, who grew up in the South Bronx, has dedicated her career to addressing environmental, social, and economic challenges in underserved neighbourhoods. She is most recognised for her innovative work with the Sustainable South Bronx (SSB) organisation, which she formed in 2001. Carter's novel method combines environmental sustainability, affordable housing, and the creation of green jobs to promote fair and sustainable development. Under her direction, SSB was instrumental in campaigning for green infrastructure projects such as the turning of a former landfill into the award-winning Hunts Point Riverside Park.

Sylvia Earle: Sylvia Earle, a renowned oceanographer, marine biologist, and environmentalist, was born on August 30, 1935, and has dedicated her life to learning about and protecting the oceans. She has earned the moniker "Her Deepness" and has become one of the most important individuals in the industry thanks to her groundbreaking work in marine research and support for ocean conservation. Earle broke down boundaries and opened the path for subsequent generations of female scientists as the nation's National Oceanic and Atmospheric Administration's (NOAA) first female top scientist. Sylvia Earle has participated in more than a hundred excursions throughout the course of her career, devoting countless hours to researching marine ecosystems and diving to the ocean's depths. Her various scientific accomplishments, notably her ground-breaking work on marine algae and the behaviour of marine species, are a testament to her passion for the ocean and its inhabitants. Beyond her contributions to science,

Earle's dedication to ocean conservation has made her a well-known spokesperson for preserving marine environments and life. She established the Mission Blue organisation to advance marine conservation initiatives, igniting a global network of "Hope Spots" to protect essential ocean habitats.

Terry Tempest Williams: A well-known American novelist, environmentalist, and supporter of public lands and the preservation of wilderness regions, Terry Tempest Williams was born on September 8, 1955. Williams has raised awareness of the nuanced interaction between people and nature through her potent words and vivid storytelling. Her writing frequently integrates ideas about feminism, environmental justice, nature, and culture, offering a distinctive viewpoint on how all life is interconnected. Williams has emerged as a key voice in the environmental movement due to her profound respect for the earth and dedication to social and environmental problems.

Terry Tempest Williams has written a few significant novels over her literary career, including "Refuge: An Unnatural History of Family and Place," which examines the effects of the Great Salt Lake and its wetlands on her family and the surrounding neighbourhood. She has also written about the difficulties encountered by native people and the requirement to safeguard public areas from being used for commercial purposes. Williams' commitment to environmental causes goes beyond her writing; she has participated in numerous environmental campaigns and held several conservation board positions. Terry Tempest Williams continues to encourage readers and environmental activists to be better stewards of the Earth and conserve the places we value with her enthusiasm for the natural world and skill for storytelling.

In their efforts to contribute to environmental conservation and activism, American women environmentalists encounter a variety of opportunities and problems. Consider the following crucial points:

OPPORTUNITIES AND CHALLENGES FOR WOMEN ENVIRONMENTALISTS IN DEVELOPED COUNTRIES

Opportunities

Advocacy and Policy affect: Advocacy efforts by American women can affect environmental policies at the local, state, and national levels. They can use their platforms and voices to promote awareness, impact public opinion, and effect positive change.

Education and Research: Environmental science, research, and education are all options for American women. They can contribute to the development of novel solutions, perform important research, and teach future generations about environmental challenges.

Grassroots Activism: American women have a long history of grassroots activism, which allows them to organise community initiatives, participate in protests, and lead grassroots environmental organisations.

Gender Diversity and Representation in Leadership Positions: There is a growing acknowledgment of the importance of gender diversity and representation in leadership positions. American women can hold positions of leadership in environmental organisations, nonprofits, government agencies, and enterprises, allowing them to create and implement sustainable practises and regulations. As per Silesbarcena (2019), despite being denied leadership positions in key environmental organisations due to existing and widespread gender inequality, women have played an important role in conserving natural resources and promoting sustainability.

Challenges

Gender Inequality: Women continue to endure gender-based discrimination and biases in the United States, notably in the environmental sector. They may face advancement difficulties, unequal remuneration, and limited access to decision-making positions. Gender variations in perception and attitude towards environmental concerns are widely documented: women express greater care about the environment and are more likely to behave pro-environmentally than males across countries. However, research into an environmental gender disparity at the elite level is scarce according to Ramstetter and Habersack (2020).

Harassment and Safety Concerns: When fighting for environmental problems, female environmentalists may encounter harassment, online abuse, and even physical threats. Keeping them safe and protected from harm is a constant challenge.

Work-Life Balance: For American women, balancing environmental action with personal and family duties can be difficult. This difficulty is exacerbated by cultural expectations, a lack of support structures, and the physically hard nature of environmental labour.

Intersectionality: Because of intersecting types of prejudice, women from varied backgrounds, such as women of colour, LGBTQ+ women, and women with disabilities, may face additional hurdles. It is critical to address intersectionality to build an inclusive and egalitarian environmental movement.

To overcome these obstacles and maximise benefits, the environmental industry must encourage gender equality, inclusivity, and diversity. This includes giving mentorship and tools, as well as campaigning for policies that promote female environmentalists. We can build a more vibrant and successful environmental movement for today and future generations by recognising the contributions of American women.

SIMILARITIES AND DIFFERENCES BETWEEN DEVELOPED, DEVELOPING COUNTRIES WOMEN ENVIRONMENTALISTS

Asian and American women environmentalists have a common enthusiasm for environmental protection and sustainability, although their tactics and challenges differ in significant ways. Both women's organisations are committed to addressing global environmental challenges such as climate change, pollution, and habitat degradation. They aggressively encourage environmental awareness and action through advocacy, grassroots activity, and leadership roles.

Asian and American women environmentalists share a commitment to grassroots engagement and community involvement. Both organisations recognise the significance of mobilising local people to follow sustainable practises and effect good environmental change. Asian women environmentalists frequently use their strong community ties and cultural expertise to drive grassroots projects. They work to address environmental concerns through educating and empowering persons at the grassroots level, encouraging sustainable practises, and cooperating with community leaders. Similarly, American female environmentalists are active members of their communities, organising local events and campaigning for sustainable legislation at the municipal and regional levels. Both parties recognise the need of community involvement in achieving long-term environmental sustainability and resilience.

There are, nevertheless, significant distinctions between Asian and American female environmentalists. Their attitudes and issues are heavily influenced by their cultural milieu. Asian female environmentalists work in a variety of cultural situations, each with its own set of societal expectations, gender conventions, and cultural values. These variables have an impact on their responsibilities and possibilities in the environmental movement. Women may confront cultural barriers and gender biases in various Asian countries, limiting their access to leadership roles and decision-making processes. American women environmentalists, on the other hand, face a variety of hurdles within the American cultural landscape, including continuing gender disparities, discrimination, and biases that limit their representation and growth in the environmental profession.

Another distinction is the focus of Asian and American female environmentalists on regional environmental issues. Due to rapid industrialisation, population growth, and urbanisation, Asian countries face unique issues such as deforestation, air pollution, water scarcity, and land degradation. Asian women environmentalists frequently prioritise solving these region-specific concerns and the social, economic, and political difficulties that accompany them. American women environmentalists, on the other hand, address issues such as fossil fuel dependency, industrial pollution, natural resource management, and the protection of sensitive ecosystems such as forests and wetlands. These geographical distinctions in environmental issues necessitate customised methods and solutions that reflect the distinct circumstances and priorities of Asian and American women environmentalists.

Current Status for the Women Environmentalists

The current situation of women leaders in environmental management is constantly changing, with some areas making progress while others face considerable hurdles. Women are increasingly taking on leadership positions in environmental organisations, government agencies, and advocacy groups, bringing their unique perspectives and talents to bear on important environmental issues. However, women continue to be underrepresented in senior leadership positions and decision-making responsibilities in the environmental sector when compared to men. Let's take a closer look at the current state of female environmental leaders:

Increasing Representation and Visibility: There has been significant growth in the number of female environmental leaders in recent years. Women are increasingly occupying major positions in environmental organizations, academia, research institutions, and government. This expanded representation contributes to a broader range of ideas and approaches to environmental problem solutions.

Environmental Advocacy and Activism: Women have been at the vanguard of environmental advocacy and activism, leading initiatives for climate action, conservation of wildlife, and equitable environmental treatment. Their participation has played a key role in promoting ecological consciousness and mobilizing people for constructive change.

Women in Science and Research: Women are increasingly pursuing occupations in environmental science and research, providing significant ideas and information to handle difficult environmental concerns. Female scientists are creating revolutionary findings and advances in sectors such as climate research, biodiversity protection, and sustainable resource management.

Challenges in Leadership roles: Despite advances, women continue to be marginalized in managerial positions in environmental management. Women's advancement to executive and decision-making positions continues to be hampered by structural impediments, gender prejudices, and stereotypes.

Gender Pay Gap: In many environmental organisations and institutions, women typically get compensated less than their male counterparts for comparable tasks and duties. This difference has an impact on job advancement and may discourage women from pursuing leadership positions.

Intersectionality and Inclusivity: Intersectionality is important in the experiences of female environmental leaders. Women from marginalised communities, such as women of color, indigenous women, and low-income women, face more obstacles and barriers to attaining leadership roles.

Mentoring and Support: There is a growing movement to mentor and support women in environmental leadership. Mentorship, skill-building, and leadership development initiatives and networks are assisting in empowering and equipping women for leadership roles.

Representation in Environmental Policy and Decision Making: Women's representation in environmental policy and decision-making bodies varies among countries and regions. Increased efforts are required to guarantee that women's voices and perspectives are heard and considered when developing environmental laws and regulations.

Representation in Policy and Decision Making: The representation of women in environmental policy-making and decision-making bodies varies among countries and regions. Increased efforts are required to guarantee that women's views and perspectives are incorporated in the development of environmental laws and regulations.

Addressing Work-Life Balance: For women leaders in environmental management, balancing family duties and job commitments can be a considerable problem. Creating supportive work environments and flexible rules can aid in the retention and advancement of women in positions of leadership.

While there has been progress in boosting the participation and visibility of female environmental leaders, there are still considerable obstacles to overcome. Addressing gender imbalances, encouraging diversity, and providing support and opportunity for women's growth are all critical steps towards creating a more varied and fair environmental leadership landscape. Efforts to level the playing field and eliminate gender prejudices would not only help women but will also increase the environmental sector's capacity to face the difficult challenges of our day. Embracing women's leadership and knowledge is critical to ensuring the planet's long-term viability and resilience.

FUTURE OF WOMEN IN THE FIELD OF ENVIRONMENTAL MANAGEMENT

The future of women in environmental management has enormous promise and opportunity, but it also faces problems that must be addressed to achieve more gender equality and inclusivity. Koohi et al. (2016) mentioned that the relationship between culture and the environment, as well as the roles of women in developing environmental culture. It highlighted the importance of women's participation in environmental management and cultural transmission to future generations. While traditional beliefs frequently limit women's participation to family and basic chores, their abilities in environmental management were clear. Here are some significant factors influencing women's future in environmental management:

Increasing Representation: The relevance of gender diversity in environmental management is becoming more widely recognised. Women's representation in leadership positions, research roles, and decision-making bodies is anticipated to expand, leading to more inclusive and effective environmental solutions.

Advocacy for Climate Action, Conservation, and Sustainable Development: As more women occupy leadership roles in environmental organisations and movements, their advocacy for climate action, conservation, and sustainable development is projected to gain importance. Women's viewpoints and expertise will help shape successful policies and strategies for dealing with environmental issues.

Education for Empowerment: Expanding access to education and training in environmental sciences and management will enable more women to enter and flourish in the sector. Supporting girls' and women's STEM (science, technology, engineering, and mathematics) education will be critical in developing a diverse and talented workforce.

Intersectionality and Inclusivity: Recognising and tackling the intersectional issues that women confront in many social, economic, and cultural contexts will be critical. Inclusion and the removal of systemic barriers will result in a more varied and representative environmental management community.

Mentorship and Support Networks: Mentorship programmes and support networks for women in environmental management will be critical in developing talent and advancing their careers. Inspiring and guiding the next generation will be enhanced by encouraging senior female executives to participate as mentors.

Gender prejudice and Discrimination: Addressing gender prejudice and discrimination in the workplace and academia will provide a more egalitarian and welcoming environment for women's professional development. The elimination of the gender pays gap and the promotion of equal opportunity will be critical first steps.

Innovation and Research: Women's contributions to innovative research and technology development in environmental management are predicted to rise in the coming years. Their various perspectives can inspire new approaches to resource management and climate adaption.

Global Collaboration: Opportunities for women to participate in global projects, bringing their knowledge to address transboundary concerns, will continue to be provided by international cooperation and collaboration on environmental issues.

Bridging the Rural-Urban Divide: Women's leadership development in both urban and rural communities will be crucial for holistic and inclusive environmental management. Recognising women's contributions to traditional and indigenous knowledge systems is also critical.

Environmental Justice Advocacy: Women's environmental management leadership frequently connects with environmental justice problems, as they advocate for vulnerable communities affected by environmental degradation. This emphasis on equitable solutions will impact environmental policy and practise in the future.

Women's future in environmental management looks bright, with more prospects for leadership, advocacy, and innovation. However, to realise the full potential of women's contributions, gender inequities, biases, and institutional hurdles must be addressed. Accepting and fostering variety will result in more effective and sustainable environmental management practises that benefit both people and the earth. Continued efforts to empower women via education and mentorship will fuel innovation and advocacy for environmentally sustainable practises. Addressing gender bias and discrimination will result in a more egalitarian workplace, encouraging women's contributions to research and policymaking. Collaboration and intersectionality will increase their impact on global environmental issues by fostering inclusive and effective solutions. Women's environmental leadership will be critical in crafting a more sustainable and resilient future for our planet.

CONCLUSION

In conclusion, while Asian and American women environmentalists share a common commitment to environmental conservation, their tactics and obstacles differ in important ways. Both organisations emphasise grassroots engagement and community involvement as critical components of their work. Cultural contexts and regional environmental challenges, on the other hand, shape their viewpoints and strategies. Recognising and comprehending these parallels and variances is critical for fostering collaboration, sharing information and resources, and constructing a diverse and inclusive global environmental movement. Asian and American women environmentalists may generate meaningful change and contribute to a more sustainable and equitable future for everybody by learning from each other's experiences and working together. There are numerous chances for female environmental leaders. First, women leaders' representation and empowerment add variety and inclusivity to decision-making processes, resulting in more comprehensive and long-term outcomes. Women leaders can fight for gender-sensitive legislation, promote social justice, and address women's and marginalised communities' special environmental issues. Their participation guarantees that the environmental movement benefits from a variety of viewpoints, experiences, and information. Women leaders also have the capacity to inspire community engagement and grassroots initiatives. They frequently have strong relationships to their communities and can effectively mobilise people in support of environmental conservation. Women leaders may bring about major change at the grassroots level by promoting local environmental initiatives and sustainable practises, thereby establishing a bottom-up movement that addresses environmental concerns from within communities.

REFERENCES

Abd Majid, N., & Jaaffar, A. H. (2023). The Effect of Women's Leadership on Carbon Disclosure by the Top 100 Global Energy Leaders. *Sustainability (Basel), 15*(11), 8491. Advance online publication. doi:10.3390u15118491

Adjei Mensah, C., Antwi, K. B., & Dauda, S. (2014). Female Traditional Leaders (Queen Mothers) and Community Planning and Development in Ghana. *Environmental Management and Sustainable Development, 3*(1), 205. doi:10.5296/emsd.v3i1.5142

Atchison, A. L., & Down, I. (2019). The effects of women officeholders on environmental policy. *The Review of Policy Research, 36*(6), 805–834. doi:10.1111/ropr.12346

Barrios, L. M., Prowse, A., & Vargas, V. R. (2020). Sustainable development and women's leadership: A participatory exploration of capabilities in Colombian Caribbean fisher communities. *Journal of Cleaner Production, 264*, 121277. doi:10.1016/j.jclepro.2020.121277

Barth, N. (2020). 10 Woman Environmentalists You Should Know About. *Greenpop*. https://greenpop.org/10-woman-environmentalists-you-should-know-about/

Birindelli, G., Iannuzzi, A. P., & Savioli, M. (2019). The impact of women leaders on environmental performance: Evidence on gender diversity in banks. *Corporate Social Responsibility and Environmental Management, 26*(6), 1485–1499. doi:10.1002/csr.1762

Buss, D., Rutherford, B., Kumah, C., & Spear, M. (2021). Beyond the rituals of inclusion: The environment for women and resource governance in Africa's artisanal and small-scale mining sector. *Environmental Science & Policy, 116*, 30–37. doi:10.1016/j.envsci.2020.10.019

De Masi, S., Słomka-Gołębiowska, A., & Paci, A. (2022). Women on boards and corporate environmental performance in Italian companies: The importance of nomination background. *Business Ethics, the Environment & Responsibility*, *31*(4), 981–998. doi:10.1111/beer.12467

Glass, C., Cook, A., & Ingersoll, A. R. (2016). Do Women Leaders Promote Sustainability? Analyzing the Effect of Corporate Governance Composition on Environmental Performance. *Business Strategy and the Environment*, *25*(7), 495–511. doi:10.1002/bse.1879

He, G., Lu, Y., Mol, A. P., & Beckers, T. (2012). Changes and challenges: China's environmental management in transition. *Environmental Development*, *3*, 25–38. doi:10.1016/j.envdev.2012.05.005

Jackson, C. (1993). Doing What Comes Naturally? Women and Environment in Development. *World Development*, *21*(12), 1947–1963. doi:10.1016/0305-750X(93)90068-K

Justin, M. S. M., Aquino, P. G. Jr, & Le, D. H. (2020). Roles and Strategies of 20th and 21st Century Women Environmentalists. In S. A. Khan (Ed.), *Global Perspectives on Green Business Administration and Sustainable Supply Chain Management* (pp. 163–180). IGI Global. doi:10.4018/978-1-7998-2173-1.ch009

Katherine. (2023). Guardians of the Planet 16 Women Environmentalists You Should Know. *Almighty Girl.* https://www.amightygirl.com/blog?p=11863

KimballA.PalmerD.MarquisC. (2012). The Impact of Women Top Managers and Directors on Corporate Environmental Performance. https://ssrn.com/abstract=2211826Electroniccopyavailableat:https://ssrn.com/abstract=2211826Electroniccopyavailableat:http://ssrn.com/abstract=2211826 doi:10.2139/ssrn.2211826

Koohi, E., Shobeiri, S. M., Koohi, E., & Meiboudi, H. (2016). Women's participation in Environmental Management and Development Promotion Culture. *International Journal of Resistive Economics*, *4*(1), 23–35.

Liu, Z., Cai, L., & Zhang, Y. (2022). Gender Diversity and Environmental Performance: New Evidence from China. *Sustainability (Basel)*, *14*(21), 13775. doi:10.3390u142113775

Lu, J., & Herremans, I. M. (2019). Board gender diversity and environmental performance: An industries perspective. *Business Strategy and the Environment*, *28*(7), 1449–1464. doi:10.1002/bse.2326

Nasrin, F. (2012). Women, environment and environmental advocacy: Challenges for Bangladesh. *Asian journal of social sciences & humanities, 1*(3), 149-172.

Piper, N. (2003). Bridging gender, migration and governance: Theoretical possibilities in the Asian context. *Asian and Pacific Migration Journal*, *12*(1-2), 21–48. doi:10.1177/011719680301200102

Ramstetter, L., & Habersack, F. (2020). Do women make a difference? Analysing environmental attitudes and actions of Members of the European Parliament. *Environmental Politics*, *29*(6), 1063–1084. doi:10.1080/09644016.2019.1609156

Shinbrot, X. A., Wilkins, K., Gretzel, U., & Bowser, G. (2019). Unlocking Women's Sustainability Leadership Potential: Perceptions of Contributions and Challenges for Women in Sustainable Development. *World Development*, *119*, 120–132. doi:10.1016/j.worlddev.2019.03.009

Silesbarcena, I. (2019). Climate change and sustainable development. *Ecofeminism*. https://sites.manchester.ac.uk/global-social-challenges/2019/06/08/why-women-as-environmental-activists-will-bring-about-necessary-political-change-for-a-sustainable-future/

Women U. N. (2023). EmPower: Women for Climate-Resilient Societies Programme Brief. UN. https://asiapacific.unwomen.org/en/digital-library/publications/2023/06/empower-women-for-climate-resilient-societies-programme-brief#:~:text=From%202018%20to%202022%2C%20UN,across%20Asia%20and%20the%20Pacific

Chapter 7
Women Leaders in Environmental Management and the Role of Media in the EMEA Region

Maybel Fernandez Vallado
https://orcid.org/0009-0000-1039-1322
University of the Philippines, Visayas, Philippines

ABSTRACT

Women leaders who are at the forefront of environmental conservation efforts, have been instrumental in mitigating the effects of climate change, protecting biodiversity, and promoting sustainable development. The EMEA Region has seen an increasing number of women leaders in various industries and they play a crucial role in shaping public opinion on environmental issues. In this Chapter, their specific contributions are examined, including awareness, grassroots mobilization, and policy and advocacy. The challenges and opportunities they face, which include gender bias, risk and safety, collaboration and partnerships, and the role of media are also identified. Women can be effective leaders in environmental management through: collaboration and teamwork; communication; problem solving and innovation; attention to detail; as well as empathy and inclusivity. Women leaders promote sustainable development, protect the environment and their work in the media inspire and empower other women to take on leadership roles and contribute towards sustainable development.

INTRODUCTION

The EMEA (Europe, Middle East, and Africa) Region has seen an increasing number of women leaders in various industries, including environmental management and media. These women have made significant contributions in promoting sustainable development and protecting natural resources. This Chapter aims to explore the role of women leaders in environmental management and media in the EMEA Region.

DOI: 10.4018/978-1-6684-5986-7.ch007

The banking industry, energy sector, governments and activism are among the key areas where women leaders have made notable contributions. Women in the media shape public opinion and create awareness of environmental issues. This Chapter highlights examples of women leaders in these areas, their specific contributions, and the challenges and opportunities they face in promoting sustainable development. Furthermore, it underscores the importance of gender equality in the context of environmental management and media. Women's voices and perspectives are often underrepresented in decision-making processes related to the environment. Therefore, highlighting the role of women leaders in these fields can inspire and empower other women to take on leadership roles and contribute towards sustainable development.

Women Leaders in Environmental Management: An Overview

The State of Climate in Europe reports Europe as the fastest warming region globally and has been experiencing double the average global warming since the 1980s. In fact, extreme heat waves, droughts, and wildfires in 2022 resulted in over 16,000 reported fatalities. The report also highlighted the urgent need for action in response to the climate crisis. (WMO, 2022, p.5)

In the United Arab Emirates (UAE) and the Arabian Peninsula, the average temperatures have risen and continue to rise. This results in the rise of surface water temperature and deep waters of the Gulf will likely to increase as well. Other seawater properties will also be affected. Seawater salinity and pH are expected to increase with water evaporation and higher absorption of CO2. The sea level in the Arabian Gulf is projected to rise due to melting polar ice caps and thermal expansion of seawater. Rainfall may also either increase or decrease. It may become less frequent, but the intensity of the precipitation events may increase. Extreme temperatures - droughts, sandstorms, and cyclones - are expected to increase in severity and frequency in the Arabian sea, which is a cause of concern for many. (UAE MOCCAE, 2021, p.8)

Average and maximum humidity will increase, due to higher evaporation. There will be further warming through an intensified greenhouse effect, lower recharge due to less rainfall, and some reservoirs will be lost because of salt intrusion from rising sea levels. (p.14)

Human activities are estimated to have caused approximately 1.0°C of global warming above pre-industrial levels, with a likely range of 0.8°C to 1.2°C. Global warming is likely to reach 1.5 between 2030 and 2050 if it continues to increase at the current rate. (IPCC, 2018, par. 4)

Women leaders at the forefront of environmental conservation efforts, and their contributions have been instrumental in mitigating effects of climate change, protecting biodiversity, promoting sustainable development and ethical practices in waste management.

The Case of Cameroon, Africa

Rural Cameroonian women interact with the environment through their numerous daily chores - cutting down and burning vegetation to create farms, fetching fuelwood and water. These activities create problems which devastate and accelerate environmental degradation and have negative consequences. Women contribute to the continuous natural resource consumption process, but they have no firm authority over the land, credit, and decision-making. The men are accomplices, as they compel the women to take a lot from the environment to satisfy the men and use for general domestic needs.

Rural Cameroonian women constitute a potential partner (56%) of rural production force. Highlighting problems created in each zone hopes to awaken public opinion, and to create awareness and envisions to propose compensatory measures, exchanges in take and give, and in farm community actions to uphold positive checks. Women can be partners in environmental management, when all these are properly focused on women's rights to land, credit, training and decision making are recognized. (Ngwa, 1995, p. 515)

Results of a random survey to evaluate innovation adoption and decision to improve the environment among rural women in North West (NW), and Far North Provinces show that 26% of women aged 40 years said they make decisions wholly in their community and do not ask other people's opinions (age, number of children and status of their husbands gave them the opportunity to make decisions); 40% of the respondents made decisions after consulting with their husband; while28% thought that men have a wonder perspective on farm matter and should decide for them; and 4% did not have anything to offer.

Among Fulbe communities in North Cameroon, notable men's wives are not expected to undertake farm activities. They have servants, called "slaves," to do everything for them. Ordinary people's wives may do a little gardening in their homes, fetch water and little bundles of wood, but they do not till the land, plant nor harvest. The future output of such neighborhoods is subject to continuous abuse, and decisions on matters involving environmental restoration are made by men, and not the women.

Men are accomplices, since they let women cut and even burn land cover to have small parcels of farmlands and bring home water for cooking and fuel wood. Exposed land surface is subject to many forms of degradation and soil erosion. Such actions have negative environmental effects on soil fertility and can no longer be held together by degraded vegetation cover. The sun's heat bakes the soil and reduces yearly expected output of these marginal lands. The responsibilities of these rural women are very limited since they exploit the land and its resources, which they have no proper control of. (p.516)

Few women know and can carry on water management for dry season vegetable irrigation and appreciate preserving the water table through agroforestry, crop mulching, and other similar activities. The majority of rural women are unaware of such environmental issues.

In South Cameroon, women perform nearly every activity, from food preparation, crop management (rice and other staples), producing plantation crops and fetching water and firewood. Men on the other hand, want a "retirement farm" and may induce their wives into deforestation and setting up such a farm. Women do their fair share of the work and mass tilling of land for crops. Thus, general moisture retention is greatly reduced in such farm communities and extensive bushfires by herdsmen aggravate the situation.

In North West Cameroon women make long ridges on any piece of land for any crop. Further south, there is a mixture of mass tilling and construction of such an earth mound. Both techniques use bush burning, which roasts both vegetation and the humus topsoil.

Negative impacts include soil erosion, laying the topsoil bare, and facilitating moisture loss. To be efficient environmental managers, women need multidisciplinary counseling to join the rest of the community, readjust their attitudes and stop the mismanagement of natural resources around them. (p.517)

Groups of rural women in the NW province of Cameroon organized into what is known as Corn-Mill Societies in the early 1960s. They learned Community Development services on how to improve their welfare and their immediate surroundings. They were taught how to raise and plant trees that later provided cover crops, fruits and firewood. "Motivated reforestation" was the term they used whereby eucalyptus trees, fruit trees, and local shrubs sprung up in and around farms and villages. This was improved upon and diffused as an Agroforestry process, and such micro-scale projects through Tuang Method were introduced in many farm groups and communities in the Western Plateau zone of Cameroon. (p.518)

The participatory role of the members of each community at the base, should be encouraged to halt the negative effects on the environment and provide alternative methodologies of work. The women should not just be told to stop the practice of laying bare any land cover for farms, fuel wood, etc. Instead, they should also be motivated and compensated for planting trees on water catchments, pasturing improvements, preserving existing woodlots, milching, and crop association. They should put in place local community organs to foresee, check and prevent the possible return to any of the negative situations that result in environmental hazards.

Local groups and NGOs should lead and uphold such initiatives where Cameroonian women can take the role as partners to take better care of our earth. (p.520)

Energy Policies in East and Southern Africa

Renewable, clean energy and gender equality are essential for sustainable development and for tackling climate change, as implied in the Sustainable Development Goals (SDGs). Central to building more environmentally sustainable and climate resilient societies (Goal 13 and 15) are gender equality and women's empowerment (Goal 5) and women's and men's equal access to secure energy services (Goal 7). (UN Women Africa, 2017, p.1)

Throughout Africa, more than 600 million people or 50 percent of the population do not have access to sustainable, clean energy resources. Furthermore, energy combustion is mostly inefficient and toxic. About 600,000 people in Africa die each year as a result of household air pollution (Africa Progress Panel, 2015), and approximately 60 percent of these victims are women (ESMAP, 2011; IRENA 2013). Dependence on firewood and biomass for household energy, contributes to climate change and environmental degradation. This exacerbates air pollution, deforestation, land degradation and desertification. However, small-scale, off-grid clean energy solutions offer the potential to bring reliable energy to the most people in need and deprived, and have limited adverse impact on the environment. (African Development Bank Group, 2012; as cited in UN Women Africa, 2017 p.1)

Women are producers and consumers of energy in both urban and rural areas. In Africa, women are responsible for producing energy mainly through collecting biomass-based fuels and consuming energy in their household activities, microenterprises and agriculture. The disproportionate effects of energy poverty on women have been well documented (Glemarec, Bayut-Reboux and Waissbein, 2016). The lack of sustainable, clean energy services has economic, social, health and security-related impacts on women's lives, which hinders their economic empowerment. In fact, as a result of deforestation in Uganda, the average distance typically walked by women and children on collecting firewood increased between 1992 to 2000 from 0.6 km to 0.9 km. (UN DESA, 2010; as cited in UN Women Africa, 2017, p.2)

On the other hand, women can be powerful agents for change in the transition to and promotion of sustainable energy, through their role as the primary energy manager in households, both in urban and rural communities (UN Women and UNDP-UNEP PEI, 2015). However, women's participation in the energy sector decision-making and policy processes, as well as in its labor force, is low. They have limited opportunities to be informed of the policies that regulate the energy sector. In Africa, as in developing countries in general, female employees are a minority in most energy enterprises, particularly in managerial and technical positions. In Botswana, for example, the number of female professionals in energy-related institutions is less than 5 percent (Botswana, 2009; as cited in UN Women Africa, 2017, p.1).

More knowledge is needed to integrate gender issues more comprehensively into energy policies. In 2015, the United Nations Agency for Gender Equality and Women's Empowerment (UN Women) for East and Southern Africa, together with the Poverty Environment Initiative of the United Nations Development Programme and UN Environment Africa, published a working paper, "Empowering Women for Sustainable Solutions to Address Climate Change," which proposes assessment of the gender responsiveness of energy policies in the Region.

Fifteen of the 22 countries in East and Southern Africa were included in the analysis: Botswana, Kenya, Lesotho, Madagascar, Malawi, Mauritius, Mozambique, Namibia, Rwanda, Somaliland, South Africa, Swaziland, Tanzania, Uganda and Zimbabwe. For the remaining seven countries in the region, no national energy policy/strategy could be obtained for the purpose of the review.

Highlights of the review of 15 energy policies show that 60 percent have nor integrated gender. Specifically, 60 percent sa the need to enhance women's participation in the sector of policy and decision-making; 60 percent also recognized that women are energy providers and consumers at the level of the household; 60 percent acknowledge the need to mainstream gender in activities, programmes and energy policy; 53 percent cited that enhancing the access of women to energy services and technologies was a means for empowerment; and, 40 percent linked empowerment of women to enhanced environmental sustainability in the sector.

Decision-makers need to go further for good practice to spread across all countries: *Mainstream gender across all aspects of policy; Generate gender specific and sex-disaggregated data on energy access and consumption patterns; Ensure that women's voices are heard in policy development and in the energy sector; Allocate the appropriate financial and human resources to implement policies on the ground* (UN Women Africa, 2017, p.1, par. 3)

These actions will inform a new narrative which clearly depicts women as active participants and agents of change in the energy sector.

The Impact of Devolution on Women and Girls in Kenya

Another research on the Assessment of *The Impact of Devolution on Women and Girls in Kenya*, highlighted the need to respect gender equality and nondiscrimination in any democracy. Although women make up half of the population in Kenya, they remain invisible in leadership and decision-making processes. The Constitution of Kenya (CoK) 2010 was envisioned to end discrimination and gender inequalities, by promoting inclusion of women and other marginalized groups (youth, the elderly and persons with disabilities. (COG, UN Women & UNCDF, 2022, p. 2)

The study used a mixed method approach that combined quantitative and qualitative methods including questionnaires, focus group discussions, semi-structured interviews, case studies and participant observation. The assessment considered the review of existing statistics, policy and legal frameworks. Primary data was collected from 14 counties. (UN Women, 2022, *Impact,* p. 3)

County governments have made efforts to empower women and girls and promote gender equality in their development. Devolution provided women and other previously marginalized groups to participate in country level planning, budgeting and to integrate gender-responsive mechanisms in development planning and budgeting. This report presented "the women and girls' voices in the story of devolution," which synthesized and documented the impact of devolution in selected sectors across the 14 counties: Busia, Garissa, Isilio, Kajiado, Lamu, Mandera, Marsabit, Narok, Samburu, Tana River, Turkana, Wajir and West Pokot. (COG, UN Women & UNCDF, 2022, p. 2)

The overall objective of the study was to assess and document the impact of devolution on women and girls in four sectors, 1) Education & Health; 2) Agriculture; 3) Water and Sanitation; and, 4) Urban Planning. The 14 select counties were those supported by the UN Women project. (p.6)

Challenges

In Busia, there was scarcity of water in urban areas; schools lack protected water sources; women and girls walked long distances in search of water for domestic use; a few schools had sunk boreholes; county water purifier broke down and had not been replaced compromising quality water being used at domestic level; and lack of clean toilets for public use. In Garissa, the average distance to the nearest water point was 25 km. However, for residents of Garissa Town, this distance had reduced considerably. Various interventions had been undertaken to mitigate against water shortages during drought, but the same remained unreliable. Part of the community still practiced open defecation. Poor supply of water contributed to poor sanitation, causing many illnesses including diarrhea among children. (COG, UN Women & UNCDF, 2022, p. 33)

In 2017, Isiolo County elected a female senator and they had 2 female County Executive Committee members. The county had no elected female MCA, all being nominated. Isiolo County officially launched its very own Gender POlicy in September 2021, to help mainstream gender in the county's policies and processes. The policy provided a clear framework for implementation with concrete measures to prevent Sexual Gender-Based Violence (SGBV), Female Genital Mutilation (FGM), and radicalization of youth extremist groups. (p.37)

Opportunities

To combat the above challenges, the county affecting women and girls, the county outlined the following strategies: advocacy for equal opportunities for all genders to be prioritized; step up initiatives against FGM and to secure productive health for all women; and, implement affirmative action for women in employment and access to other opportunities.

The high female illiteracy levels at a staggering 85% hindered the ascendance of women to decision-making positions. The county prioritized civic education programmes to promote girl child education and encouraged women to take up political leadership and decision-making positions. (p.37)

THE ROLE OF WOMEN LEADERS IN THE EMEA REGION

The role of women in government is important for sustainable and inclusive policies. Women bring unique perspectives and experiences to the table that can lead to more comprehensive and effective solutions.

Women have traditionally been underrepresented in government, but as more women are elected to office, there has been an increase in policies that prioritize environmental protection and sustainable development. For example, studies have shown that female politicians are more likely to prioritize issues such as climate change and environmental protection.

Gender should not be the only determining factor in deciding who should lead in environmental preservation and crisis management. Both men and women can be effective leaders in these roles, and the most important factor is selecting a leader who has the necessary skills, experience, and passion for environmental protection, and can work collaboratively with others to develop and implement solutions to environmental challenges.

Women play a critical role in environmental management, as primary caregivers in their families and communities, and they are likely to be impacted by environmental issues such as air pollution, water contamination, and food security. As decision makers, women in government can work to address these issues and ensure that policies are inclusive and consider the needs of all members of society.

WOMEN TRAILBLAZERS IN EMEA

Several women have been helpful in environmental protection. According to Wangari Maathai, "We cannot tire or give up. We owe it to the present and the future generations of all species to rise up and walk!"(The Green Belt Movement, 2023, par. 1)

Wangari Muta Maathai (+) (1 April 1940 - 25 September 2011), Kenyan social, environmental and political activist and 2004 Nobel Peace Prize Laureate. She was the founder of the Green Belt Movement (GBM), an environmental non-government organization focused on planting trees, conserving the environment and advocating women's rights. They have planted over 50 million trees in Kenya. She wrote: *The Green Belt Movement; Unbowed: A Memoir; The Challenge for Africa;* and, *Replenishing the Earth.* She was featured in several books, and the Green Belt Movement was the subject of a documentary film, *Taking Root: the Vision of Wangari Maathai.* (Marlboro Productions, 2008 as cited in *Biography*, The Green Belt Movement, 2023, par. 1)

The United Nations (UN) Secretary General named Professor Maathai "UN Messenger of Peace" in December 2009, with a focus on the environment and climate change. In 2010, she was appointed to the Millenium Development Goals (MDG) Advocacy Group, to galvanize worldwide support for the achievement of the MDGs. She was a trustee of the Karura Forest Environmental Education Trust, established to safeguard the public land. In partnership with the University of Nairobi, she founded the Wangari Maathai Institute (WMI) for Peace and Environmental Studies. This brought together academic research (in land use, forestry, agriculture, resource-based conflicts, and peace studies) with the Green Belt Movement approach, and members of the organization. (The Green Belt Movement, 2023)

Dr. Gro Harlem Brundtland, is the first woman Prime Minister of Norway and the former Director-General of the World Health Organization (WHO). Dr. Brundtland champions health as a human right, and puts sustainable development in the international agenda. In 1981, she became Norway's first woman Prime Minister; 1998-2003, served as Director General of the WHO; 2007, Founding member of The Elders; 2007-2010, UN Special Envoy on Climate Change; 2013-2018, Deputy Chair of The Elders; 2018, appointed Co-Chair of the Global Preparedness Monitoring Board (GPMB); 2020, appointed member of the UHC (Universal Health Coverage) Movement Political Advisory Panel. (The Elders, 2023). Dr. Brundtland did not want to be an environmentalist or politician; but to be like her father, who was a specialist in rehabilitation medicine which was in demand after World War II. She also inherited her passion for political activism from her father. At an early age of 7, she was enrolled as a member of the Norwegian Labor Movement's children section and she has been a member ever since. She worked on children's health issues, including breastfeeding, cancer prevention, and other diseases (after attending Harvard School of Public Health). She returned to Oslo and the Ministry of Health in 1965, and became Director for Health Services for Oslo's schoolchildren. In 1983, the United Nations Secretary-General invited her to establish and chair the World Commission on Environment and Development - best known for developing the broad concept of sustainable development - and published its report *Our Common Future* in April 1987. The Commission's recommendation led to the Earth Summit in 1992, the United Nations Conference on Environment and Development (UNCED) in Rio de Janeiro. Dr. Brundtland is acting Chair of the Elders. (UN Foundation, n.d.)

Women bring unique perspectives and strengths to the table that can help create more sustainable future outcomes. Their leadership is essential for creating inclusive solutions to environmental challenges. The role of the media is also important. Media coverage of these women have impacted nations and have resulted in positive actions towards climate change mitigation efforts and sustainable development, especially in the MEA Region.

Impact of Governance Structures on Environmental Disclosures in MEA

Kilincarslan et al. (2020) investigated the impact of corporate governance structure on environmental disclosure practices in the Middle East and Africa (MEA), using a panel dataset of 121 listed (non-financial and non-utility) firms from 11 MEA countries over the period 2010-2017. They used alternative dependent variables and regression techniques, and applied these to various subgroups to improve robustness. Their empirical results strongly indicated that MEA firms with high governance disclosures tend to have better environmental disclosures practices. Characteristics of gender diversity, size, CEO chairperson duality and audit committee size of the board had a positive impact on MEA firm's voluntary environmental disclosures. The board independence however, had a negative influence. Because of the positive impact of board gender diversity on firms' environmental disclosures, there should be policy reforms to increase female directors. MEA corporations that aim to be more environmentally friendly should recruit females to top managerial positions. The study provided insights from the efficiency and legitimation perspectives of neo-institutional theory. It explained the relationship between MEA's firms' governance structures and environmental disclosures, but was limited to firms whose data were available from Bloomberg. The results have important practical implications for MEA policy makers and regulators.

CHALLENGES AND OPPORTUNITIES

Women face various challenges, such as (1) Gender inequality: Women often have a limited access to resources, including education, finances, and decision-making power, which limit their ability to participate in environmental efforts; (2) Gender-based violence: Women are often subjected to violence and harassment when they try to speak out on environmental issues; (3) Cultural societal norms: Women are often expected to take on traditional roles and may not be given equal opportunities to participate in decision-making processes related to environmental protection; and, (4) Lack of representation: Women are often underrepresented in leadership positions in environmental organizations and agencies, which limit their influence and impact.

Women also have opportunities to share, such as (1) Unique perspectives: Women can bring unique perspectives and experiences to environmental protection efforts, which can lead to more inclusive and effective solutions; (2) Collaboration and partnerships: Women are skilled collaborators and can work together with other stakeholders to create meaningful change; (3) Grassroots organizing: Women are often at the forefront of grassroots movements and can mobilize communities to take action on environmental issues; (4) Advocacy and policy: Women can use their voices to advocate for policies and regulations that promote environmental protection and sustainability. Overall, there is a need for greater gender equity in environmental protection efforts, and women have a critical role to play in leading the way.

ENVIRONMENTAL MANAGEMENT PRACTICES

The Dubai, United Arab Emirates Experience

Women Leaders in Dubai also played important roles.

Firstly, Her Excellency Razan Khalifa Al Mubarak, Managing Director of the Environment Agency Abu Dhabi (EAD), the Mohamed bin Zayed Species Conservation Fund, and current president of the International Union for Conservation of Nature. She also serves as UN Climate Change High-Level Champion. (Razan Al Mubarak, n.d.) Recently, Razan Al Mubarak urged decisive action by the private, local governments and civil society at the Bonn Climate Change Conference, held from 5-15 June, 2023. This event is a part of the formal United Nations negotiations on climate and marks an important stop on the road to the UN Climate Change Conference (COP 28) which will take place in the UAE in December 2023. (Environment Agency Abu Dhabi, 2021)

Secondly, Her Excellency Dr. Aisha Bint Buti Bin Bashr (Director General of the Smart Dubai Office), responsible for implementing Dubai's smart city vision. Under her leadership, Smart Dubai has launched several initiatives to promote sustainability, including the Dubai Paperless Strategy, which aims to reduce paper usage in government offices.

Dr. Aisha Bin Bashr leads the creation of "The Smart City Index" - the first-ever benchmark for smart city implementations across the globe in cooperation with ITU and the United Nations. She also serves as a member of the World Economic Forum's Global Future Councils; The Fourth Industrial Revolution's smart cities readiness index team; and the World Happiness Council where she heads "The Council of Happy Cities." (World Government Summit, 2023)

Thirdly, Her Excellency Mariam Bint Mohammed Saeed Almheiri (Minister of State for Food and Water Security), has been working to ensure food and water security in the UAE through sustainable agriculture, water conservation, and the use of renewable energy.

H.E. Mariam Almheri was appointed as Minister of State for Food and Water Security in July 2020. Previously, she was Minister of State for Food Security in October 2017. She is currently representing the UAE in the Food and Agriculture Organisation of the United Nations (FAO) as well as the International Center of Biosaline Agriculture (ICBA). She is a member of the Board of Directors in Abu Dhabi Agriculture and Food Safety (ADAFSA) and a member of the Board of Trustees of Khalifa International Award for Date Palm and Agricultural, the Dubai Future Foundation (DFF) and the World Economic Forum's Global Future Council for Food Systems Innovation. (United Arab Emirates University, 2022).

Young Women at the Forefront of Climate Action in Europe

Women and girls are powerful leaders and change-makers in the design and implementation of climate action. Their leadership, knowledge and participation in climate responses today, realized solutions for a sustainable planet and gender-equal world. Across the region, they are advancing feminist climate justice and leading the charge on climate change adaptation, mitigation and response by mobilizing local, national, regional and global climate movements and harnessing the transformative power of feminist leadership to face the unprecedented challenges of our times.

International Women's Day on 8 March 2022, highlighted some of the young women champions driving feminist climate action across Europe with the theme "Gender equality today for a sustainable tomorrow".

Gabriela Isac (Moldova, Europe), 29, is an environmental activist, co-founder of the *Seed It Forward*, a volunteer agroforestry initiative and a project coordinator at the EcoVisio grass-roots ecological non-profit in Moldova. With the *Seed It Forward* team, she organizes tree-planting events, consults civil society organizations, local public authorities, schools and general public on environmental issues, and educates them through informational materials on trees, composting and permaculture. They have planted over 50,000 trees and bushes, while their recent environmental campaign reached more than 1.5 million people online.

Ania Sauku (Albania, Europe), 19, is an active voice for gender equality, climate action and youth empowerment in Albania, and is one of the incumbent Albanian Youth Delegates to the United Nations, where she advocates for climate issues and sustainable development and shares the perspective of youth in her country. She raises awareness on climate change and feminism and how they are inextricable from one another. Sauku believes that for many people in Albania, climate change is still not an issue, and that gender equality and climate are not related. Together with her team, she organizes movie nights on the environment, protests and marches for climate justice, and other educational initiatives to raise awareness about climate change and intersectional feminism.

Sanne Van de Voort (The Netherlands, Europe), 27, is Advocacy Officer for *Women Engage for a Common Future* (an international ecofeminist network), and an NGO representative on the Dutch Delegation *Commission on the Status of Women* (CSW) in 2022. As an Advocacy Officer, she believes each individual is affected differently by climate change and they can lend their unique experiences to finding solutions and that feminist climate justice recognizes the intersectionality of climate and environmental issues. She works to ensure that Dutch and international decisions taken on climate and environmental issues reflect the needs, perspectives and solutions of women and feminists across the world, especially from the Global South. She contributes to preparations and priority-setting in the Dutch Government's CSW delegation alongside other Dutch civil society organizations, as a Dutch NGO representative to CSW. (UN Women, 2022, March 11)

The Plastic-Free Coastlines Initiative

British environmentalist Emily Penn, leads *The Plastic-Free Coastlines Initiative*. The project aims to reduce plastic pollution in the world's oceans by working with educating local communities on the impacts of plastic pollution, implementing plastic recycling programs, and promoting sustainable tourism practices. She established *The Plastic-Free Coastlines Initiative* in 2016, an ongoing project that has implemented plastic waste reduction strategies in coastal communities in countries such as Indonesia, the Philippines and Vanuatu. The project continues to work towards reducing plastic pollution in the world's oceans and has also promoted sustainable tourism practices and educated local communities on the impacts of plastic pollution.

Overall, these projects have demonstrated the importance of innovative and sustainable approaches to environmental management and protection, and their impact can be seen in the positive changes they have brought in their respective regions.

In October 2019 Emily's organization eXXpedition embarked on a 2 year global circumnavigation involving 300 amazing women to study and find solutions to for the ocean plastic issue. (Penn, n.d.)

The Bonobos of Congo (Central Africa)

Claudine André was born in Belgium and moved to the Democratic Republic of Congo (DRC) with her family when she was three years old. While living in the capital Kinshasa with her husband Victor, years later, she ran a luxury boutique while raising her five children. At that time, there was political unrest and looting, and extreme poverty became the norm. The animals at the Kinshasa Zoo were neglected and at risk of starving, so Claudine volunteered and she learned that the zoo had taken in a sick, baby bonobo. People were hunting adult bonobos for meat due to starvation because of the war and leaving bonobo babies as orphans.

Baby bonobos cannot survive without their mother, because bonobos need their mother's love and would rather die without it.

Claudine's gift for saving bonobos spread, and people from different places began giving orphaned bonobos to her. She founded the nonprofit organization Amis des Bonobos du Congo (ABC) in 1994 to get support from other people who want to help orphaned bonobos. In 2002, she opened *Lola ya Bonobo* sanctuary.

In 2009, after so many years of planning and conservation efforts, Claudine finally found the right place and she led the first successful rewindling of Bonobos and more than 30 bonobos are back in the wild. Now, more than 60 Bonobos of all ages wander, eat, and play on 75 acres of Lola ya Bonobo Sanctuary.

For Claudine, it has been a gift to dedicate her life to conservation which should be the work of everyone else. She believes we are all responsible for our one and only beautiful planet. (Friends of Bonobos, 2023)

WOMEN LEADERS IN THE BANKING SECTOR

The banking sector plays a crucial role in promoting sustainable development and addressing environmental challenges. Women leaders and board members in the banking sector in the EMEA region have been instrumental in driving change and promoting sustainable practices.

Contributions in the Banking Sector

The women leaders in the banking sector initiated a lot of developmental programs and environmental protection projects to mitigate climate change and ensure that the next generation enjoy the beauty of our God-given resources. These initiatives are not limited to these efforts, namely, (1) Developing green financing: Women leaders and board members in the banking sector have led efforts to create new financial instruments and products that support sustainable development, such as green bonds, sustainability-linked loans, and renewable energy financing. (2) Integrating Environmental, Social and Governance (ESG) considerations: Women leaders and board members have been instrumental in integrating ESG considerations into the core business strategies of banks. They have helped to ensure that environmental risks and opportunities are identified, assessed, and managed appropriately. (3) Promoting sustainable practices: Women leaders and board members have been encouraging their institutions to adopt sustainable practices in their operations, such as reducing carbon emissions, improving energy efficiency, and reducing waste.

(4) Advocating for policy changes: Women leaders and board members have been engaging with policymakers and regulators to promote adoption policies that encourage sustainable practices, such as tax incentives for renewable energy projects and regulatory frameworks that support green finance.

Birindelli, et al. (2019) analyzed the impact of women leaders on environmental performance and included a sample of 96 listed banks in the EMEA region from 2011 to 2016, and explored gender diversity in leadership positions. This was done through women in the board of directors, chief executive officer gender, and the interaction between these two aspects. Results suggested that mainly, there was a nonlinear relationship between women directors and the environmental performance of banks. The female executive officers played a strategic role in shaping the relationship, and it can be concluded that leader-gender diversity was an important impetus for environmental sustainability in banks. These banks were increasingly involved in environmental issues, through their lending activity in the process, either directly or indirectly. (p.1485)

THE ROLE OF WOMEN LEADERS IN MEDIA

Women leaders in the media use their platform to raise awareness, educate the public, and advocate for sustainable solutions. They can help to shape public opinion and influence policy decisions related to environmental issues. They can bring attention to environmental challenges and highlight the impact on communities and individuals, especially those who are marginalized and often disproportionately affected by environmental degradation. They can shed light on the solutions and innovations that are being developed to address these challenges. In addition, women leaders in the media can help amplify the voices of other women and marginalized groups who are working on environmental issues. They can provide a platform for these individuals to share their stories and perspectives, which can lead to a more diverse and inclusive approach to environmental management.

Women in Media and their Contributions to Environmental Management

Women in the media who have promoted environmental solutions:

First, Christiane Amanpour (international correspondent and anchor), has covered environmental issues such as climate change, deforestation, and pollution. She has highlighted the urgency of addressing these issues and interviewed experts and policymakers to discuss potential solutions. She is CNN's chief international anchor of "Amanpour," the network's award-winning flagship global affairs program which also airs on PBS in the United States. Fearless and uncompromising, she has reported the aftermath of many humanitarian crises including the devastating 2012 earthquake in Haiti, the 2011 Japanese tsunami, and Hurricane Katrina, where she visited a community center which had been converted to a makeshift morgue for victims of the storm. Aside from being an anchor and reporter, Amanpour is an active rights campaigner. A board member of the Committee to Protect Journalists, the Centre for Public Integrity and the International Women's Media Foundation, she has used her profile to raise awareness of key global issues and journalists' rights. She has interviewed educational rights activist Malala Yousafzai for CNN on several occasions — bringing focus to her courage and international advocacy work. In May 2014 she used an appearance on BBC television to raise awareness of the plight of the 200 Nigerian schoolgirls abducted by Boko Haram - asking British Prime Minister David Cameron to join the #BringBackOurGirls campaign.

Amanpour received major television journalism awards, including twelve News and Documentary Emmy Awards, four Peabody Awards, two George Polk Awards, three duPont-Columbia Awards and the Courage in Journalism Award. She has received nine honorary degrees, has been named CBE and inducted into the Cable Hall of Fame. She is an honorary citizen of Sarajevo and a UNESCO Goodwill Ambassador for Freedom of the Press and the Safety of Journalists. She graduated summa cum laude from the University of Rhode island with a Bachelor of Arts in Journalism. (CNN, 2023)

Second, Lucy Siegle (journalist and author), has written extensively on sustainable fashion and the impact of the fashion industry on the environment. She has highlighted the need for sustainable fashion practices and has called on consumers to make more conscious choices when it comes to their clothing. She is dubbed as *"a journalist who writes about ethical living."*

Siegle advocates for extending the lifespan of garments, selling, and reusing them, which is critical in the fight to bring sanity in the fashion industry. (Siegle, 2022)

Freedom of information requests reveal that, on average, 11% of rubbish collected for recycling is incinerated (Siegle, 2021, par. 2).... In England, more waste is burned than recycled - 11.6 million tonnes was incinerated in 2019, while 10.9 million was sent for recycling. (par. 3)

One of Siegle's articles, also discussed the importance of using reuse, recycling and better design for industrialized nations to become circular economies to tackle the problem of pollution. (Siegle, 2021)

Third, Mariella Frostrup (journalist and broadcaster), has highlighted the need for action on climate change and has interviewed climate experts to discuss potential solutions. She is a broadcaster and journalist based in the UK who is an advocate for equal rights and women's causes, Save the Children's Gender Ambassador and Founder Trustee of The Great Initiative. She is also a Patron of the Shark Trust and supporter of Comic Relief. She has written for almost every mainstream UK Newspaper and various magazines: *The Times*, *Sunday Times*, *Daily Mail*, *Mail Sunday*, *Harper's Bazaar*, *Vanity Fair*, *Observer* newspaper's erstwhile Agony Aunt and contributor for *Family Traveller* magazine.

In one of her documentaries "Britain's Novel Landscapes with Mariella Frostrup," she sets off to discover some of our most impressive and dramatic landscapes that helped inspire Britain's best loved female authors to write their great novels.

In the series, she featured breathtaking landscapes, shared incredible insights and had access to fascinating buildings and places revealing the real role of different regions and landscapes in our literary history. (Frostrup, 2023).

Women Environmental Activists in Media

Here are women activists in the media advocating changes in environmental issues in the EMEA Region:

First, Hindou Oumrou Ibrahim is an environmental activist from Chad, known for her advocacy on climate change and indigenous rights. She is the founder of the Association for Indigenous Women and Peoples of Chad (AFPAT), which works to promote the rights and voices of indigenous peoples in Africa. She has been involved in various international forums, including the United Nations Framework Convention on Climate Change (UNFCCC), where she advocates for the inclusion of indigenous peoples in the decision-making process regarding climate change. She is also an expert in the adaptation and mitigation of indigenous peoples to climate change, is a member of the Mbororo pastoralist people in Chad and President, Association for Indigenous Women and Peoples of Chad (AFPAT). As an advocate for the greater inclusion of indigenous people and their knowledge and traditions in the global movement to fight the effects of climate change, she received the Pritzker Emerging Environmental Genius Award and was appointed as a United Nations Sustainable Development Goals Advocate. She serves as a Member of the United Nations Sustainable Development Goals Advocate; a Member of the United Nations Permanent Forum for Indigenous Issues; Member of the Indigenous Peoples of Africa Coordinating Committee (IPACC); Member of the Advisory Committee to the Secretary-General's 2019 Climate Action Summit; and Conservation International Senior Indigenous Fellow. She was listed by Time Magazine as one of 15 women championing action on climate change in 2019. (WEF, 2023).

Second, Isatou Ceesay is a Gambian environmental activist, known for her work on plastic pollution. She is the founder of Women's Initiative Gambia (WIG), which focuses on waste management and recycling, and the organization has trained over 100 women in recycling techniques and has helped reduce waste in Gambia while providing economic opportunities for women. She has worked for the U.S. Peace Corps office in Gambia, the Swedish organization Future in Our Hands, and as a consultant for development organizations. Her recycling project began in 1998, even amidst much resistance and taunting, it is still thriving today. She was awarded a TIAW "Difference Maker" award in Washington, DC in 2012. She is humbled by the fact that there is a book about her and hopes it will inspire others to join her cause or be a changemaker in their own communities. Her book: **"***One Plastic Bag: Isatou Ceesay and the Recycling Women of the Gambia"* tells the inspiring story of five women who creatively dealt with their village's plastic trash problem. Despite limited resources and ridicule, Isatou and her friends persevered for more than a decade, eventually realizing economic empowerment through their recycled plastic purse project. "The Bags" idea of making recycled plastic bag purses may seem like a simple idea - but it was life-changing to the villagers of Njau, Gambia in the late 1990s. The project continues to save livestock, prevent greater outbreaks of disease, and make the soil healthier. In the early years of the project, many of the women and girls who joined the co-op did not know how to read or write. Eventually, these resources provided classes and business opportunities. The women have also been educated about Fair Trade principles and have gained a significant amount of confidence in themselves, their business skills, and their respect for the environment. They are always trying out new designs - from recycled cell phone and passport holders to dolls and ornaments. (Paul, 2014)

Third, Greta Tintin Eleonora Thunberg FRSCGS is a Swedish environmental activist, 18 years old internationally recognized climate activist, who emerged as the voice of the youth. Her voice outside the Swedish parliament resonated with numerous youths around the world in 2018, to initiate the urgent climate action from political leaders and inspired youth-led climate strikes to address the climate crisis. She was nominated for the Nobel Prize in 2019 and also listed as Time's Person of the Year the same year. Apart from the protests and awareness campaigns, she donated all her winnings of $1 Million Euro to climate change charities, for the betterment of human civilization and to protect the environment. (Volunteerfdip, 2023).

Thurberg's activism began when she persuaded her parents to adopt lifestyle choices that reduced their own carbon footprint. In August 2018, at age 15, she started spending her Fridays outside the Swedish Parliament to call for stronger action on climate change by holding up a sign reading *Skolstrejk for klimatet* (School Strike for Climate). Thunberg initially gained notice for her youth and her straightforward and blunt speaking manner. (Nordstrom, 2019) Soon other students engaged in similar protests in their own communities. Together they organized a school climate change strike movement under the name *Fridays for Future*.

After Thunberg addressed the 2018 United Nations Climate Change Conference, student strikes took place every week all around the world. In 2019, there were multiple coordinated multi-city protests involving over a million students each. (Haynes, 2019) To avoid carbon-intensive flying, Thunberg sailed on a yacht to North America, where she attended the 2019 UN Climate Action Summit. Her speech there, in which she exclaimed "How dare you?" was widely taken up by the press and incorporated into music. (Clark, 2019). She speaks fluent English, and most of her public interactions are in English. Her sudden rise to world fame made her both a leader in the activist community (Wallace-Wells, 2019) and a target for critics, (The New York Times, 2020) especially due to her youth. Her influence on the world stage has been described by *The Guardian* and other newspapers as the "Greta effect." (Watts, 2019). She received numerous honors and awards, including an honorary Fellowship of the Royal Scottish Geographical Society, inclusion in Time's 100 most influential people, being the youngest *Time* Person of the Year, inclusion in the Forbes list of the World's Most Powerful Women in 2019 (Forbes, 2022); and, nominations for the Nobel Peace Prize in 2019, 2020, 2021, and 2022 (Read, 2021; Solsvik, 2020). She is also included in the 100 Most influential people in climate policy 2022/23. (Apolitical, 2022)

Environmental conservation campaigns led by celebrities having huge fan following and their involvement can draw attention to the public, and they have played a significant role in contributing towards environmental issues, conducting charities, making donations and supporting the funding of different organizations that narrates the role of combating climate change. (Volunteerfdip, 2023).

Overall, these examples demonstrate that women have been effective in managing environmental issues in the EMEA Region. Their contributions have been instrumental in promoting sustainable development, combating environmental degradation, and creating economic opportunities for women.

Challenges and Opportunities

Here are some of the challenges that women leaders in media face:

Gender Bias and Stereotyping: Women leaders in media often face gender bias and stereotyping, which can limit their opportunities and hinder their professional growth. They face additional scrutiny and criticism for their actions, which can undermine their authority and credibility.

Lack of Representation: Women leaders are often underrepresented in the media industry, especially in leadership positions, which can lead to a lack of diversity of perspectives and ideas. This may result in women leaders being overlooked for promotions or not being taken seriously in decision-making processes.

Discrimination and Harassment: Women leaders in environment management may face discrimination and harassment, which can create a hostile work environment and negatively impact their work. They may face harassment or discrimination from colleagues or superiors, which can create a hostile work environment.

Limited Access to Resources: Women leaders in media may face challenges in accessing resources, including funding, networks, and mentorship, which can limit their ability to pursue innovative projects and achieve their goals. They may struggle to access funding or resources for their environmental projects due to bias or limited networks.

Balancing Work and Family: Women leaders in environmental management may struggle to balance work and family responsibilities, which can lead to burnout and limit their professional opportunities. This can be particularly challenging because they face social pressure to prioritize caregiving over their careers.

Where there are challenges, there are also opportunities for women leaders in media:

Promoting Diversity and Inclusion: Women leaders can help promote diversity and inclusion in the media industry by advocating for more equitable hiring practices, creating mentorship opportunities for underrepresented groups, and challenging bias and discrimination. They can create mentorship programs or networks for underrepresented groups, such as women of color or LGBTQ+ individuals, to increase diversity in the media industry.

Fostering Innovation: Women leaders can bring new perspectives and innovative solutions to environmental challenges, which can help create more sustainable and effective approaches to environmental management in the media industry.

Building Networks: Women leaders can build strong networks of support and collaboration, which can help amplify their voices and increase their impact in the media industry.

Leading By Example: Women leaders can serve as role models for other women and girls, inspiring them to pursue careers in environmental management and the media industry.

Creating Opportunities for Others: Women leaders can create opportunities for other women and girls, including mentorship, training, and access to resources, which can help increase diversity and build a stronger and more resilient media. They can create opportunities for others by advocating more equitable hiring practices or funding for environmental projects led by underrepresented groups.

ANALYSIS AND DISCUSSIONS

Among the key areas where women leaders have made notable contributions towards sustainable development include the Banking industry, Energy sector, Governments, and Activism. Additionally, women in the media also play a crucial role in shaping public opinion and awareness on environmental issues.

There is a positive trend at the policy level in integrating gender into energy policies. In fact, 60 percent of the policies acknowledge the need for enhancing women's participation in policy and decision-making in the sector; while 55 percent note the need to enhance women's access to energy services and technologies as a means of empowerment; and, 40 percent of the policies highlight the link between women's empowerment and enhanced environmental sustainability in the sector.

The remaining challenge is that gender is considered a stand-alone chapter of policy documents and women are still mainly considered vulnerable, rather than recognized as key agents of change towards sustainable energy solutions. In Kenya's and Malawi's policies, the needs of women, youth and the physically challenged are lumped together without reference to the different vulnerabilities and most seriously, two policies in the region do not include any reference to gender, women or men.

Socially structured gender roles, identities and underlying power dynamics affect how women and men access and use energy and participate in decisions and investments.

Overall, women leaders and board members in the banking sector in the EMEA region have made significant contributions to environmental management. Their efforts have been critical in promoting sustainable practices, creating new financing options, and advocating for policy changes that support sustainable development.

Moreover, women activists and their contributions have been instrumental in promoting sustainable development, combating environmental degradation, and creating economic opportunities for women.

In addition, women in the media have played a significant role in raising awareness about environmental issues and advocating for sustainable development. Reporting on environmental abuses, issues on policies and showcasing the importance of environmental protection and the need to mitigate climate change, informed viewers and listeners on the importance of preserving our natural resources and contributing to help save our planet for future generations.

CONCLUSION: LOOKING TO THE FUTURE

Women leaders in environmental management in the EMEA region play a critical role in addressing environmental challenges in the banking sector, energy sector, governments, activism, and the media industry. Despite facing numerous challenges, including bias and limited access to resources, women leaders have the potential to foster innovation, promote diversity and inclusion and create opportunities for others.

The notable contributions of these women leaders - Wangari Muta Matthai, Gro Harlem Brundtland, Gabriela Isac, Ania Sauku, Sanne Van de Voort, Emily Penn, Claudine André, Christiane Amanpour, Lucy Siegle, Mariella Frostrup, Hindou Oumrou Ibrahim, Isatou Ceesay, Greta Tintin Eleonora Thunberg - highlight the important work that women are doing in the EMEA region to address environmental issues and advocate for more sustainable and equitable practices.

It is clear that women leaders in the media have the power to shape the conversation around environmental issues and influence policy and decision-making in the industry. As we continue to face urgent environmental challenges, it is important to recognize and support the important work of women leaders in environmental management in various sectors, especially, in the media industry.

REFERENCES

Africa Progress Panel. (2015). *Africa progress report: power people planet - seizing energy and climate opportunities.* Africa Progress Panel.

African Development Group (2012). *Annual development effectiveness review.* African Development Group.

Apolitical.co. (2022). *100 Most influential people in climate policy 2022/23.* Apolitical.co. https://apolitical.co/list/en/most-influential-climate-100-2022

Birindelli, G., Iamuzzi, A. P., & Savioli, M. (2019, November). The impact of women leaders on environmental performance: Evidence on gender diversity in banks. *Corporate Social Responsibility and Environmental Management, 26*(6), 1485-1499. doi:1002/csr.1762

Botswana. (2009). *National energy policy for Botswana - Final Draft.* Botswana.

Clark, R. (2019, October 15). What I learned about writing from reading GretaThunberg's speech to the UN. *Poynter.*

CNN (Cable News Network). (2023) "Christiane Amanpour: Chief international anchor." *Cable News Network.* https://edition.cnn.com/profiles/christiane-amanpour-profile

COG. UN Women & UNCDF (Council of County Governors, UN Women and the United Nations Capital Development Fund). (2022). The Impact of Devolution on Women and Girls in Kenya. *Assessment of the Impact of Devolution on Women and Girls Report.* Council of Governors (CoG) in partnership with UN Women and the United Nations Capital Development Fund. www.maarifa.cog.go.ke

Environment Agency Abu Dhabi. (2021). *H.E. Razan Khalifa Al Mubarak.* EAD. https://www.ead.gov.ae/en/About-Us/Our-Leadership/Leaders/Razan-Khalifa-Al-Mubarak

ESMAP (2011). *Improved cookstoves to conserve biomass resources.* ESMAP.

UN Foundation. (n.d.). Gro Harlem Brundtland (Norway). UN Foundation. https://unfoundation.org/who-we-are/our-board/gro-harlem-brundtland-norway-vice-chair/

Friends of Bonobo. (2023). *Our founder, Claudine André: A hero for bonobos.* Bonobos. bonobos.org/our-founders-story.

Frostrup, M. (2023). Mariella Frostrup. *MariellaOnline.* https://mariellafrostrup.co.uk/britains-novel-landscapes-with-mariella-frostrup-which-airs-on-thursdays-at-9pm-from-3-february-on-more4/

Glemarec, Y., Bayut-Renoux, F., & Waissbein, O. (2016). *Removing barriers to women entrepreneurs' engagement in decentralized sustainable energy solutions for the poor.* Research Gate.

Haynes, S. (2019, May 24). Students from 1,600 cities just walked out of school to protest climate change: It could be great Thunberg's biggest strike yet. *Time.*

IPCC (Intergovernmental Panel on Climate Change). (2018, October 6). Summary for Policymakers. In: Masson-Delmotte, V., P. Zhai, H.-O.Pörtner, D. Roberts, J. Skea, P.R. Shukla, W. Moufouma-Okia, C. Péan, R. Pidcock, S. Connors, J.B.R. Matthews, Y. Chen, X. Zhou, M.I. Gomis, E. Lonnoy, T. Maycock, M. Tignor, and T. Waterfield (eds.) Global Warming of 1.5oC. An IPCC Special Report. Cambridge University Press.

IRENA (International Renewable Energy Agency) (2013). *Renewable energy and jobs, annual review.* IRENA.

Kilincarslan, E., Elmagrhi, M. H., & Li, Z. (2020) Impact of Governance Structure on Environmental Disclosures in the Middle East. *The International Journal of Business in Society.*

MacGrath, M., Jones, N., & Burho, E. (2022). The world's 100 most powerful women. *Forbes.* https://www.forbes.com/lists/power-women/?sh=6c438baf5a95

Ngwa, E. N. (1995, April). The role of women in environmental management: An overview of the Cameroonian situation. *GeoJournal, 35*(4), 515–520. https://www.jstor.org/stable/41146448. doi:10.1007/BF00824366 PMID:12291368

Nordstrom, Louise (2019, January 25). The Swedish teen holding world leaders accountable for climate change. *France 24.*

Paul, M. (2014) Isatou Ceesay. *One Plastic Bag.* http://oneplasticbag.com/the-women

Read, R. (2021) Opinion: This year's Nobel Peace Prize should've gone to Greta Thunberg. *The Independent.*

Siegle, L. (2021). Revealed: why hundreds of thousands of tonnes of recycling are going up in smoke. *The Guardian.* https://www.theguardian.com/environment/2021/mar/07/revealed-why-hundreds-of-thousands-of-tonnes-of-recycling-are-going-up-in-smoke

Siegle, L. (2022, February 25). Can we enjoy fast fashion without destroying the planet? *The Guardian.* https://www.theguardian.com/environment/2022/feb/25/rent-clothes-fashion-sustainable-climate

Solsvik, T. (2020, February 26). Climate activist Thunberg heads growing field of Nobel Prize candidates. *Reuters.*

The Elders. (2023). Gro Harlem Brundtland. *The Elders.* https://theelders.org/profile/gro-harlem-brundtland

The Green Belt Movement. (2023) Wangari Matthai. Biography. *The Green Belt Movement.* https://www.greenbeltmovement.org/wangari-maathai/biography

The Guardian. (2023). Lucy Siegle. *The Guardian.* https://www.theguardian.com/profile/lucysiegle

The New York Times. (2020). Greta Thunberg joins climate march on her last day in Davos. *The New York Times.*

UAE MOCCAE (United Arab Emirates Ministry of Climate Change and Environment). (2021). *The UAE State of Climate Report: A Review of the Arabian Gulf Region's Changing Climate & Its Impacts.* UAE MOCCAE. www.moccae.gov.ae

UN DESA (United Nations Department of Economic and Social Affairs). (2010). *The world's women 2010, trends and statistics.* UN.

UNDP-UNEP PEI (United Nations Development Programme - UN Environment Programme Poverty-Environment Initiative) (2015). *Mainstreaming Environment and Climate Change for Poverty Reduction and Sustainable Development - A Handbook.* UN.

UNFCCC. (2023, June 19,). Razan Al Mubarak. razanalmubarak.com. https://www.ead.gov.ae/en/About-Us/Our-Leadership/Leaders/Razan-Khalifa-Al-Mubarak

United Arab Emirates University. (2022, November 21). H.E. Mariam Bint Mohammed Saeed Hareb Almheiri. *Board of Trustees.* United Arab Emirates University. https://www.uaeu.ac.ae/en/about/board-of-trustees/mariam-almheiri.shtml

Wallace-Wells, D. (2019, September 17). It's Greta's world. *Intelligencer.*

Watts, J. (2019, April 23). The Greta Thunberg effect: At last MPs focus on climate change. *The Guardian.*

WEF (World Economic Forum). (2023). *Hindou Oumarou Ibrahim*. World Economic Forum. https://www.weforum.org/people/hindou-oumarou-ibrahim

WMO (World Meteorological Organization). (2022). WMO's state of the climate in Europe Report for 2022 urges immediate action on Europe's climate crisis. *State of Climate in Europe 2022*. WMO. https://community.wmo.int/en/news/wmos-state-climate-europe-report-2022-urges-immediate-action-europes-climate-crisis

Women U. N. (2022). *Impact of devolution on women and girls in Kenya*. UN.https://africa.unwomen.org/sites/default/files/2022-12/IMPACT%20OF%20DEVOLUTION11%20eVersion.pdf

Women U. N. (2022, March 11). *Five young women on the forefront of climate action across Europe and Central Asia*. Europe and Central Asia. https://eca.unwomen.org/en/stories/news/2022/03/five-young-women-on-the-forefront-of-climate-action-across-europe-and-central-asia

Women U. N. (2022). *Impact of devolution on women and girls in Kenya*. UN. https://africa.unwomen.org/sites/default/files/2022-12/IMPACT%20OF%20DEVOLUTION11%20eVersion.pdf

U. N. Women Africa (2017). Gender, Energy and Policy - A Review of Energy Policies in East and Southern Africa. UN Women Africa.

KEY TERMS AND DEFINITIONS

Advocacy and Policy: Women can use their voices to advocate for policies and regulations that promote environmental protection and sustainability.

CGGCF: This stands for Commission for Global Governance and Commission of the Future.

Collaboration and Partnerships: Women are skilled collaborators and can work together with other stakeholders to create meaningful change.

Cultural Societal Norms: Women are often expected to take on traditional roles and may not be given equal opportunities to participate in decision-making processes related to environmental protection.

Devolution: The transfer or delegation of power to a lower level, especially by central government to local or regional administration, allowing women to have an opportunity to be part in the decision-making process

Gender Inequality: Women have limited access to resources, including education, finances and decision-making power, which limit their ability to participate in environmental efforts.

Gender-Based Violence: Women are often subjected to violence and harassment when they try to speak out on environmental issues.

Grassroots Organizing: Women are often at the forefront of grassroots movements and can mobilize communities to take action on environmental issues.

Lack of Representation: Women are often underrepresented in leadership positions in environmental organizations and agencies, which limit their influence and impact.

UNGA: This stands for United Nations General Assembly.

Unique Perspectives: Women can bring unique perspectives and experiences to environmental protection efforts, which can lead to more inclusive and effective solutions.

UNFCCC: This stands for United Nations Framework Convention on Climate Change.

Chapter 8
Women Leadership in Fulfilling the Sustainable Development Goals in India

Somya Khare
Jagran Lakecity University, India

Rohit Mishra
https://orcid.org/0009-0000-4847-860X
Jagran Lakecity University, India

ABSTRACT

Rio in 1992 known as United Nations Conference on Environment and Development focused mainly on issues relating to environmental resources and intergenerational concerns of global warming, ozone layer depletion collectively termed as 'green agenda.' However, with the development of sustainable development goals and growing awareness of 'brown agenda' which is directly connected to the development process and environmental problems, it becomes quintessential to talk about gender equality.

INTRODUCTION

Goals to "promote gender equality and empower women and to ensure environmental sustainability" were included in the Millennium Development Goals (MDGs) adopted at the United Nations Millennium Summit in New York in 2000. According to the Rio Declaration of 1992, it was asserted that the involvement of women in environmental management and development is of utmost importance, as their complete participation is crucial for the attainment of sustainable development. The Millennium Development Goals acknowledge the imperative of advancing gender equality and women's empowerment, as well as the necessity of mitigating poverty and fostering sustainable environmental conservation. One of the objectives pertaining to the aim of achieving environmental sustainability is to reduce by half, by the year 2015, the percentage of individuals lacking sustainable access to safe drinking water and basic sanitation (United Nations, 2019).

DOI: 10.4018/978-1-6684-5986-7.ch008

This was a major focus of India's Ten-Year Plan (2002-2007). The National Policy for the Empowerment of Women was given top priority by being translated into action with a three-fold strategy focusing on economic empowerment, social empowerment, and gender justice. However, the Millenium Development Goals were replaced with Sustainable Development Goals (SDGs) which were established during the United Nations Conference on Sustainable Development in Rio de Janeiro in 2012. The aim was to generate a collection of universally applicable objectives that effectively address the pressing environmental, political, and economic challenges confronting our global community (Patel, 1996). In contrast to the Millennium Development Goals (MDGs), which exclusively focused on developing nations, the Sustainable Development Goals (SDGs) encompass all countries, regardless of their economic status, including high-income, middle-income, and low-income nations. The Sustainable Development Goals (SDGs) are characterised by a nationally owned and country-led approach, allowing each country the autonomy to develop a national framework for attaining the SDGs (Gwagwa, 1991).

The Sustainable Development Goals (SDGs) represent a resolute commitment to effectively conclude ongoing endeavours and address the most urgent global challenges currently encountered. The interconnection of all 17 Goals implies that the achievement of one goal has a direct impact on the achievement of others. The management of our delicate natural resources is influenced by the challenge posed by climate change. Additionally, the eradication of poverty is facilitated by the attainment of gender equality and improved health outcomes. Furthermore, the promotion of peace and inclusive societies contributes to the reduction of inequalities and the enhancement of economic prosperity. In summary, this presents a significant opportunity to enhance the quality of life for forthcoming generations which has been incorporated in India.

The Five-Year Plans (FYPs) provide a comprehensive framework and direction for India's policies, programmes, and projects. Annual plans, commonly referred to as APs, are utilized to implement the objectives outlined in five-year plans. The Twelfth Five-Year Plan from 2012 to 2017 was a comprehensive and centralised national economic blueprint. The subtitle of the Twelfth Plan articulates the comprehensive vision and aspirations that it seeks to accomplish, which are centred around achieving 'faster, sustainable, and more inclusive growth.

Presently, India does not have an operational Five-Year Plan. Subsequently, the government of India made the decision to cease the implementation of the Five-Year Plan system and instead embraced a more extensive development strategy known as the NITI Aayog. NITI Aayog is the apex think tank of the government of India which functions as the primary public policy research institution of the Government of India. It is responsible for driving economic development and promoting cooperative federalism by engaging State Governments in the process of economic policy formulation, employing a bottom-up approach to move away from a bargaining-based federalism model. Two of the many objectives of this think tank is "to create a knowledge, innovation and entrepreneurial support system through a collaborative community of national and international experts, practitioners and other partners" and "To maintain a state-of-the-art resource centre, be a repository of research on good governance and best practices in sustainable and equitable development as well as help their dissemination to stakeholders." (United Nations, n.d.)

WOMEN AND ENVIRONMENTAL MANAGEMENT

The way we conceptualise the environment is likely to shape our perception of environmental issues. It will also have an impact on the types of strategies or management programmes that are developed for the purpose of intervention.

Definition of Sustainable Development

"Sustainable development is development that meets the needs of the present without compromising the ability of future generations to meet their own needs." (Khatibi & Indira, 2017, p.3).

The International Council of Local Environmental Initiatives provides further insight into the matters that must be tackled in order to attain sustainability, as outlined in their definition of sustainable development:

"The process of development that ensures the provision of fundamental environmental, social, and economic services to every member of a community, while simultaneously safeguarding the long-term viability of the natural, built, and social systems that are essential for the provision of these services."(Greenly, 2023, p. 5)

The transition from perceiving the environment as a collection of separate elements, primarily focused on the physical aspects, to a more comprehensive and integrated comprehension that encompasses the spaces where individuals reside and work, as well as their means of sustenance, carries significant implications for examining the connection between women and the management of the environment.

A planetary emergency and the pandemic have depleted the world's natural resources and widened poverty and inequality. Unless decisive action is taken to advance the Sustainable Development Goals (SDGs), it is projected that approximately one billion individuals may be subjected to extreme poverty by the year 2030 (GLOPP, 2008).

Since the United Nations Conference on Environment and Development (UNCED), there has been an increasing recognition of the 'brown agenda'. This agenda encompasses environmental issues that are intricately connected to the process of development which are:

- The absence of a secure water supply, adequate sanitation facilities, and effective drainage systems.
- Insufficient management practises for solid and hazardous waste.
- Unregulated emissions stemming from motor vehicles, industrial facilities, and the utilisation of low-quality domestic fuels such as paraffin and wood.
- Accidents that are associated with traffic congestion and overcrowding:
- The utilisation and deterioration of ecologically vulnerable territories
- The insufficient and inadequate allocation of housing resources for individuals with low socioeconomic status.

There is an increasing recognition of the inter-relationship among these issues. The revised environmental agenda acknowledges the interconnectedness among poverty cycles, resource depletion, and environmental degradation. As a result, environmental concerns have become an integral part of the political economy. The implementation of this emerging environmental agenda disproportionately affects women from lower socioeconomic backgrounds. The survival of the family is contingent upon the women's dependence on the environment.

The existing body of research and policy has predominantly concentrated on examining the correlation between poverty and environmental degradation, highlighting the dual role of the impoverished population as both recipients and contributors to environmental deterioration. The individuals in question can be regarded as victims due to their higher likelihood of residing in ecologically susceptible regions. Additionally, they can be considered agents as they may lack alternative choices but to exploit environmental resources, thereby contributing to the deterioration of the environment. To foster environmental sustainability, it is imperative to prioritise social conditions and poverty alleviation, owing to the heightened consciousness surrounding this issue (Puhazhendhi & Jayaraman, 1999).

Women Leading the Way in Environmental Management: Empowerment and Sustainability

The management of the environment plays a pivotal role in our collective endeavour to address the complexities associated with climate change and the exhaustion of natural resources on a global scale. Throughout the course of time, women have emerged as significant contributors in the realm of environmental preservation and sustainable development, actively participating in these endeavours.

1. Women's Integral Role in Environmental Conservation: Throughout the world, women have emerged as prominent figures in the realm of environmental conservation and the pursuit of sustainability. They have demonstrated active engagement in diverse capacities, including but not limited to researchers, policymakers, community leaders, educators, and activists. The perspectives of women frequently incorporate a distinctive amalgamation of conventional wisdom and contemporary expertise, resulting in inventive and comprehensive approaches to environmental obstacles.
2. Bridging the Gap in Environmental Sciences: Throughout history, the discipline of environmental sciences has predominantly been characterised by a significant gender imbalance, with males occupying most positions and exerting influence within the field. Nevertheless, it is worth noting that there has been a significant increase in the presence and prominence of women in positions of representation and leadership within this field in recent times. Efforts aimed at promoting the engagement of girls and women in STEM (Science, Technology, Engineering, and Mathematics) education have proven instrumental in addressing the disparity between genders within the field of environmental sciences. The increasing participation of women in the field has contributed to the enrichment of environmental research and management through the introduction of diverse viewpoints and approaches.
3. The power of Women's Environmental Network: The proliferation of women's environmental networks and organisations on a global scale has facilitated the establishment of platforms that foster collaboration, knowledge exchange, and collective efforts. These networks play a significant role in promoting various environmental initiatives, encompassing sustainable farming practises as well as advocacy for addressing climate change. The efficacy of these networks resides in cultivating a collective consciousness and agency, empowering women to magnify their influence and engender constructive transformation within their respective communities.
4. Women as Agents of Change in Sustainable Development: Frequently, women assume the role of change agents within their communities, actively advocating for sustainable development and

the adoption of environmentally conscious practises. Women have spearheaded initiatives at the grassroots level to tackle various concerns such as waste management, the adoption of clean energy, and the conservation of biodiversity. These region-specific endeavours not only serve to protect the natural surroundings but also enhance the socio-economic development and overall welfare of communities.

5. Empowering Women for Environmental Sustainability: The empowerment of women in the field of environmental management is not solely a matter of achieving gender equality, but also a pragmatic approach to bolstering sustainability endeavours. There are multiple strategies that can be implemented to enhance gender equality within the environmental industry such as:
    - Promoting Gender Parity: Advocating for the equitable inclusion of women in influential positions within environmental organisations and policymaking entities.
    - Educational Opportunities: Facilitating the provision of high-quality educational and training programmes to women who express interest in pursuing professional paths within the fields of environmental sciences and management.
    - Funding and Support: Establishing funding opportunities and support networks that are specifically designed to cater to environmental initiatives led by women.
    - Challenging Gender Stereotypes: Addressing the Impact of Stereotypes and Biases on Women's Advancement in the Field.

Women and the Conservation, Use, and Management of Natural Resources

Women holds a rich warehouse of technical knowledge of environment, which can be used in the problem-solving process. In addition to their profound knowledge of fragile ecosystem management and sustainable resource use, women have a high ability to collusively and childcare. Women also have a powerful influence over future insight towards the environment.' (Shashikala, 2007). Women are organised into collectives with the overarching objective of attaining gender equality and promoting sustainable and comprehensive community development (United Nations Development Programme, 2001).

In 2001, the United Nations underscored the imperative of incorporating women into sustainable development efforts, asserting that the long-term viability of such endeavours would be compromised without their active involvement. Women play a crucial role in various dimensions of sustainable development, including social, economic, and environmental aspects. The significance of women's involvement in environmental management, particularly in the domain of water resource management, was underscored. International efforts, such as the International Drinking Water Supply and Sanitation Decade and the United Nations Conference on Environment and Development (UNCED), have played a significant role in advocating for the involvement of women in the preservation of natural resources (United Nations Department of Economic and Social Affairs, n.d).

There is an assumption that women residing in rural areas have greater levels of environmental engagement compared to their urban counterparts. Women play a crucial role in maintaining the sustainability of the community through their direct connection with the environment. Women play a significant role in agricultural activities, encompassing tasks such as cultivating crops, traversing extensive distances to procure firewood and water, and tending to livestock, among other responsibilities. In most rural regions, women assume the primary role of producers, cultivating crops not solely for household consumption, but also for commercial purposes. Rural women often face the necessity of engaging in extensive labour throughout the day to fulfil both the productive and reproductive demands of their households.

In the context of sustainable development, which seeks to provide essential environmental, social, and economic services to all members of a community, it is evident that the relationship between women and the environment is unsustainable. Women are experiencing exclusion from various aspects of social and economic life, while also facing an excessive burden of environmentally related responsibilities. There exists a prevailing assumption that women residing in rural areas possess a greater degree of interaction with their surrounding environment in comparison to their urban counterparts. There is an argument suggesting that urban women participate in similar activities as their rural counterparts. Low-income urban women, specifically those residing in informal settlements, are required to allocate significant amounts of time to gather wood for fuel, procure water, and maintain market gardens for subsistence purposes.

The challenges faced by urban women are exacerbated by the issues related to the 'Brown agenda' and the limited availability of natural resources in densely populated urban areas. The lack of secure tenure experienced by women residing in informal settlements contributes to a sense of dislocation and displacement, hindering their ability to make substantial investments in their living environments.

The planners responsible for designing urban environments frequently neglect the requirements of women, thereby exacerbating the marginalisation of women and impeding their ability to participate fully in society. The initial step in analysing the connection between women and the environment involves recognising the gendered responsibilities, labour, and work routines that are characteristic of both urban and rural systems.

In terms of environmental interaction, it is important to note that divisions exist not only between women and men, but also among women themselves, based on age and status. In many households, younger women and daughters-in-law often bear the burden of additional and more demanding responsibilities, primarily because of their subordinate position and corresponding lack of authority. Moreover, female children are often allocated a greater share of domestic and environmentally-oriented tasks compared to their male counterparts. Females frequently assume the responsibility of gathering fuel and water, whereas males are typically assigned the duty of herding (Jackson, 1995).

The utilisation of natural resources by women and their significant reliance on these resources for their daily activities, particularly in roles related to sustenance that contribute to the well-being of their families and communities, has led to the emergence of prevailing viewpoints that emphasise women's unique connection with the environment, positioning them as either users or stewards of natural resources. This approach has gained recognition within development circles under the acronym WED, which stands for women, environment, and development. The Women-Environment-Development (WED) perspective posits that women possess a distinct connection with the natural environment and asserts that the interests of women and the environment are mutually reinforcing. This perspective suggests that actions and policies that benefit women also have positive implications for the environment, and conversely, environmental initiatives can contribute to the well-being of women.

Nevertheless, one could contend that the notion of women possessing a distinct "special relationship" or "affinity" with the environment is not inherent, but rather a result of societal structures that have imposed this relationship upon women. Therefore, it is by means of this compelled correlation between women and the environment that women acquire an augmented understanding of the environment. However, from a policy perspective, the connection between women and the environment has frequently been used as a rationale for advocating for greater involvement of women in the management of environmental affairs.

There may be significant variations in environmental perceptions, values, and priorities both within and among rural and urban communities, particularly when considering the differences between women and men. The relationship between women and the environment, whether in rural or urban systems, is characterised by a multitude of interconnected and intricate factors. Nevertheless, despite the absence of a unanimous agreement regarding the involvement of women in environmental management, it is worth noting that women are the primary users and managers of environmental resources. Therefore, any policy intervention aimed at influencing environmental management should duly acknowledge this fact.

Women Empowerment and Their Role in Fulfilling Sustainable Development Goals 2030

The promotion of women's empowerment is an essential element in the attainment of the Sustainable Development Goals (SDGs) within the timeframe of 2030. Gender equality is not solely a fundamental entitlement of individuals, but also an essential condition for a future that is both sustainable and inclusive.

The 2030 Agenda for Sustainable Development was unanimously adopted by all Member States of the United Nations in 2015. This comprehensive framework delineates 17 interrelated objectives aimed at the eradication of poverty, preservation of the environment, and promotion of inclusive prosperity for all individuals. At the core of this agenda lies the pursuit of gender parity and the empowerment of women and girls, as outlined in Sustainable Development Goal 5. The unique capabilities and contributions of women make them crucial in the promotion and advancement of sustainable development. This paper examines the significant role that women play in the achievement of the Sustainable Development Goals (SDGs) and addresses the obstacles they face in their efforts towards sustainable development.

Figure 1. Sustainable development goals (SDGs)

1.1. Education and Healthcare: The Significance of Education and Healthcare in Promoting Women's Empowerment and their Contribution to the Attainment of Multiple Sustainable Development Goals.

1.2. The Importance of Economic Empowerment: Examining the Role of Women's Economic Participation and Financial Independence in Promoting Sustainable Development and Alleviating Poverty (SDGs 1 and 8).

2. Women and Climate Action (SDG 13): The imperative role of women in addressing climate change and advancing sustainable environmental practises.

2.1 Climate-Induced Challenges for Women: Examining the unequal effects of climate change on women and the necessity of implementing gender-responsive policies.

3 Gender Equality and Inclusive Governance (SDG 16):

3.1 Political Representation: The Significance of Women's Political Involvement and Decision-Making in Attaining Sustainable and Inclusive Governance.

3.2 Eradicating Violence and Discrimination: Tackling gender-based violence and discriminatory practises as fundamental requirements for achieving sustainable development.

4. Women's Role in Ensuring Zero Hunger and Sustainable Agriculture (SDG 2):

4.1 **The Significance of Women in Agriculture:** Acknowledging the Vital Role of Women in Food Production, Agriculture, and Ensuring Food Security.

4.2 The Importance of Ensuring Women's Land and Property Rights for the Promotion of Sustainable Agricultural Practises.

5. Women's Health and well-being (SDGs 3 and 5):

5.1 Reproductive Health: The topic of reproductive health pertains to the examination and resolution of women's health concerns, as well as the facilitation of unrestricted entry to reproductive healthcare services.

5.2 Family Planning and Education: The Role of Family Planning and Education in Empowering Women: Enhancing Women's Empowerment through Family Planning and Educational Initiatives.

6. Empowering Women in Urban and Rural Settings (SDG 11):

6.1 Urbanization and Gender Equality: Examining the impact of urban development on women's empowerment.

6.2 Women in Rural Areas: Identifying challenges faced by rural women and strategies for their empowerment.

The successful achievement of the Sustainable Development Goals by 2030 is contingent upon the prioritisation of women's empowerment and the promotion of gender equality. To effectively attain the Sustainable Development Goals (SDGs), it is crucial to acknowledge and tackle the distinct obstacles encountered by women and girls. It is imperative for policymakers, governments, civil society, and international organisations to collaborate in order to guarantee gender equality by providing women with equitable opportunities, unrestricted access to resources, and the ability to participate in decision-making processes. The construction of a sustainable, resilient, and inclusive world for all necessitates the establishment of authentic gender equality.

EMPOWERMENT OF WOMEN THROUGH SELF HELP GROUPS (SHGs)

"Empowerment involves people in an active role, not only in decision making, but also an understanding of the factors that shape a situation, and the nature of oppression itself." (Purushothaman, 1998, p. 10).

Empowerment is characterised by heightened engagement in the decision-making process, wherein individuals perceive themselves as competent decision-makers with the corresponding entitlement to exercise this capacity. There exists a clear and reciprocal relationship between involvement and empowerment, wherein each factor exerts an influence on the other. The active involvement of empowered women in projects has the potential to contribute significantly to the achievement of sustainable development outcomes.

Two crucial processes have been identified as significant for the attainment of empowerment. One key factor is social mobilisation and collective agency, as impoverished women frequently face a dearth of fundamental capabilities and self-assurance to effectively address and confront prevailing inequalities and obstacles. Frequently, the presence of change agents is required to deliberately initiate social mobilisation. Furthermore, it is imperative that the process of social mobilisation is accompanied and supplemented by the provision of economic security. The inability of marginalised individuals to address their economic deprivation and livelihood insecurity hinders their capacity to engage in mobilisation efforts (International Consultation, 1993).

Self Help Groups in India and Their Importance

The term "Self-Help Groups" (SHGs) refers to a collective of poor women, typically ranging from 10 to 20 members, who voluntarily join forces to pool their savings through thrift and self-management. The primary objective of these groups is to establish their creditworthiness to financial institutions. The concept is predicated upon the principles of mutual trust and assistance, collective ownership, peer influence, and the promotion of group cohesion and unity (SIDA, 1996).

Currently, Self-Help Groups (SHGs) in India are actively engaged in the process of poverty alleviation by empowering women. Self-help groups (SHGs) in India are autonomous collectives comprised of individuals who voluntarily unite to collectively tackle shared social, economic, or developmental concerns. These collectives generally consist of individuals who share similar socio-economic backgrounds and share a common goal of self-empowerment through reciprocal support, financial aid, and the enhancement of skills. Self-Help Groups (SHGs) have emerged as a pivotal force in the context of poverty alleviation, women's empowerment, and rural development within the Indian context. The following are salient aspects concerning self-help groups (SHGs) in India:

1. **Empowerment of Women:** The empowerment of women has been facilitated by the establishment of self-help groups in rural and semi-urban regions of India. The organisation offers a forum for women to convene, exchange personal narratives, and offer mutual assistance across a range of domains, encompassing financial affairs, well-being, and educational pursuits.
2. **Micro finance and Savings:** The promotion of savings habits among members is a key objective of Self-Help Groups (SHGs) in the realm of microfinance. Individuals make regular, modest contributions to a shared pool of funds, which are subsequently utilised to offer interest-free loans to fellow members who are experiencing financial hardship. The implementation of this microfinance approach has proven to be instrumental in facilitating credit accessibility for numerous individuals, enabling them to initiate and establish small-scale entrepreneurial ventures.
3. **Livelihood Enhancement:** Self-help groups frequently participate in income-generating endeavours such as animal husbandry, agriculture, handicraft production, and small-scale enterprises. Through

engaging in these activities collectively, individuals have the potential to enhance their household income and overall economic welfare.

4. **Government Support:** The Indian government has been actively engaged in promoting and facilitating the establishment of self-help groups through a range of programmes and initiatives. The National Rural Livelihood Mission (NRLM) and the Deendayal Antyodaya Yojana-National Rural Livelihood Mission (DAY-NRLM) are prominent endeavours designed to foster the growth and advancement of Self-Help Groups (SHGs) by providing them with comprehensive support.
5. **Skill Development and Capacity Building:** Self-help organisations place a strong emphasis on helping members develop their abilities and talents. Training programmes are held to provide participants with the information and abilities needed for a variety of income-generating occupations.
6. **Social Issues and Advocacy:** SHGs deal with social issues like education, health, sanitation, and community development in addition to economic ones. They serve as a support system and fight for the rights and welfare of its members.
7. **Community Development:** SHGs contribute to the growth of their communities by taking part in endeavours that are of benefit to the community. They do it by contributing to initiatives for regional development and by taking part in the processes of communal decision-making. They are actively engaged in development initiatives that prioritise watershed management, a critical aspect of environmental governance. There is a correlation between women's empowerment and environmental management, highlighting the potential for women to serve as catalysts for advancing environmental sustainability.
8. **Urban Self-Help Groups:** Self-help groups are sometimes thought of as being more prevalent in rural areas; nevertheless, it is possible to find them in metropolitan settings as well. Urban SHGs are primarily concerned with resolving problems that are unique to low-income urban populations, such as housing, sanitation, and access to fundamental services.

Self-help groups have been shown to be beneficial in improving the socio-economic position of their members, particularly women, and have made substantial contributions to the growth of grassroots organisations in India. The members of these organisations develop a sense of empowerment, self-reliance, and solidarity with one another, which can lead to beneficial changes on both the individual and the community level.

Given the extensive quantity of Self-Help Groups (SHGs) in India, it is unfeasible to furnish a comprehensive enumeration of all. However, there are several prominent self-help groups and organisations that have exerted considerable influence in various regions of India which are mentioned below:

1. **SEWA (Self-Employed Women's Association):** SEWA, established in 1972 by Ela Bhatt, is a trade union and women's cooperative that is dedicated to the empowerment of women workers operating within the informal sector.
2. **Kudumbashree:** Kudumbashree is a community-based poverty reduction programme with a focus on women, which has been implemented by the State Poverty Eradication Mission (SPEM) in the state of Kerala. The SHG initiative in India is notable for its substantial size and impressive level of success.
3. **Velugu**: Velugu is a rural development initiative implemented in the state of Andhra Pradesh, aimed at fostering the establishment of Self-Help Groups (SHGs) and facilitating women's empowerment through diverse income-generating endeavours.

4. **Mahila Arthik Vikas Mahamandal (MAVIM):** MAVIM is a women's development corporation located in Maharashtra, India, dedicated to promoting the socio-economic empowerment of women through the establishment and support of Self-Help Groups (SHGs).
5. **NRLM SHGs:** The National Rural Livelihood Mission (NRLM) has facilitated the establishment of a considerable number of Self-Help Groups (SHGs) in various states with the objective of tackling rural poverty and promoting women's empowerment.
6. **Pudhu Vaazhvu Project:** The project, which has been implemented in Tamil Nadu, seeks to eliminate poverty, and enhance the empowerment of women by establishing Self-Help Groups (SHGs) and implementing diverse livelihood initiatives.

Role of Non-Governmental Organisations (NGOs) in Supporting SHGs

In India, several Non-Governmental Organisations (NGOs) are actively engaged in initiatives aimed at empowering women through Self-Help Groups (SHGs). These non-governmental organisations (NGOs) help small and medium-sized groups become more sustainable and effective by providing capacity building, financial assistance, training, and other support services. The following is a list of well-known NGOs that have been active in providing support to SHGs:

1. **PRADAN (Professional Assistance for Development Action):** PRADAN engages in extensive collaboration with Self-Help Groups (SHGs) in rural regions of India, with the aim of empowering women and fostering the development of sustainable livelihoods.
2. **MYRADA (Mysore Resettlement and Development Agency):** MYRADA has been actively engaged in collaborating with Self-Help Groups (SHGs) to facilitate rural development, watershed management, and the promotion of sustainable livelihoods.
3. **NABARD (National Bank for Agriculture and Rural Development):** NABARD, being a financial institution, assumes a pivotal role in facilitating the support of non-governmental organisations (NGOs) and self-help groups (SHGs) through the provision of financial resources and technical expertise for initiatives aimed at rural development.
4. **Access Development Services:** Access Development Services engages in partnerships with multiple Self-Help Groups (SHGs) to offer them technical proficiency, market connections, and opportunities to avail financial services.
5. **CARE India:** CARE India provides support to Self-Help Groups (SHGs) through a range of initiatives encompassing livelihood programmes, women's empowerment, and facilitating access to health and education services.
6. **Rashtriya Mahila Kosh (RMK):** RMK facilitates the empowerment of women's Self-Help Groups (SHGs) through the provision of financial aid and the enhancement of their capacities.
7. **Hand in Hand India:** This organisation collaborates with Self-Help Groups (SHGs) to facilitate the advancement of livelihood opportunities, entrepreneurship, and the empowerment of women.

Self-Help Groups (SHGs) have emerged as a potent approach for addressing poverty, fostering human development, and promoting social empowerment. These groups, championed by non-governmental organisations (NGOs), have gained recognition from financial institutions and the Government of India. They are acknowledged as influential mechanisms for empowering women through their active involvement and the generation of employment opportunities (Puhazhendhi & Jayaraman, 1999). The participation of Self-Help Groups (SHGs) in management initiatives, facilitated by training programmes, represents an environmental concern by fostering the inclusion of women as active participants in environmental governance.

According to an evaluation conducted by the World Bank, the inclusion of women in a project was found to significantly enhance its effectiveness, with a reported increase of six to seven times compared to projects where women were not involved. The inclusion of women in the design and planning phases yields several outcomes, including the mitigation of corruption, enhancement of management transparency, improvement of financial management, and the provision of empowerment to women through exemplification.

A significant portion of India's Gross National Product (GNP), approximately 37%, is attributed to the agricultural sector. This assertion is supported by data from the National Sample Survey organisation, which indicates that agriculture provides employment to around 70% of the working population in India. Furthermore, it is noteworthy that approximately 84% of economically active women in the country are engaged in agricultural activities (International Household Survey Network, 1991). Among the estimated population of 168 million individuals living below the poverty line, it is observed that a substantial proportion of 60-80 million consists of women. In India, women who are economically disadvantaged face a triple burden characterised by poverty, social disadvantage, and gender-based discrimination. Therefore, these NGOs help the vulnerable class to have the opportunity to get rid of the economic disadvantage (GLOPP, 2008).

Women exhibit a higher level of engagement in environmentally sustainable activities and environmental management compared to men. Hence, the engagement of women in development initiatives via Self-Help Groups (SHGs) can significantly enhance societal consciousness towards the promotion of environmental sustainability.

CONCLUSION

The absence of women from decision-making processes concerning the protection of natural resources and environmental management is a notable concern, given the existing relationship between women and the environment, as well as their specialised knowledge in this domain. To effectively support these interventions, it is imperative to establish and enhance the institutional frameworks that enable women to voice their concerns and protect their rights. The exclusive focus on the Ministry of Environmental Affairs and Tourism as the primary agent for reshaping the connection between women and the environment is not advisable. Various governmental departments, such as those responsible for health, education, economics, housing, and land reform, possess significant implications for women's access to resources both at the national and regional levels.

While the facilitation of women's access to decision-making structures may enhance their participation, it does not necessarily ensure that bureaucracies will effectively incorporate the interests of women in environmental management. In rural areas, this would additionally involve the process of altering tribal structures and questioning various cultural beliefs. This argument posits that the inclusion of women and the environment in decision-making structures necessitates a transformation of said structures, shifting them away from a patriarchal and dominant framework towards one characterised by equality. One potential strategy for accomplishing this objective involves increasing the representation of women in leadership roles pertaining to decision-making processes. It is imperative to provide women with training opportunities in various fields related to natural resource management, including but not limited to land-use planning, farming and agriculture, forestry, marine science, and environmental law.

The current government delivery mechanism is not only insufficient, but also unsuitable for facilitating the active involvement of individuals in institution-building and promoting long-term development and fair distribution of advantages.

However, the empowerment of women through women's self-help groups (SHGs) can be a great opportunity for non-governmental organisations (NGOs) to involve women in environmental management in order to achieve sustainable development. This can be done through training, sensitization, and mobilisation using Participatory Rural Appraisal (PRA) methodologies, as well as community development (Khatibi & Indira, 2017).

It is imperative for environmental management policies to acknowledge the inter-relationship between women and the environment, while refraining from making broad generalisations about the inherent affinity between women and the environment. There exists a significant concern regarding the potential consequence of including 'environment' as an additional responsibility within the extensive array of women's roles, thereby exploiting women as a means of acquiring inexpensive and unremunerated labour. It is imperative to ensure that the time afforded by the provision of these services is effectively utilised to enhance the overall well-being of these women.

In order to facilitate the advancement of sustainable development, it is imperative for policies to guarantee gender parity across all societal domains. This encompasses the facilitation of women's literacy, education, training, nutrition, and health, as well as their active involvement in pivotal decision-making roles and environmental management. Of particular significance is the equitable allocation of resources and benefits to women (United Nations, n.d). Enhancing access to diverse credit options, particularly within the informal sector, and implementing measures to secure women's access to property rights, agricultural resources, and tools can effectively facilitate the objectives.

The initiation of this transformation can be found within the domain of local government, as it assumes a crucial role in addressing the specific 'practical' requirements that arise from women's interconnected roles as reproducers, producers, and community managers (Gwagwa, 1991). Nevertheless, notwithstanding the potential exhibited by local government in effecting changes in gender and environmental dynamics, there remains a substantial amount of work that must be undertaken in order to fully realise this potential (Patel, 1996).

The involvement of women in environmental management is crucial for the achievement of sustainability objectives. By acknowledging and empowering women as leaders and catalysts for change in this field, we can cultivate a more inclusive and efficient approach to the preservation of the environment and the attainment of a more promising future for future generations. The incorporation of women's perspectives, knowledge, and dedication to environmental issues will undeniably contribute to the establishment of a more resilient and sustainable global community.

REFERENCES

GLOPP. (2008). *DFID's Sustainable Livelihoods Approach and its Framework*. GLOPP. http://www.glopp.ch/B7/en/multimedia/B7_1_pdf2.pdf

Greenly. (2023). *The International Council for Local Environmental Initiatives*. Greenly. https://greenly.earth/en-us/blog/company-guide/the-international-council-for-local-environmental-initiatives-iclei

Gwagwa, N. (1991). Women in local government: Towards a future in South Africa. *Environment and Urbanization*, *3*(1), 70–78. doi:10.1177/095624789100300107

International Consultation. (1993). *To Advance Women in Ecosystem Management and to Promote Co-operation, Co-ordination and Strategic Planning Among Multilateral, Regional and Bilateral Agencies and Organisations*. Washington D.C. https://digitallibrary.un.org/record/241092

International Household Survey Network. (1991). *National Survey 1991*. IHSN. https://catalog.ihsn.org/index.php/catalog/4672/study-description

Jackson, C. (1995). From conjugal contracts to environmental relations: Some thoughts on labour and technology. *IDS Bulletin*, *26*(1), 33–39. doi:10.1111/j.1759-5436.1995.mp26001005.x

Khatibi, F. S., & Indira, M. (2017). Empowerment of Women through Self Help Groups and Environmental Management: Experiences of NGOs in Karnataka State, India. *Journal of Human Ecology (Delhi, India)*, *61*(3), 213–224.

Patel, Z. (1996). Women and Environmental Management. Agenda. *Empowering Women for Gender Equity*, *29*, 66–71.

Puhazhendhi, V., & Jayaraman, B. (1999). Increasing Women's Participation and Employment Generation among Rural Poor: An Approach through Informal Groups. *Indian Journal of Agricultural Economics. Indian Society of Agricultural Economics*, *54*(3), 287–295. doi:10.22004/ag.econ.297680

Purushothaman, S. (1998). The Empowerment of Women in India: Grassroots Women's Networks and the State. *Sage (Atlanta, Ga.)*, *8*(1), 109–111. doi:10.1177/1018529119980107

Shashikala, S. (2007). *India: Promoting Urban Social Development through Self Help Groups in Karnataka. A Report to the Operations Evaluation Department for the Project Performance Evaluation Report of Karnataka Urban Infrastructure Development Project in India*. Asian Development Bank (ADB). https://www.adb.org/sites/default/files/evaluation-document/35856/files/karnataka-urban-infrastructure-developmentproject.pdf

SIDA. (1996). Promoting Sustainable Livelihoods: A Report from the Task Force on Poverty. Stockholm: Swedish International Development Cooperation Agency (SIDA).

United Nations. (2019). *Sustainable Development Goals Report 2019*. UN. https://digitallibrary.un.org/record/478154?ln=en

United Nations. (n.d.). Rio 1992: *United Nations Conference on Environment & Development*. UN. https://www.un.org/en/conferences/environment/rio1992

United Nations Department of Economic and Social Affairs. (n.d). *2030 Agenda for Sustainable Development: Sustainable Development Goals*. UN. https://social.desa.un.org/2030agenda-sdgs

United Nations Development Programme (UNDP). (2001). *Annual Report 2001*. UNDP. https://www.undp.org/publications/annual-report-2001

Chapter 9
Role of Women Leaders to Deal With Environmental Crisis:
Failure and Success

Katyayani Singh
Jagran Lakecity University, India

Rohit Mishra
https://orcid.org/0009-0000-4847-860X
Jagran Lakecity University, India

ABSTRACT

Human consciousness is the pre-requisite for equality and equality postulates the involvement of women leadership in the environment in every sphere. Though the 21st century is characterized as the century of globalization and new technology and we have witnessed science and technology grow by leaps and bounds, the gender disparity in society persists till date. The spirit of patriarchal mindset seems to be looming over society at large which makes it difficult for women to come out as the frontrunners. Hence, the major challenge women face as leaders is primarily of getting the recognition as a leader. In this chapter the socio-political thoughts of the state and society, the perception about women leaders would be explored with a comparative aspect to understand that whether the challenges women face in environmental management are different from the challenges they must face in other spheres. Further, the chapter would explore the role and success of women leaders in environmental management to educate and train the society about environmental concerns.

INTRODUCTION

Human evolution has not only led to physical changes but also transformed human thoughts and attitudes. This transformation is reflected in the shift from an age of divine rule to an era of democracy, where 'individual will' guides human life. The Age of Enlightenment played a significant role in securing individual rights and promoting rationality, equality, liberty, and justice. Democracy has become the

DOI: 10.4018/978-1-6684-5986-7.ch009

defining feature of modern governments. However, challenges like widespread illiteracy persist, limiting people's understanding of political and social issues. This issue was highlighted during the Brexit referendum by biologist Richard Dawkins, who argued against letting uninformed individuals make complex decisions. He states "You might as well call a nationwide plebiscite to decide whether Einstein got his algebra right, or let passengers vote on which runway the pilot should land." (Dawkins, 2017, p. 1). To this argument Yuval Noah Harari responded by saying that for democracy feelings are more important than what one thinks. Human feelings reflect free will and this "free will is the ultimate source of authority, and that while some people are more intelligent than others, all humans are equally free" (Harari, 2019, p. 3). If rationality were the sole criteria for running a successful democracy, it would likely lead to the dominance of elites, silencing the common people. Rationality is effective if we have all necessary facts and analytical abilities.

However, overreliance on rationality while neglecting emotions like compassion and empathy has contributed to issues like global warming. The pursuit of economic and scientific progress often came at the expense of the environment. Early warnings of environmental catastrophes were often dismissed. Steven Pinker shares this perspective, suggesting that environmental concerns may be exaggerated, and renewable energy sources can support scientific advancement without harming the environment. He proposes in his book *Enlightenment Now* that science and technology have the solution to every problem that humankind faces and if currently there aren't any there is a high probability of developing them if future (Pinker, 2018). Indeed, his hypothesis is quite an optimistic one but when we observe environmental catastrophe in the form of: heat waves, floods, earthquakes, tsunami, cyclones on regular intervals there stands much reason for us to raise question on such optimism.

In the recent years the world has seen an escalating impact of environmental crisis such as global warming, biodiversity loss, pollution and deforestation. Addressing these challenges requires effective leadership and efforts from society to tackle these emerging issues in the economy. This chapter explores the role of women leadership in dealing with environment crises and examining their failures and successes. Women have made significant contributions to the environmental movement, policy making and sustainable goals and practices. With these contribution to the diverse sectors they have also faced obstacles in their efforts. By exploring both the sides of the coins we can gain a deeper knowledge (Mies & Shiva, 1993).

The chapter discusses the achievements of women leaders in environmental conservation and sustainability, highlighting their strategies and accomplishments. It emphasizes the importance of women's leadership in fostering community engagement for environmental management. However, it also acknowledges the challenges and limitations these leaders face, including systemic barriers. The chapter underscores the need to address these gaps and support women leaders. It explores how women leaders from diverse backgrounds identify environmental issues and how their experiences can inform more effective strategies, policies, and support systems for dealing with environmental crises and building a sustainable and equitable world.

RISE OF WOMEN AS LEADERS

Women have had always played a key figure in maintaining and nurturing the role of family and marriage however, that's not their sole contribution to the civilizations. Upon studying history, one can also find that women have consistently demonstrated their exceptional capacity to stimulate innovation, encourage community participation, and fervently champion policies that prioritize the responsible management of the environment. Despite the remarkable progress achieved, it's crucial to recognize that gender disparities persist in leadership positions across various sectors focused on sustainable environmental efforts. To address this barrier and promote gender equality in leadership organizations, governments of modern liberal state had implemented various programs, mentorship opportunities, policies for the solace of women leadership and create inclusive environment for equal opportunities and promotions. It is noteworthy here that not all governments in the modern society have taken this initiative rather only the governments which observe democracy, liberty, and equality as their defining feature. Despite these steps taken by the liberal societies, the challenges for the women to come out as leaders have not ceased to exist. All sorts of challenges are there for women while they try to assert themselves as leaders in different areas of private and public sectors.

One of the major impediments for women to rise as leaders is being subjected to sexist comments even while they have proved themselves to be efficient leaders. Perception about women as leaders is an important factor and this was the finding of Reykjavík Index for Leadership (Women Political Leaders, 2020). This index was established to quantify levels of comfort in society with the prospect of female leadership. This index was formally launched in 2018 for the G7 countries: Canada, France, Germany, Italy, Japan, United Kingdom and USA and was repeated in 2019. In 2020 and 2021 the G7 countries undertook it again along with India and with the addition of two African countries: Kenya and Nigeria.

The methodology used for building the index was by exploring a question: "For each of the following sectors or industries, do you think men or women are better suited to leadership positions?" This question allowed responses of 'men', 'women' or 'both equally' for 23 different sectors. A response of 'both equally' scores a point while a response of 'men' or of 'women' scores zero points. On the scale from 0 to 100 U.K. and Canada scored the highest being 81 amongst the G7 countries. In this index 23 sectors were explored where the overall perception of the G7 countries was given a score of 73. From the index it was concluded that only in the field of media entertainment, natural sciences, economics and political science, banking and finance, food and drinks manufacture and pharmaceutical and medical research there was optimism with regards to women leadership.

Further exploring the sector such as that of education, government, and politics, which are the key areas for raising awareness about environment and also for making policies related to it we find the women leaders have been positively accepted. But the important point to note here is that this is the status for the G7 nations, implying it is the situation in developed countries only.

While exploring the lesser developed countries such as Kenya and Nigeria the overall perception for women as leaders is quite low, i.e., 53 for Kenya and 47 for Nigeria. In this report it was concluded that despite the increase in the comfort level of having a woman as a leader that in no group of the 23, in any country is there a complete comfort of women being accepted as the leader.

The Reykjavík Index for Leadership helps us to comprehend that despite living in a modern democratic state, democracy in true sense has still not been achieved. The patriarchal mindset has been detrimental not just to the growth of females but also has damaged the ecosystem around us. For patriarchy is about power and dominance in society. A society that is patriarchal in nature will thus have a militaristic approach for progress wherein the place of women will always be secondary because of their feminine qualities. The historical and cultural connections between the subjugation of women and the exploitation of nature have been highlighted by Mies and Shiva in their book Ecofeminism. In this book it is argued that patriarchal societies have perpetuated a worldview that objectifies and dominates both women and the environment. It is because of this oppressive mindset that the world is dealing with depletion of natural resources, environmental degradation, and social injustices.

According to empirical research, it has been observed that women in leadership positions tend to prioritize and emphasize certain qualities and behaviours such as teamwork, empathy, and relationship-building. They also encourage active engagement from the community, with an emphasis on empowering local stakeholders, has been noted to lead to the creation of environmentally sustainable initiatives. Studies show that women leaders leverage a range of viewpoints to formulate creative solutions for intricate environmental issues. Women leaders prioritize inclusive environmental policies, leading to positive outcomes. Their initiatives are resilient and adaptable in changing conditions, fostering proactive problem-solving. They highlight the link between economic and environmental goals for sustainable development. Women leaders connect effectively with younger generations and promote environmental education, cultivating a sense of responsibility for future resource sustainability, even in male-dominated industries (Balabantaray, 2023).

On the basis of these findings, we conclude the following traits of women leadership: -

Inclusivity: - In today's complex world where the diversity in thoughts is important in tackling the problems and innovative ideas the women leadership plays a important role for the collective decision making process.

Emotional Intelligence: - for understanding the emotions and experience of others women have been praised for their higher level of emotional intelligence. In modern world where the employees play a important role as factors of production, leaders with empathy can foster positive response in the working environment.

Adaptive leadership: - The traditional women have shown the flexibility in the adoption of changes and embrace new approaches, this attire in modern world is crucial in the leadership for the decision making and handling the dynamic business environment.

Conflict resolution and building relationship: - The problem of conflict resolution is very critical in every organization and being as a leader it's important to deal with that very carefully. Women leader traditionally knows for their communication skill and conflict management which help the organization to create harmony among the various stakeholders.

Long term vision and sustainable practices: - research suggest that women leader are exhibit more long term vision and inclined towards sustainable practices in business decision making. This approach is important in the face of environmental challenges and social responsibilities faced by today's world.

Resilience: - overcoming the barriers and biases, the women as leader have historically demonstrated resilience and perseverance. This quality in today's dynamic environment of business can play a vital role.

The new FAO report offers a comprehensive view of women's roles in agri-foods systems, emphasizing their participation and benefits. It highlights the differential impacts of shocks like COVID-19 on different genders and their coping methods. The report underscores the importance of gender equality and women's empowerment in achieving sustainable agri-foods systems, showing how transforming these systems can promote gender equality and women's empowerment (FAO, 2023). Studies by different organs of U.N. have emphasized that women should be included in different environmental areas such as: energy studies, technology, there should be gender sensitive response to climatic changes and there should be strong empowerment and investment in women so as to combat the effects of desertification which is major environmental concern. Furthermore, a 2006 study of 141 natural disasters by the London School of Economics found that when economic and social rights are fulfilled for both sexes, the same number of women and men die in disasters. At the same time, when women do not enjoy economic and social rights equal to men, more women than men die in disasters. This gender discrepancy has come to light in a range of major disasters, including the Asian Tsunami; Hurricane Mitch, Hurricane Katrina, and other storms in the Americas; European heat waves; and cyclones in South Asia, etc. In some cases, such as in Central America, more men than women have died in natural disasters as they placed themselves at risk while helping their families.

Environmental Concerns

World today is at the brink of global warming because of pollutions level reaching its peak in every area, i.e., air, water, and soil. Women are reportedly said to be the direct victims of this environmental crisis as they gather water, fish, or farm lands which are affected by floods. Due to such extreme situation developing in the world the Intergovernmental Panel on Climate Change (IPCC), during the 25th Session of the Conference of the Parties (COP25) of the United Nations Framework Convention on Climate Change (UNFCCC) held in Madrid, Spain in 2019, warned that global temperature would increase to 4 degrees Celsius if the greenhouse emission was not checked in time. Further the UN, in preparation of COP26 in Glasgow in 2021 categorically stated that "human activity is changing the climate is changing the climate in unprecedented and sometimes irreversible ways." The warnings were not taken seriously by the world leaders and as a consequence to that it is being said that the world leaders are now steadily losing the fight against climate change.

The emission of greenhouse gases is continuously on the rise making it difficult to achieve the global target of maintaining the global temperature well below 1.5 degrees Celsius. Primarily seven industries are responsible for this (Howell, 2022):

Energy industry emitting 15.83 billion tons of greenhouse gas;
Transport emitting 8.43 billion tons;
Manufacturing and construction: 6.3 billion tons;
Agriculture: 5.79 billion tons;
Food retail: 3.1 billion tons;
Fashion: 2.1 billion tons;
Technology: 1.02 billion tons

This is the yearly emission of these industries which is proving to be dangerous for human survival. In this challenging situation some scholars believe that incorporating or mainstreaming eco-feminists or female environmentalists actively into climate change discourse will successfully reinvigorate the fight against climate change (Abdulai, 2022). Abdulai explains the generic factors that make eco-feminist or female environmentalists highly relevant in tackling the climate change menace. It has been stated that due to certain qualities like childbirth, caretakers, homemakers, etc., women are essentially nearer or closer to nature than men. Hence, environmental related ill health or the fatal impact of the climate changes such as earthquakes, floods, tsunami, etc. are disproportionately felt by women and their children. A study conducted by an organisation called Women Deliver (2021) indicated that increasing temperatures, heat waves, erratic rainfall patterns, and droughts due to climate change affect maternal health greatly.

The report further came up with the finding that directly relate to women. It states that climate change issues have negative impacts on maternal health and create conditions that result in increases in gender-based violence, including harmful practices such as child marriage (Women Deliver, 2021). Based on these findings, it has been concluded by Abdulai those female sentiments regarding the environment, especially climate change, reflect more than half of global environmental sentiments. For this reason, Abdulai opines that co-feminists or female environmentalists can successfully tackle the climate crisis when they are actively incorporated or mainstreamed into the fight against climate change. Further the author is of the view that by actively collaborating with the female world leaders the problem related to climate change can be better tackled.

The Beijing Platform for Action in 1995 echoed similar sentiments, outlining three key strategic objectives for governmental action in the environmental realm. These objectives encompass actively involving women in decision-making processes related to the environment at all levels, incorporating their perspectives and concerns into policies and programs, and establishing mechanisms for evaluating the impact of development and environmental policies on women. Despite significant progress over the two decades since this report's release, there is still much work to be done to bolster women's participation in decision-making and to ensure a brighter future for everyone. To address this, the United Nations is actively striving to place women at the forefront of sustainable development initiatives and endeavours aimed at mitigating the impacts of climate change.

Despite these suggestions, as authors we are of the view rather than imposing all the responsibility of dealing with the climate crisis women need to collaborate with the male counterparts as well if they really want to be successful in this endeavour. Indeed, there have been various women leaders who have been recognized world-wide for their contribution in raising environmental awareness yet they have had support from the male counterparts without whose help the initiatives taken by these women probably would have been lost.

Contribution of Women in Environmental Education and Awareness

In the field of environmental management, a lot of women leaders have emerged to prove their efficiency whose contribution would be explored in this chapter. These include: Christiana Figueres, from Costa Rica who served as the Executive Secretary of the United Nations Framework Convention on Climate Change (UNFCCC) from 2010 to 2016; Vandana Shiva, an Indian environmental activist and scholar known for her work in promoting biodiversity, sustainable agriculture, and social justice; Wangari Maathai, from Kenya founded the green belt movement that focused on tree planting, environmental conservation, and empowering women to become active agents of change; Marina Silva, from Brazil, served as Minister of the Environment of Brazil from 2003 to 2008 and has been actively involved in advocating for environmental policies and indigenous rights; Jane Goodall, a British primatologist and anthropologist, renowned for her advocacy for wildlife conservation and habitat protection. These women belong to different parts of the world and have been picked up from developed nations, developing nations and the poor nations in terms of economy.

Christina Figueres had an instrumental role to play in the Paris Agreement of 2015 (United Nations, 2015) Her leadership at the UNFCCC is said to be crucial in the successful outcome of the Paris conference. The goal of Paris Agreement was to limit global warming to well below 2° Celsius, preferably to 1.5° Celsius, compared to pre-industrial levels. It was a legally binding treaty adopted by 196 countries. In her book titled *The Future We Chose* she describes this decade as the most consequential in human history because of the possession of capital and technology to bring impactful changes. Through this book she describes the existential threats that the humankind faces due to the economic crisis. She gives suggestions such as that individuals becoming citizens and not consumers.

Another change to mitigate the environmental crisis is to create gender equality in the society. According to her gender equality is needed to empower women to come to the decision-making table. The women according to her have the tendency to be more collaborative and they tend to think much more in long terms compared to men (Carrington, 2020). She calls for global co-operation in reducing the emission of harmful gases. Further speaking at different conferences, summits, and events she has been able to rally support for climate action. After leaving her position at UNFCCC, she co-founded Global Optimism, an organization that was aimed to drive climate action. She has been working on reduction of emission for long. First in 2002 she had proposed a Sectoral Clean Development mechanism (CDM). This was basically a mechanism introduced to reduce the emission of greenhouse gases and ensure sustainable development (Figueres, 2023). Taking this initiative further she had launched Mission 2020 initiative, which aimed to accelerate efforts to reduce greenhouse gas emissions.

Vandana Shiva, an Indian environmentalist has helped in raising environmental concerns through her rigorous activism. Shiva through her work has highlighted the role of traditional knowledge in sustaining ecological balance and since women in India specifically have always been the kea bearers of traditional values, they can help in providing an insight into sustainable living. One such example is that of *Tulsi,* known as holy Basil. The Tulsi plant is a revered plant in Indian tradition. In general, almost all traditional households one is deemed to find a Tulsi plant in their backyard which is worshipped regularly. This plant is basically a medicinal herb which can be used for various purposes such as respiratory issues, providing stress relief, developing immunity system, digestive system and is extremely useful for skin due to its antimicrobial and anti-inflammatory properties. The modern science today has installed various manufacturing industries that produces medicines and beauty products to deal with issues relating to our body whether it is about healing our physical body or beautifying it. In doing so large-scale pollution is caused as a by-product. Now if traditional knowledge had been used for healing and beautification purpose herein a lot of damage to the environment could have been easily controlled. Similarly, there are various other plants which are worshipped in India and these include: Banyan tree, *Neem* Tree, *Peepal* Tree, or Sacred Fig. Worshipping of tress first of all denotes that there is great reverence shown to them and when one truly reveres something or someone causing harm to them would be the last thing on their mind. When the world today is witnessing problems related to deforestation it is because the reverence given to them in the olden times has been lost. All these trees like *Tulsi* are not ornamental trees proving for shade only. They all have some certain healing properties and those who know of them find it better to cure themselves through it rather than relying on medicines from the market which sometimes have side effects as well.

Further Vandana Shiva had been associated with the green movement in different continents such as: Africa, Asia, Latin America, and Europe. The Green movement is a political ideology that is rooted in the ideas of environmentalism. This political ideology is primary concerned with conservation and promulgates Gandhian ideas to sustain and conserve the environment. She founded the Research Foundation for Science, Technology and Ecology in 1982 that led to the creation of Navdanyaa national movement to protect the diversity and integrity of living resources, especially native seed, the promotion of organic farming and fair trade (Navdanya International, 2023). Not only has she extensively written with respect to environmentalism she has also been associated with governments, and NGO's such as: International Forum on Globalization, the Women's Environment & Development Organisation and the Third World Network. A unique concept that has found the support of Shiva is Ecofeminism. This is basically an approach to build more sustainable and productive agriculture by reinstating the system of farming in India that is more centred on engaging women. She believes that ecological destruction caused primarily by the capitalist mindset joined by the industrial catastrophes have threatened daily life, and the maintenance of these problems have become the onus of women (Mies & Shiva, 1993).

Wangari Maathai was a key was a Kenyan environmentalist and is credited with the founding of the Green Belt movement, a movement that was focused on the planting of trees, environmental conservation, and women's rights (The Green Belt Movement, 2023). She has made significant contribution in the field of reforestation, women empowerment and sustainable development. Her contribution in the aforesaid field bestowed upon her the Nobel Peace prize in 2004; in fact she was the first African women to have been awarded this prestigious prize. The Green Belt movement was to protect the interest of the Kenyan women who reported about their hard survival due to problems such as: drying up of streams, inadequate food supply, depletion of resources like firewood to attain which they often had to cross long distances. After beginning her work Mathai realized that there were deeper issues behind the problems relating to environmental degradation, deforestation, and food insecurity and they were: disempowerment, disenfranchisement, and a loss of the traditional values by which the communities used to protect the environment.

Consequently, the movement fought against land grabbing and encroachment of agriculture into the forests. It was in the later part of her life that she started to focus on waste production. In 2006 she spearheaded the United Nations Billion Tree campaign a project which aimed to plant one trillion trees world-wide. Continuing her legacy after her death in 2011, the Green movement in recent years has reached internationally to campaign and advocate on climate change, the importance of Africa's rainforests in the Congo, to initiate the *mottainai* campaign—an effort to instil the notions of "reduce, reuse, recycle" in Kenya and around the world—and has partnered with the United Nations Environment Programme (UNEP) in its Billion Tree Campaign (The Green Belt Movement, 2023). It was in the later part of her life that she started to focus on waste production.

Marina Silva, is a Brazilian politician and environmentalist. She is serving as Brazil's Minister of the Environment and Climate Change since 2023 and has been working for environment conservation for long. Her contribution has won for her, number of awards from US and international organizations. These include the Goldman Environmental Prize for South & Central America for helping establish a two-million-hectare reserve managed by Indigenous communities; the Sophie Prize, 2009; in 2007 she was named as one of the Champions of the Earth by the United Nations Environmental Programme. She is a strong proponent of clean and renewable energy such as solar power and wind power and stands in opposition to Brazil's nuclear programme.

In her initial years she started her environment activism by working as rubber tapper and had led demonstration called *empates* to warn against deforestation and the outplacement of forest communities from their traditional locations along with Chico Mendes, another environmentalist from Brazil. Silva played a crucial role in raising awareness about the importance of the Amazon rainforest and its biodiversity. In 2003 she was appointed as Environment Minister. Her administration established new management of public forests, created a forest service and a biodiversity institute, and several funds for nature, including the largest international effort to preserve the rainforest, the Amazon Fund. Under her leadership the deforestation decreased by 59% from 2004 to 2007 (Schueman, 2023). Silva had been a grassroot worker and working from the bottom she is more connected with the people as well as the nature. For her environmental initiatives she had faced death threats too. Brazil and Colombia have seen the highest number of deaths of the environmentalists, yet she was not discouraged in her endeavours.

Jane Goodall through her sixty years of groundbreaking work had shown the world the urgent need for the protection of chimpanzees from extinction. When she entered Gombe Stream National Park, Tanzania the world knew very little about chimpanzees, and even less about their unique genetic relations to humans. She immersed herself in their habitat and rather than being a distant observer she studied them closely not just as some species but as individuals with emotions and long-term bonds. In her long-term study of chimpanzee behaviour and social interaction she discovered in 1960 that chimpanzees can make and use tools. This was a significant discovery and is considered as one of the greatest achievements of twentieth-century scholarship.

Later in 1977, she founded the Jane Goodall Institute to aid the research in Gombe and increase protection of chimpanzees in their habitats. However, this was only part of the solution to a much bigger, rapidly growing problem of deforestation and declining chimpanzee populations across Africa (Jane Goodall Institute USA, 2023). Through her research she has become deeply involved in wildlife conservation and environmental advocacy. Her institute Jane Goodall Institute is a global non-profit organization dedicated to wildlife conservation, research, and community centred conservation efforts. The institute operates in various countries wherein it supports projects that protect endangered species, promote reforestation, and empower local communities to protect their natural resources. She has also initiated, Roots & Shoots program, a youth-led environmental and humanitarian initiative that empowers young people to act in their communities (Roots & Shoots., 2023). It is a worldwide programme inspiring young leaders to address local environmental challenges and engage in meaningful conservation projects.

Traditionally Contribution of Women in Biodiversity

Women in many endemic and local communities have a deep knowledge of biodiversity and possess the traditional knowledge through generations (Shiva, 1992). Some of the attires of their knowledge as discussed by Shiva are:

Plants and herbs: - women in traditional communities have ambiguous knowledge of medicinal plants and herbal remedies. They know which plant have healing properties and which have their use in various health conditions.

Wildlife behaviour and migration: - based on the environmental cues' women can predict the movement of animals and that can be useful for wildlife conservation.

Crop Diversity: - women farmers have deep knowledge of local crops varieties and practices.

Plant propagation: - In traditional farming system women were often responsible for seed saving and plant propagation, they preserve seeds ensuring the local plant varieties to be continued.

Water management: - In some communities, women were responsible for the constructions and maintenance of irrigation systems and water storage structures.

Weather forecasting: - traditional knowledge passed down through generations to generations and enables women to predict weather patterns based on natural indicators such as cloud formation, wind directions, animal behaviour etc.

Challenges to the Environment Conservation Projects

Briefly going through the contribution made by these prominent women leaders we could also grasp the challenges they had to face in their respective initiatives. Two things that have come as a major challenge are: patriarchal mindset and the scientism of certain section of the society ranging from the common populace to the academic intellectual. If the common populace is having a patriarchal mindset thereby destroying the ecosystem by exercising the power construct, the academic intellectuals have come out with the theories of power and scientism that has destroyed the community building approach by decrying the traditional values that shows reverence to nature. It is worth to be noted that in other spheres also women had to face the onslaught of patriarchy and that has been one of the major impediments in their rise as leaders.

The problem of scientism is a more recent problem of the modern world. It can be described turning of science into a fundamentalist, i.e., when science closes its doors to enquiry and questions and assumes an authoritarian nature while addressing the issues of the world it is called as scientism. As can be observed a lot of the environmental crisis occurred because of the loss of traditional values which have often been described by thinkers such as Steven Pinker to be obstacles in development of society. Interestingly these thinkers don't have the answer to some of the major global crisis; they know that human interference in nature is the cause of the problem but rather than adopting the traditional values they want to wait till science comes out with some alternate solution. Clearly, they have downplayed the environmental crisis and overcoming their ignorance is a great challenge for women environmentalists.

CONCLUSION

The chapter started with the notion that human consciousness is important for ascertaining liberty and equality, however, it is not limited to that alone. The same human consciousness is required to understand all the problems around us from inequality and injustice at individual level to the environmental problems at global level. Had human only been conscious about the harm they are causing all the people, things, and the nature around them, the problem that we are facing today could have been avoided. Nevertheless, consciousness towards the environment was raised by various women leaders from different parts of the world with respect to deforestation; use of clean and renewable energy; greenhouse gas emission; conservation of the environment; protection of different species that are on the verge of extinction; sustainable development for peace and progress of the society. The path shown by these women leaders, however, is still not sincerely being followed due to patriarchal mindset and also because they are seen as threats to scientific progress. However, despite these challenges we can positively assert that women can serve as efficient environmentalists who can manage protection of nature along with sustainable development.

REFERENCES

Abdulai, A. M. (2022). Could Eco-Feminists or Female Environmentalists Boost the Fight Against Climate Change. *African Journal of Climate Change and Resource Sustainability*, *1*(1), 107–114. doi:10.37284/ajccrs.1.1.1039

Balabantaray, S. R. (2023). Women's Leadership and Sustainable Environmental Initiatives: A macroscopic investigation from Ecofeminism framework. *International Journal of Multidisciplinary Research and Growth Evaluation*, *4*(4), 1039–1046.

Carrington, D. (2020). Christiana Figueres on the climate emergency: 'This is the decade, and we are the generation. *The Guardian*. https://www.theguardian.com/environment/2020/feb/15/christiana-figueres-climate-emergency-this-is-the-decade-the-future-we-choose

Dawkins, R. (2017). *Richard Dawkins: We need a new party – the European Party*. New Stateman. https://www.newstatesman.com/politics/2017/03/richard-dawkins-we-need-new-party-european-party

FAO. (2023). *The status of women in agrifood systems*. FAO.

Frank, A. (2023). What is scientism, and why is it a mistake? *Bigthink*. https://bigthink.com/13-8/science-vs-scientism/

Harari, Y. N. (2019). *21 Lessons for the 21st Century*. Vintage.

Howell, B. (2022). The 7 Biggest Polluters by Industry in 2022, as Ranked in New Research. *EP Online*. https://eponline.com/articles/2022/10/14/the-7-biggest-polluters.aspx

Jane Goodall Institute USA. (2023). *About Jane*. Jane Goodall Institute. https://janegoodall.org/our-story/about-jane/

Jane Goodall Institute USA. (2023). *Our Story*. Jane Goodall Institute. https://janegoodall.org/our-story/about-jane/#conservationist

Mies, M., & Shiva, V. (1993). *Ecofeminism*. Fernwood Publications. doi:10.5040/9781350219786

Navdanya International. (2023). *Vandana Shiva*. Navdanya International. https://navdanyainternational.org/our-staff/vandana-shiva/

Pinker, S. (2018). *Enlightenment Now: The Case for Reason, Science, Humanism, and Progress*. Penguin Books.

Roots & Shoots. (2023). *About Roots & Shoot*. Roots & Shoots. https://rootsandshoots.org/about-us/

Schueman, L. J. (2023). Climate Hero: Marina Silva. *One Earth*. https://www.oneearth.org/climate-hero-marina-silva/

Shiva, V. (1992). *Women's Indigenous Knowledge and Biodiversity Conservation*. India International Centre Quarterly, 205-2014.

The Green Belt Movement. (2023). *Our History*. Green Belt Movement. http://www.greenbeltmovement.org/who-we-are/our-history

United Nations. (2015). *The Paris Agreement*. UN. https://unfccc.int/process-and-meetings/the-paris-agreement

Women U. N. (2015). The Beijing Declaration and Platform for Action Turns 20. New York: United Nations Entity for Gender Equality and the Empowerment of Women. UN Women.

Women Deliver. (2021). The link between climate change and sexual and reproductive health and rights: An evidence review. *Women Deliver.*

Women Political Leaders. (2020). *The G7 Reykjavík Index for Leadership*. Kantar.

Women Watch U. N. (2009). *Fact Sheet: Women, Gender Equality and Climate Change.* UN. https://www.un.org/womenwatch/feature/climate_change/

Chapter 10
Fisher Women Find Opportunities to Preserve the Coastal Environment and Face the Challenges:
Climate Change and Environment

Md. Shafiqul Islam
International University of Business Agriculture and Technology, Bangladesh

Ummul Fatiha Binte Shafiq
Jatiya Kabi Kazi Nazrul Islam University, Bangladesh

ABSTRACT

Coastal environment is a complex term in relation to definition, context, and importance. Fisher women do not get recognition and importance in the social realm. This study was conducted with aims to explore the role of fisher women in preserving coastal environment and what challenges they face. Both qualitative and quantitative data were used in this study. Direct engagement of women in coastal fishing is hard to find, but the indirect involvement of women is remarkable. They are involved in multiple tasks, including fish drying, cleaning, sorting, storage, packing, net maintenance, grading, and collection. Several livelihood activities get narrowed, and fisher women suffer more. They face several problems and live with disasters in the coastal environment. The poor fisher women often settle in the risky areas. Fishing activities are often low-income, but they stopped different livelihood activities for environmental shocks. They initiated new livelihood options switching from fishing.

DOI: 10.4018/978-1-6684-5986-7.ch010

INTRODUCTION

Social-ecological resilience is an essential part of the fisher women to cope with and adapting coastal disasters and environment. This study explores the ways through how the fisher women use their knowledge and build capacity and take actions. Natural hazards are the creeping phenomenon and a part of human history (Adger et al., 2011). Asia and the Pacific is in the sixth position of the World's ten highest risk countries as mentioned by the World Risk Index. It has been mentioned that 23 percent of people (1.2 billion) live within the vicinity of the coast globally and 50 percent are going to be at risk by 2030 (Zimmerman et al., 2004). Around fifty percent death toll from disasters is caused by natural hazards in Asia and the Pacific in the 21st century (UNESCAP, 2012). Another study mentioned that global damages from disasters have increased from USD 14 billion to USD140 billion over the 10 years period from 1976-1985 and 2005-2014 (GFDR, 2016). The scale and frequency of natural disasters are increasing due to climate change and anthropogenic factors (Campanella, 2006). Bangladesh is the country of coastline and very much vulnerable to disasters. The coastal fisher women are exposed to coastal disasters including, cyclones, tidal surges, floods, and other natural disasters due to geographical settings of Bangladesh. The coastal fisher women also have limited access to modern technology for warning and responses to disasters in Bangladesh. It is assumed that 50 million people are expected to be at risk by 2080 for increasing population density and climate change issues (Amin, 2004). Jones (2017) mentioned that resilience enhances people's understanding to respond to climate change risks. Societies become stuck in a series of problems including loss, depletion, disappearance and collapse even death (Elliott, 2018).

A study mentioned that climate change impact will increase and affect economic sectors and many aspects of natural and human wellbeing (Jones, 2019). The coastal fisher women are trying to fight with disaster by using their local knowledge/indigenous knowledge from generation to generation. The degree of adaptive response is not enough to make them resilient to coastal environment. In this connection more adaptive and coping measures essentially required in managing future disaster impacts and preservation of coastal environment. Social ecological network is interdependencies between social and ecological factors (Barnes et al., 2019). Local resilience is the capacity of linked socio-ecological systems to absorb and respond to the recurrent disturbance and impacts as to retain essential processes and feedback to sustain (Moss & Townsend, 2004). Mutahara et al. (2017) reported that Bangladesh faces several environmental extreme events including floods, droughts, cyclones, tsunami, tidal surges, riverbank, and coastal erosion and is intermingled to salinity intrusion. The coastal area is frequently affected due to the country's funnel shaped coast situated in the Bay of Bengal. The area is also translated as the high poverty zone that makes local community more vulnerable (Hossain et al., 2011). Environment and climate change impacts surrogates more disasters and increase extreme events in Bangladesh as well as Globally (IPCC, 2012). Onset coastal districts of Bangladesh are more vulnerable due to its geographical location and characteristics. The setting of Bangladesh coastal environment describes very low elevation with the terrain topography that makes it more vulnerable to extreme environment (Agrawala et al., 2003). Several studies found that south coastal regions have been affected by an increased occurrence of typical shocks including coastal floods, erosion, floods, storm surges and riverbank erosion as the influencing factors of environment and climate changes (Chowdhury, 2003; Huq et al., 2015). The area is affected by tidal flooding, inundation, and movement of saline water (ground and surface) during the dry season (Dasgupta et al., 2014a). Climate change increases river salinity leading to shortages of drinking water and irrigation, and significant changes in the aquatic ecosystems in the Southwest coastal areas of Bangladesh during dry season by 2050 (World Bank Report, 2016). Sea level rise, salinity intrusion, limited access to the resources and livelihood options with several environment and climate extreme events make resource poor communities more vulnerable to environment related shocks, even they are not well addressed in terms of their adaptation measures, local knowledge, and their resiliency to changing environment. Social action and interaction are essential for social ecological resilience and adaptation. This study will explore local knowledge, social resiliency of adaptation measures of vulnerable coastal fisher women (communities) to coastal disasters and climate extreme events in Bangladesh.

BACKGROUND

In Bangladesh, the coastal districts are more vulnerable to environmental change. Among 64 Districts 19 Districts are situated in the coastal areas of Bangladesh and home of 50 million people, are vulnerable to climate extreme events. In addition, the intense frequency of extreme events also exacerbates the conditions of the resource poor vulnerable coastal communities (fisher women, land less, wage labor, van pullers etc.). Poverty prevails in Bangladesh; particularly coastal poverty is more acute. Around eighty percent, including the resource poor population, stay and keep in touch with rural areas and their vicinity. They depend on nature-based activities and tasks as the smallholdings, wage and agricultural laborers, fisher women and other occupations. According to Climate Change Vulnerability Index (2017), Bangladesh stands on sixth as the most vulnerable country to climate change on earth. Evidence tells us the incidence of tropical cyclones in Bangladesh, a total of 61 cyclones hit Bangladesh during the period of 1961-2013 in which 32 hit Bangladesh from 1978 to 2007 (GoB, 208a). It was mentioned that 28 percent distress south-western zones and 16 percent of cyclones distress south-central coastal zones (Quadir & Iqbal, 2008). Haque et al. (2016) reported that more than 45 devastating cyclones hit Bangladesh affecting coastal livelihoods (dependent on fisheries, forestry, and agriculture). The coastal resources, especially natural resources are spiraling with environmental changes and extreme events thus affecting coastal livelihood (DMB, 2010). To make resource poor people resilient to coastal environment extremes and to stop out-migration from coastal Bangladesh (Kartiki, 2011) and switching occupation, there is an urgent need to explore local adaptation measures and their social ecological to environmental changes. Similarly, Awal et al. (2016) mentioned the necessity of exploration of adaptive capacity and resilience of resource poor coastal communities on present and future vulnerabilities to climate extreme events. The chronological perspectives of coastal disasters in Bangladesh are given in Table 1.

MAJOR ENVIRONMENT EXTREME EVENTS AND IMPACTS OF COASTAL ENVIRONMENT

The occurrence of coastal disasters is very frequent and unpredictable in Bangladesh. These coastal disasters and their devastating effects are provided chronologically (Table 1).

Table 1. Coastal disasters chronological perspectives

Occurrence date	Affected area	Nature of extreme event	Loss and damage
October, 1872	Cox's Bazar	Cyclonic storm	270 people died
December, 1909	Cox's Bazar	Cyclonic storm	Damage data is not available
May, 1926	Cox's Bazar	Cyclonic storm	606 people died
May, 1963	Noakhali-Cox's Bazar coast	Severe cyclonic storm W=201/kmh, s=5m and t=0.3m at Chittagong	11520 people killed
December 1965	Cox's Bazar- Teknaf coast	Severe cyclonic storm W=200kmh, s=4.0m, t=0.2m	12000 people killed
December 1966	Cox's Bazar	Cyclonic storm	Damage data is not available
October,1967	Chittagong –Cox's Bazar coast	Cyclonic storm W=160km/h, s=2m	128 people killed
May, 1970	Chittagong-Teknaf coast and landfill Cox's bazar	Cyclonic storm W=148km/h, s=1.3m, t-0.2m	18 people killed
November 1974	Cox's bazar-Chittagong offshore line	Severe cyclonic storm, w=161km/h, s=3.1, t=0.2m	20 people killed
November, 1983	Chittagong –Teaknaf coast and land fill Chittagong and cox's bazar	Severe cyclonic storm, w=136km/h, s=2.5m	300 fishermen with 50 boats missing
May 1985	Noakhali-cox's Bazar coast	Severe cyclonic storm W=154km/h, s=3.2m, t=2.8m	11069 people killed
April 1991	Patuakhali-Cox's Bazar coast	Most severe cyclonic storm, w=255km/h, s=5.8m, t=1.7 m	145000 people killed
May 19, 1997	A cyclone lashed the coastal line	W=230Km/h	606 people died
Cyclone Sidr (November 15, 2007)	Cyclone Sidr hit the country's coastal belt with a wind speed of 223 km per hour	Very Severe Cyclonic Storm W=260Km/h	Killed 3663 people
Cyclone Aila (May 25, 2009)	Cyclone Aila hit offshore 15 districts of southwestern part of Bangladesh with wind speed of 120 km per hour.	Very Severe Cyclonic Storm, W= 120 km/h	Killed 150 people and 200000 houses were damaged
Cyclone Mahasen (May 16, 2013)	Cyclone Mahasen hit near Chittagong with a wind speed of 85 km per hour.	Cyclonic Storm w=85 km/h	Killed 17 people
Cyclone Roanu (May 21, 2016)	Cyclone Roanu made landfall near Chittagong	Cyclonic Storm 110 km/h	Killed 26 peoples and damaged 40000 houses
Nov 19, 2019, Cyclone Bulbul	Cyclone Bulbul hit Patukakhali and Khulna	W=120Km/h	

Compiled from T.S Murty et al., 1986; Khalil, M.G., 1992 and Murty, T.S. and El-Sabah, M.I., 1992 and Dhaka Tribune 2013.

The study explores how the fisher women find the opportunities to preserve coastal environment and what challenges are the face? It also elicits the local knowledge, innovations, adaptation measures of resource poor vulnerable rural fisher women to environmental changes and climate extreme events. The vulnerability and resilience of coastal fisher women are heavily reliant on the process present to past. Access to information and new technologies on disasters are limited to the coastal fisher women. The increased frequency and scale of disasters make them more vulnerable to coastal environment. The local resilience and response capacity to disasters are not well documented and examined. There is an urgent need to explore and document the local resilience and response capacity to preserve coastal environment.

Objectives

The main objective of this study is to explore local adaptive response and opportunities of fisher women to preserve coastal environment. The study also addresses the following objectives:

To know the impacts and forms of coastal disasters
To find the opportunities for fisher women to preserve coastal environment.
To explore the adaptation measures and the challenges the fisher women face.

METHODOLOGY

The study was conducted on the coastline to assess the social ecological knowledge of the fisher women on coastal environment over the period of three months. Both primary and secondary data were collected in exploring opportunities for fisher women. Households survey, Focus Group Discussions and Key Informants' interviews were carried out in getting information for this study. A total of 95 households surveys were carried out to explore their local knowledge. Secondary information and data were gathered through literature review and review of documents available for this study. Journal articles and books, newspapers and project documents were collected for this study. The knowledge and capacity of fisher women were assessed with the following questions (Table 2) to explore their opportunities and knowledge to preserve coastal environment. These questions helped in exploring the opportunities, adaptation measures, challenges of the fisher women to preserve coastal environment.

Table 2. List of resilience capacity related questions used in this study

#	Resilience capacity	Question	References
1	Understanding capacity	How can you perceive coastal disasters?	Islam, 2018
2	Capacity of absorption	Can your household back to home after challenges?	Bahadur et al.,2017
3	Transformation Capacity	Do you have any alternative options during hardship?	Kates et al., 2015
4	Adaptive capacity	Do you get prompt solutions when your family feel threats?	Bahadur et al., 2017
5	Finacial capacity	Do you get financial support during hardship?	Mayunga, 2007
6	Social capacity/capital	Can you rely on your friends or any one for support when you need?	Aldritch, 2012
7	Political capacity	Can you rely on politician for their help support when you need?	Magis, 2010
8	Learning capacity	Can you use learning from previous hardships that will you help?	Cutter et al., 2008
9	Prophylactic capacity	Are you fully prepared for future disaster?	Bahadur et al., 2017
10	Early warning capacity	Do you receive any useful early warning information on future risks	Aldritch, 2012

RESULTS

The Study Area

The study concentrated in Ward no. 1 under Cox's Bazar SadarUpazila and very close to the heart of Cox's Bazar District, the longest sea beach in the world. Cox's Bazaar is a District for tourism, fishing, and other business activities in the southeastern part of Bangladesh. It is also termed as the "largest natural unbroken sea beach" in the world. The district covers lots of beauty and socioeconomic activities around us. It is the more popular tourist destination place in Bangladesh. The study sites were composed of Kutubdia Para, NazirarTek, Padnad Del and Samity Para. The population of the study area came from Kutubdia Island almost 40 years ago after severe life-threatening riverbank erosion disasters. They lost their shelter, land, assets, and other belongings. A few of them even lost their father/mother/sister/brother or relatives. After losing everything, they migrated to Cox's Bazar SadarUpazila to survive with the new hope. The most interesting things are they engaged themselves in the same profession and tried to tie them in the same community. The study place is very close to the beach and in risky areas.

Persuasive and Adaptation Measures

Persuasion is the individual belief of social, political, or religious. Persuasion is also the ability or acts to influence others to perform adaptation to change their mind. Several perceived causes are mentioned by the respondents (Table 3) It also depends on awareness-raising training, level of knowledge, warning systems on extreme events, education, adaptation skills, organizing skills, and environmental education. The common attributed causes of drought and the sources of information (knowledge) that mentioned by respondents and participants of Focus Group Discussions. The fisher folks perceived the causes for coastal disasters to include punishment, sin, lack of sacrifices, reckless behavior & immorality, natural occurrence, industrialization, oil spill in the deep sea. The sources of information (knowledge) included generational knowledge transfer, religious leaders, teachers, Mass media (radio, television, and newspaper), NGOs and government sources (Table 3). Eventually this information was collected from informal discussion at the tea stall during fieldwork. Generally, the above-mentioned causes had a strong logic on social realms, especially sacred reasons, and religious disobedience.

Table 3. Perceived general cause of coastal extreme environmental events

#	Perceived causes of drought	Sources of information
1	Retribution of God (Allah)	Aged persons, religious leaders like Imam, Priest
2	Irresponsible and wicked behaviour	Aged person, teacher, and religious leaders
3	No or less sacrifices to divine beings	Religious leader
4	As a test of faith to God (Allah)	Religious leaders and aged person
5	Natural occurrence	General people (Farmers, housewives etc)
6	Human induced activities (cutting of trees, use of chemicals, pond filling etc)	Farmers, housewives, teachers etc
7	Sign of *Akherat* (Day of judgement)	Religious leader and aged person
8	Commercialization and industrialization	Mass media, NGOs, and general people
9	Switch for even more hazardous events	Religious leader

Source: Field work (Discussion, Interviews, and Focus Group Discussion)

Allah (God) may bring disasters in the area for several reasons, including immoral conduct, failure to perform religious customs, and unwillingness to give *Zakat*. *Zakat* is a Muslim custom to pay the needy at least once a year. People believed that the generation is not practicing religious customs and culture as religious bylaws. The prime aim of these measures is to make people conscious and build resilience capacity. It also helps people in rebuilding social ecological resilience of fisher folks to coastal disasters. Ecological education and adaptation to environmental extreme events training is a keen need for primary education and adult groups in long term strategies. It should be started in academic institutions formally and adult education in non-formal education. It will reduce hazardous impacts on their livelihood; enhance knowledge and skills on social ecological resilience and effective adaptation measures. Several mock drills and training will improve the knowledge base and reduce the negative impacts of fishing, processing, product diversification or value chain, supply chain, storage, and marketing. The campaign events and training should be continued and repeated and extended to school, college, and universities. This will ensure rebuilding of social ecological resilience to coastal disasters. Active participation of family members and community people can enhance the resilience of the tested adaptation measures. These can be ensured through field day, rally, ecology and disaster fair, community gathering and sharing on value for life, environment, water, land, fish resources, forest, social cohesion, norms, ethics, morality, and other consequences. The adaptation measures and more effective measures should be documented, translated, and disseminated accordingly. In persuasion, the more effective practices should be discussed with other fishers at the time of sharing and caring. In addition, the following will harvest good results in building the resilience of the fisher folks:

Sharing of adaptation measures and their effectiveness to disasters in relation to preparedness, response, and rehabilitation

Specific case base effective coping and adaptation measures for safeguarding, shelter, food, livelihood options for recovery and sustainability

Creating more volunteers and enhanced their skills for SER.

Forming a core community for extension

Community working groups including schools and college for disaster warning, environmental management and stopping illegal fishing.

Public awareness and community practices can make people more resilient in extreme events management. Weather forecast, preparedness, warning system (miking), response to hazards is very effective for building resilience. Fishers usually go fishing on a day, and on a weekly basis. Thus, they reported that the highest interest in continuing fishing in the deep sea than normal hours and day. Eventually, they stay in the dull weather in the sea avoiding weather forecasts and warn for more fishing. Few people don't have the trust in weather signals and warning systems; it delays the process of making decisions to leave houses or fishing for shelter. This type of decision makes them more vulnerable than others. Traditional methods and knowledge applied in fishing, trade and responding to the coastal hazards. Usually, they build their houses in a safe area from vulnerable areas. In a few extents, they can predict weather and indication of good fishing in their own ways. So, it needs good persuasion moving forward them in a timely manner.

They predict and forecast weather and environment extreme events using local and indigenous knowledge in the following ways (Table 4).

Table 4. Means of weather prediction using local and indigenous knowledge

Key indication	Statement	Explanation
Observation of sea wave	If they find plain wave from deep sea to the coast shore; means that no wave movement in the sea and it indicates possibility of cyclone. When they can see large waves from deep sea towards coast then they usually go fishing. It means that the weather will be fine for next week. They go five miles from the shore towards sea and it takes 10-15 hours journey for fishing and after two weeks come back again.	The sea wave behaves very peculiarly and provides indication for safety to the fisher folks. The plain wave in the sea indicate normal weather and fisher can go for safe fishing
Observation of bright sun in the east sky	They assumed that bright sun in the east sky indicate better weather it means that sunny weather and there are no risks with fishing and other relevant activities.	Community uses nature's indication. The bright sun light in the east sky denotes bright and normal weather without having any trouble or risks
Observation of wind direction	They also used wind direction in fishing portfolio. If wind blow from south to north, it indicates fishing time and opportunity for more fishing and if it is from north to south it indicates lean period / time for fishing.	This knowledge is inherent among the respondents over decade from generation to generation. They believed that wind direction is the vital instrument in predicting bad weather.
Observation of churi fish	Another indicator they used in predicting environment change extreme event and it is very popular to the fisher folks. If the churi fish start jumping frequently in the sea it is indication of cyclone.	As a notion of bad weather Churi fish try to come out from typical conditions.
Catching whiter icha	If they can catch more whiteicha fish through their net it indicates immense event of cyclone.	White Icha try to secure shelter during extreme weather, and they trapped more
Observation of environment	When trees, stones and sands removed from coastal areas there will be landslides or coastal erosion	When trees removed and sand collect the soil become fragile and landslides and erosion occurs

In safeguarding, they need to understand the value of their own life and need quick moves towards safer places or cyclone/flood center. They also have attachments with their belongings, boats, and nets, houses. The most convenient response to the coastal hazard was to take shelter in the safest places (cyclone shelter, school and mosque and well build structure) after getting warnings. 10.3 percent of respondents tried to solve it in their own ways. 8.4 percent of respondents used traditional methods in responding to coastal disasters and the other 8.4percent of respondents were not able to respond with coastal disasters. Few respondents rebuilt their housing structure and used a tent in a suitable location in tackling the situation. Sometimes they removed fallen trees from the structures and road. Fisher folks used natural resources in responding to coastal hazards long back ago. Forest resources including timber, fuelwood, wild food, and other consumable resources they used as they required in the interim period. Rivers, ocean, land, forest, aquatic resources especially fish are the great natural resources for them. They are blessed with these resources for their livelihood options. It was strongly mentioned that nature is being skeptical and punished them accordingly due to human actions against nature. In responding to coastal hazards and environmental extreme events, well-connected road communication is essentially important. Cyclone and flood shelters, especially for the fishing communities, should be constructed in the coastal areas.

In addition to these, embankment, tree plantation, appropriate housing structures (low-cost resilient building), hospitals and school structures can reduce coastal disaster impacts. Dry food, pure drinking water and riverbank erosion control measures have tailored them in responding to coastal hazards. The fishers have natural resources including land, plants, animals, marine animals, fish resources, aquatic resources, sand, stone, and freshwater sources. These resources are interconnected and affect each other due to environmental change. Few equipment and tools are essential for them. These include fishing gear, boats, other nets, and infrastructures for storage, marketing, drying and transportation such as trucks, boats which depend on connecting roads, waterways, railways; other warning equipment and facilities including mobile phone, radio, electricity, fuel etc. Poor coastal communities bear more shocks and are vulnerable to disasters due to low resilience, low capacity, lower economic status, and low level of social protection. Observation of wave, sky, sun, weather, clouds, moon, animals, plants, and insects allow them to predict storms, cyclone, tidal surges, heavy rainfall, flood, drought, and other extreme events; and have prepared themselves for the reduction and mitigate impacts. Fishers practice rituals and religious-based respect to nature to stop cyclones and storms. The customary conservation law of the environment and management of natural resources are essential in preventing and mitigating extreme events. A combination of local and indigenous knowledge with science might be an effective strategy for resilience to coastal disasters. Such strategies will enhance and develop community resilience of predicting and dealing with hazards, climate environmental change and disaster impacts.

It is well known to all that coastal fishers suffer more from disasters than other people. Coastal fishers help others at the very beginning of disasters before coming to outside assistance. Thus, local, and indigenous knowledge linkage improves disaster risk preparedness, mitigation, and prevention as the national policy.

Table 5. Means of adaptation using local and indigenous knowledge to preserve coastal environment

Key indication or tasks	Description	Course for action
Food storage and preservation	Dig a hole on a suitable place and put all dry food and other utensils to ensure food security during crisis period of cyclone and storms	This is natural way to preserve food from bad weather
Making Killah for livestock	People make high place with mud and sand to keep their livestock safely during floods and tidal surges	Usually flood and tidal surges washed away all assets during the crisis, high place (*killah*) protect livestock from water and wave during the period.
Colonization and planting of Keya plants Sacrificing with quantity and quality of diets	Nature responds to survive itself during the crisis. People also plant *keya* for anchoring and stabilizing the soil in the coastal areas. Respondents in the study areas consume low dietary foods and less quantity to cope with the disasters	Keya plants can spread roots through soil and sands, and they can anchor soil in the coast or newly disturbed or formed soil. Usually, they cut off their diets quality and then cut off their meals from three to two meals in a day. They believe that the system will not hamper their health for a short period.

It was mentioned that they build an informal structure in the risky areas spending less money. They believed that huge investment in the risky areas will make them more fragile with the environment change extreme events. Fisher women usually consumed 3 meals in a day. While onset disasters come, they compromised with meals (Table 5). It was found that 22.4 percent of respondents consumed meals two times a day. Only one percent of respondents cut off meals two times. 29 percent of respondents mentioned that the quality of food is not good enough and even not sufficient during a crisis. Most of the respondents opined that food quality declined at the time of disasters and after an immediate crisis. They consumed less quantity and diverse food during detrimental impacts of disasters. They sacrificed with diet and changed it based on reality and other perspectives. These sacrifices started with women, then adult men. Children, disabled, pregnant mother and elderly people got empathy for meal at the crisis period. Few limitations in adaptation strategies towards coastal hazards were mentioned by the respondents. Few of these are beyond their capacity and within their capacities. Lack of eagerness and awareness makes them fragile to coastal hazards. On its other side, funds, facility, and technological innovation held up on the adaptation strategy. They also lay off their works and switch to other works for better livelihood. They don't have modern technology or infrastructure for fish processing.

Till now they have been using sunlight to follow natural processes to dry fish. They stored dry food inside polythene bags and damped it in high places to meet food demand during crises. They believed that the traditional knowledge is good but not enough in responding to hazards. They need new initiates that have a simple and friendly application. Responsible tourism and sustainable harvesting systems can contribute to preventing frequent coastal hazards impacts on the earth. They are willing to live within natural disasters risk zone due to the long back heritage profession and passed it to them inherently over the generation. Faith-based fishers often report that faith-based beliefs, knowledge, and practices have made them resilient and inner enabled to move forward with coastal disasters. It was mentioned by the respondents that local and indigenous knowledge is their great assets even they can ignore it over the generations. A study also mentioned that local and indigenous knowledge is the body of science that can enhance social ecological resilience of the fisher folks. Poor people benefit less from protection against hazards. There is a tendency in investment in the wealthier community, this decision intensifies the exposure gap between the poor and well of communities and creates a pocket of risks. Poor people are more often fall in at risk. It was found that non well fishers affected by natural disasters. Poor people invest less in preparedness and risk reduction. Obviously, the poor fishers invest fewer resources in preventing and mitigating adverse effects of natural hazards and environmental change extreme events. Early warning system is helpful, but the system is biased against wealthy people and against poor people.

DISCUSSION

They study intimate that fisher women are more vulnerable to several coastal disasters and mostly lost their assets. A study also mentioned that the coastal fisher women lose their assets when they affect by terrible natural hazards (Jongman et al., 2015). Poor coastal fisher women are more vulnerable and have less capacity dealing shocks than non-poor households due to their limited access in savings, borrowing cash or kinds and social networks and protections. Similar results were mentioned by Highfield et al. (2014) in adapting flood hazards. Some evidence is available in the literature that shows how environmental change and human activities devastated human life, livelihood, and their assets. In 1991, one terrible tropical cyclone caused 100000 deaths and dislocation of millions of people from coastal areas of Bangladesh. The people are living with disasters for natural resources on the coast. Fleisher and Chen (1997) reported that people stay in risky areas for getting better opportunities with industries and natural resources in coastal areas. Some credentials are required in recognition and external support can help the vulnerable fisher communities to enhance their capacity. If not to ensure improved capacity to them, future environment extreme events may trap poverty to coastal fishers (Hallegatte et al., 2016). Land, sea, plants, animals affect people. It was found that high vulnerability, inadequate coping measures, hazards on livelihood activities and low or no saving and investment create more risks to the fisher communities. Similarly, Hallegatte et al. (2016) reported that five reasons are responsible making poor people more vulnerable to natural disasters including overexposure; higher vulnerability; low coping ability; long term impacts on health, education, and livelihood; and effects on saving and risky investment. Poor fishers often must settle in high-risk areas for diverse opportunities in those areas. Similarly, Loayza et al. (2012) mentioned that risky areas often become more attractive despite risks, when such places provide a wider window of tangible benefits, opportunities, more income, productivity, easy access to natural resources (fish, water, land), cheap transport and other facilities (health facility). Few other favors including cheaper accommodation (Husby et al., 2018) and housing structure opportunity make people attractive though it often more fragile (World Bank and Australia Aid, 2014). Recognition of local knowledge is very important in managing coastal disasters and its impacts on the fishing communities.

One study mentioned that inclusion of local and indigenous knowledge in disaster management and climate change adaptation yet not well recognized in Indonesia (Winsemius et al., 2018). The sustainability of local communities depends on the level of interactions between human and natural environment. Social ecological resilience is linked up among human-nature relationships, practice domain and sustainability (Wilkinson et al., 2012). The study found that most informal settlements occur in the risky area for earning more. Risky areas provided more opportunities for employment and health services. Similarly, Lal and Deichmann (2010) reported that informal settlements occurred in risky areas. One study reported that people are interested in living in disaster-prone areas due to several opportunities offered by the areas including employment, education, and health services (Patankar, 2015 cited in Winsemius et al., 2018). In fact, social ecological system linked between social and natural systems (Berkes, 1993). It was found that poor people are affected more than non-poor people during coastal disasters. Akhter and Mallick (2013) expressed similar results that 25 percent of poor communities affected by Aila, whereas only 14 percent from non-poor households in 2009. The ecological resilience is diminishing within the communities due to the fate of frequent disasters. The loss of ecological resilience that occurs within society is due to the nasty surprises and unforeseen effects of the changing environment (Berkes., 2001). Holling et al. (1995) defined resilience in two ways: firstly, it concentrates on stability at a presumed steady state; resistance to disturbance and speed of return to the equilibrium position. Berkes et al. (2019) (cited in Cinner and Bernes, 2019) mentioned that numerous interdependencies are linked between social and ecological nodes to scale up social ecological approach. Resilience falls one of the important focuses on intertwined social ecological ecosystem upon which society depends and that helps in adaptation measures to disasters (Folke et al., 2010). Policy support is also essentially important to make the fisher women capable to preserve coastal environment.

CONCLUSION

The term social ecological resilience is essentially very strong terminology. It does not need further clarification. Still, fishing communities are performing, responding, and preserving coastal environment, natural hazards in their own ways and to some extent, with external support. The capacity is enhancing with the changing environment, but they are becoming reliant on external help. The use of local and indigenous knowledge is becoming less in practice. The status of social tie, mutual trust, social cohesion, and their social ecological resilience is decreasing. The study is also showed the way women fighting with natural disasters on the coast. Illegal fishing practices made them more vulnerable to environmental change extreme events. The disaster intense practical and research should be included in the curriculum and science books. It was recommended that fishing (both legal and illegal) should be stopped during environmental extreme events. Policy support in favor of fisher women should be enacted. Fisheries activities in the mangrove and coastal areas should not be extended. Otherwise, the fisher women will lose resilience, but Goldstein (2018) explained "let go and learn to love other terminology for resilience".

REFERENCES

Adger, W. N., Barnett, J., Chapin, F. S. III, & Ellemor, H. (2011). This Must Be the Place: Under representation of Identity and Meaning in Climate Change Decision-Making. *Global Environmental Politics*, *11*(2), 1–25. doi:10.1162/GLEP_a_00051

Adger, W. N. (2000). Social and ecological resilience: Are they related? *Progress in Human Geography*, *24*(3), 347–364. doi:10.1191/030913200701540465

Agrawala, S., Ota, T., Ahmed, A. U., Smith, J., & Van Aalst, M. (2003). *Development and climate change in Bangladesh: focus on coastal flooding and the Sundarbans*. OECD.

Ahern, J. (2011). From fail-safe to safe-to-fail: Sustainability and resilience in the new urban world. *Landscape and Urban Planning*, *100*(4), 341–343. doi:10.1016/j.landurbplan.2011.02.021

Akter, S., & Mallick, B. (2013). The poverty–vulnerability–resilience nexus: Evidence from Bangladesh. *Ecological Economics*, *96*, 114–124. doi:10.1016/j.ecolecon.2013.10.008

Aldrich, D. P. (2012). *Building resilience: Social capital in post-disaster recovery*. University of Chicago Press. doi:10.7208/chicago/9780226012896.001.0001

Awal, M. A., Rashid, M. H. A., Islam, A. F. M. T., Imam, M. F. & Sarkar, M. A. R. (2016). Measuring Climate Change Vulnerability and its Adaptive Capacity: Policies and Planning for Bangladesh. *British Journal of Environment & Climate Change*.

Bahadur, A., & Pichon, F. (2017). *Analysis of resilience measurement frameworks and approaches*. Overseas Development Institute.

Barnes, M. L., Bodin, Ö., McClanahan, T. R., Kittinger, J. N., Hoey, A. S., Gaoue, O. G., & Graham, N. A. (2019). Social-ecological alignment and ecological conditions in coral reefs. *Nature Communications*, *10*(1), 1–10. doi:10.103841467-019-09994-1 PMID:31053708

Berkes, F. (1993). Traditional ecological knowledge in perspective. *Traditional ecological knowledge: Concepts and cases, 1*.

Berkes, F. (2001). *Managing small-scale fisheries: alternative directions and methods*. IDRC.

Bloor, M., Frankland, M., Thomas, M., & Stewart, K. (2000). Focus groups on social research. *Sage (Atlanta, Ga.)*.

Campanella, T. J. (2006). Urban resilience and the recovery of New Orleans. *Journal of the American Planning Association*, *72*(2), 141–146. doi:10.1080/01944360608976734

Choudhury, A. M., Neelormi, S., & Quadir, D. A. (2003). Socio-economic and physical perspectives of water related vulnerability to climate change: Results of field study in Bangladesh. *Science and Culture*, *23*, 225–238.

Chowdhury, Z. (2001). *Final Report on Network/ Political Analysis Study of the Shrimp Component of the Social Feasibility Studies for the Fourth Fisheries Project*. University of Dhaka and Bangladesh Centre for Advance Studies.

Cinner, J. E., & Barnes, M. L. (2019). Social Dimensions of Resilience in Social-Ecological Systems. *One Earth*, *1*(1), 51–56. doi:10.1016/j.oneear.2019.08.003

Cinner, J. E., Adger, W. N., Allison, E. H., Barnes, M. L., Brown, K., Cohen, P. J., & Marshall, N. A. (2018). Building adaptive capacity to climate change in tropical coastal communities. *Nature Climate Change*, *8*(2), 117–123. doi:10.103841558-017-0065-x

Coser, L. A., & Rosenberg, B. (Eds.). (1965). Sociological theory: A book of readings (No. 301.082 C834 1964). Macmillan.

Cutter, S. L., Barnes, L., Berry, M., Burton, C., Evans, E., Tate, E., & Webb, J. (2008). A place-based model for understanding community resilience to natural disasters. *Global Environmental Change*, *18*(4), 598–606. doi:10.1016/j.gloenvcha.2008.07.013

Dasgupta, S., Kamal, F. A., Khan, Z. H., Choudhury, S., & Nishat, A. (2014a*). River Salinity and Climate Change: Evidence from Coastal Bangladesh* (Policy Research Working Paper No. 6817). Development Research Group, World Bank. https://documents.worldbank.org/en/publication/documents-reports/documentdetail/522091468209055387/river-salinity-and-climate-change-evidence-from-coastal-bangladesh

Haugen, S. E. (2009). *Measures of labor underutilization from the current population survey* (Working Paper No. 424). Bureau of Labor Statistics. https://www.bls.gov/osmr/pdf/ec090020.pdf

DMB. (2010). *National Plan for Disaster Management 2010-2015*. Disaster Management Bureau, Bangladesh. https://ddm.portal.gov.bd/sites/default/files/files/ddm.portal.gov.bd/page/332124ba_20b1_42cc_816b_d97d7eb4f478/npdm_final.pdf

Elliott, R. (2018). The sociology of climate change as a sociology of loss. *European Journal of Sociology. Archives Européennes de Sociologie*, *59*(3), 301–337. doi:10.1017/S0003975618000152

Fleisher, B. M., & Chen, J. (1997). The coast-noncoast income gap, productivity, and regional economic policy in China. *Journal of Comparative Economics*, *25*(2), 220–236. doi:10.1006/jcec.1997.1462

Folke, C., Carpenter, S. R., Walker, B., Scheffer, M., Chapin, T., & Rockström, J. (2010). Resilience thinking integrates resilience, adaptability, and transformability. *Ecology and Society*, *15*(4), 1–9. doi:10.5751/ES-03610-150420

Friedmann, J. (2008). The uses of planning theory: A bibliographic essay. *Journal of Planning Education and Research*, *28*(2), 247–257. doi:10.1177/0739456X08325220

GFDR (Global Facility for Disaster Reduction and Recovery). (2016). *The Making of a Riskier Future: How Decisions Are Shaping Future Disaster Risk*. World Bank.

Giddens, A., & Sutton, P. W. (2013). *Sociology* (8th ed.). Polity.

GoB. (2008a). *Cyclone Sidr in Bangladesh: Damage, Loss and Needs Assessment for Disaster Recovery and Reconstruction*. GFDRR. https://www.gfdrr.org/sites/default/files/2275_CycloneSidrinBangladeshExecutiveSummary.pdf

Goldstein, J. (2018). *Planetary Improvement: Cleantech Entrepreneurship and the contradictions of green Capitalism*. MIT Press.

Hallegatte, S., Bangalore, M., Bonzanigo, L., Fay, M., Kane, T., Narloch, U., & Vogt-Schilb, A. (2015). *Shock waves: managing the impacts of climate change on poverty*. The World Bank.

Hallegatte, S., Vogt-Schilb, A., Bangalore, M., & Rozenberg, J. (2016). *Unbreakable: building the resilience of the poor in the face of natural disasters*. World Bank Publications.

Haque, M. A., Rahman, D., & Rahman, M. H. (2016). The importance of community-based approach to reduce sea level rise vulnerability and enhance resilience capacity in the coastal areas of Bangladesh: A review. *Journal of Sustainability Science and Management, 11*(2), 81–100.

Highfield, W. E., Peacock, W. G., & Van Zandt, S. (2014). Mitigation planning: Why hazard exposure, structural vulnerability, and social vulnerability matter. *Journal of Planning Education and Research, 34*(3), 287–300. doi:10.1177/0739456X14531828

Holling, C. S., Schindler, D. W., Walker, B. W., &Roughgarden, J. (1995). Biodiversity in the functioning of ecosystems: an ecological synthesis. *Biodiversity loss: economic and ecological issues*, 44-83.

Hossain, M. A., Reza, M. I., Rahman, S., & Kayes, I. (2011). Climate Change and its Impacts on the Livelihoods of the Vulnerable People in the Southwestern Coastal Zone in Bangladesh. *Climate Change Management*, 237–259

Huq, N., Hugé, J., Boon, E., & Gain, A. K. (2015). Climate change impacts in agricultural communities in rural areas of coastal Bangladesh: A tale of many stories. *Sustainability (Basel), 7*(7), 8437–8460. doi:10.3390u7078437

Husby, T., de Groot, H. L., Hofkes, M. W., & Filatova, T. (2018). Flood protection and endogenous sorting of households: The role of credit constraints. *Mitigation and Adaptation Strategies for Global Change, 23*(2), 147–168. doi:10.100711027-015-9667-7 PMID:30093827

IPPC. (2012). *Managing the Risks of Extreme Events and Disasters to Advance Climate Change Adaptation*. IPPC.

Islam, M. S. (2018*). Sustainable approaches to Drought Management: Exploratory Study on selected regions of Northwest in Bangladesh* [Unpublished PhD Thesis, University of Dhaka].

Jones, L. (2019). Resilience isn't the same for all: Comparing subjective and objective approaches to resilience measurement. *Wiley Interdisciplinary Reviews: Climate Change, 10*(1), e552. doi:10.1002/wcc.552

Jones, L., & D' Errico, M. (2019). *Whose resilience matters? like-for-like comparisons of objective and subjective measures of resilience*. London School of Economics and Political Science, LSE Library.

Jones, L., & Tanner, T. (2017). 'Subjective resilience': Using perceptions to quantify household resilience to climate extremes and disasters. *Regional Environmental Change*, *17*(1), 229–243. doi:10.100710113-016-0995-2

Jongman, B., Winsemius, H. C., Aerts, J. C., De Perez, E. C., Van Aalst, M. K., Kron, W., & Ward, P. J. (2015). Declining vulnerability to river floods and the global benefits of adaptation. *Proceedings of the National Academy of Sciences of the United States of America*, *112*(18), E2271–E2280. doi:10.1073/pnas.1414439112 PMID:25902499

Kartiki, K. (2011). Climate change and migration: A case study from rural Bangladesh. *Gender and Development*, *19*(1), 23–38. doi:10.1080/13552074.2011.554017

Kates, R. W., Travis, W. R., & Wilbanks, T. J. (2012). Transformational adaptation when incremental adaptations to climate change are insufficient. *Proceedings of the National Academy of Sciences of the United States of America*, *109*(19), 7156–7161. doi:10.1073/pnas.1115521109 PMID:22509036

Khalil, G. M. (1992). Cyclones and storm surges in Bangladesh: Some mitigative measures. *Natural Hazards*, *6*(1), 11–24. doi:10.1007/BF00162096

Lall, S. V., & Deichmann, U. (2010). *Density and disasters: economics of urban hazard risk*. The World Bank. doi:10.1596/1813-9450-5161

Loayza, N. V., Olaberria, E., Rigolini, J., & Christiaensen, L. (2012). Natural disasters and growth: Going beyond the averages. *World Development*, *40*(7), 1317–1336. doi:10.1016/j.worlddev.2012.03.002

Amin, M. (2004). Introduction: Real-time Systems. In R. Zimmerman & T. Horan (Eds.), *Digital Infrastructures: Enabling civil and environmental systems through information technology* (pp. 116–140). Routledge.

Magill, R. S. (1993). Focus groups, program evaluation, and the poor. *Journal of Sociology and Social Welfare*, *20*(1), 103. doi:10.15453/0191-5096.2059

Magis, K. (2010). Community resilience: An indicator of social sustainability. *Society & Natural Resources*, *23*(5), 401–416. doi:10.1080/08941920903305674

Mahmood, S. M., & Ansary, B. S. (2013). Shrimp Fry Collection as Alternative Livelihood: A Case Study on the Southwest Coastal Region of Bangladesh. *ASA University Review*, *7*(2), 139–149.

Mayunga, J. S. (2007). Understanding and applying the concept of community disaster resilience: a capital-based approach. *Summer academy for social vulnerability and resilience building, 1*(1), 1-16.

Moser, S. C., Melillo, J. M., Jacobs, K. L., Moss, R. H., & Buizer, J. L. (2016). Aspirations and common tensions: Larger lessons from the third US national climate assessment. *Climatic Change*, *135*(1), 187–201. doi:10.100710584-015-1530-z

Moss, M. L., & Townsend, A. M. (2004). Moving information in the twenty-first century city. In R. Hanley (Ed.), *Moving People, Goods, and Information in the 21st Century: The Cutting-Edge Infrastructures of Networked Cities* (pp. 69–84). Routledge.

Murty, T. S., & El-Sabh, M. I. (1992). Mitigating the effects of storm surges generated by tropical cyclones: A proposal. *Natural Hazards*, *6*(3), 251–273. doi:10.1007/BF00129511

Murty, T. S., & Neralla, V. R. (1992). On the recurvature of tropical cyclones and the storm surge problem in Bangladesh. *Natural Hazards*, *6*(3), 275–279. doi:10.1007/BF00129512

Mutahara, M., Warner, J. F., Wals, A. E. J., Khan, M. S. A., & Wester, P. (2017). Social learning for adaptive delta management: Tidal River Management in the Bangladesh Delta. *International Journal of Water Resources Development*, 1–21.

Olsson, L., Jerneck, A., Thoren, H., Persson, J., & O'Byrne, D. (2015). Why resilience is unappealing to social science: Theoretical and empirical investigations of the scientific use of resilience. *Science Advances*, *1*(4), e1400217. doi:10.1126ciadv.1400217 PMID:26601176

Patankar, A. (2015). *The exposure, vulnerability, and ability to adapt of poor households to recurrent floods in Mumbai*. World Bank. https://documents1.worldbank.org/curated/en/418931467991989631/pdf/WPS7481.pdf

Quadir, D. A., & Iqbal, A. (2008). *Tropical Cyclones: Impacts on Coastal Livelihoods. Investigation of the Coastal Inhabitants of Bangladesh*. IUCN.

Schipper, E. L. F., & Langston, L. (2015). *A comparative overview of resilience measurement frameworks. Analyzing Indicators and Approaches*. ODI. https://odi.org/en/publications/a-comparative-overview-of-resilience-measurement-frameworks-analysing-indicators-and-approaches/

Sultana, P., Thompson, P. M., & Ahmed, M. (2002). *Women-led fisheries management-a case study from Bangladesh*. World Fish Center. https://digitalarchive.worldfishcenter.org/

UNESCAP (United Nations Economic and Social Commission for Asia and the Pacific). (2012). *Statistical Yearbook for Asia and the Pacific 2012*. UNESCAP. https://www.unescap.org/stat/data/syb2012/index.asp

Weber, E. U. (2010). What shapes perceptions of climate change? *Wiley Interdisciplinary Reviews: Climate Change*, *1*(3), 332–342. [REMOVED HYPERLINK FIELD] doi:10.1002/wcc.41

Wilkinson, C. (2012). Social-ecological resilience: Insights and issues for planning theory. *Planning Theory*, *11*(2), 148–169. doi:10.1177/1473095211426274

Winsemius, H. C., Jongman, B., Veldkamp, T. I., Hallegatte, S., Bangalore, M., & Ward, P. J. (2018). *Disaster risk, climate change, and poverty: assessing the global exposure of poor people to floods and droughts*. LSE. https://eprints.lse.ac.uk/87536/1/Bangalore_Disaster%20risk_2018.pdf

World Bank and Australian Aid. (2014). *Where are we during flooding?' A qualitative assessment of Poverty and Social Impacts of Flooding in selected neighborhoods of HCMC*. World Bank. https://documents1.worldbank.org/curated/en/669141603288540994/pdf/People-in-Harms-Way-Flood-Exposure-and-Poverty-in-189-Countries.pdf

World Bank Group. (2016). *Impact of Climate Change and Aquatic Salinization on Fish Habitats and Poor Communities in Southwest Coastal Bangladesh and Bangladesh Sundarbans*. World Bank. https://elibrary.worldbank.org/doi/abs/10.1596/1813-9450-7593

Zimmerman, R., & Horan, T. A. (Eds.). (2004). *Digital Infrastructures: Enabling civil and environmental systems through information technology*. Psychology Press. doi:10.4324/9780203357125_chapter_ONE

Chapter 11
Opportunities and Challenges for Indian Women Leaders in Environmental Management:
A Woman in a Patriarchal Society

Priyadarshini Nidhan
Mumbai University, India

ABSTRACT

This chapter examines the opportunities and challenges faced by Indian women leaders in the field of environmental management. As India undergoes rapid development and faces growing environmental concerns, the need for effective leadership in environmental management becomes increasingly crucial. The chapter highlights the progress made in promoting women's involvement in this field, such as increasing awareness, government policies, education and skill development, networking opportunities, and global collaborations. However, it also identifies the challenges that hinder women's progress, including gender biases, limited representation, work-life balance, resource access, and cultural constraints.

INTRODUCTION

This chapter examines the opportunities and challenges faced by Indian women leaders in the field of environmental management. As India undergoes rapid development and faces growing environmental concerns, the need for effective leadership in environmental management becomes increasingly crucial. The chapter highlights the progress made in promoting women's involvement in this field, such as increasing awareness, government policies, education and skill development, networking opportunities, and global collaborations. However, it also identifies the challenges that hinder women's progress, including gender biases, limited representation, work-life balance, resource access, and cultural constraints.

Despite these challenges, the chapter proposes strategies for empowering women leaders in environmental management. These strategies include the creation of inclusive policies and frameworks that promote gender equality and provide support for women leaders. Enhancing educational opportunities

DOI: 10.4018/978-1-6684-5986-7.ch011

for women in environmental management is also crucial, ensuring access to quality education and skill development programs that equip them with the necessary knowledge and expertise. Moreover, encouraging mentorship and networking opportunities allows aspiring women leaders to connect with experienced professionals and gain valuable guidance and support.

Promoting women's representation and leadership is another key strategy. This can be achieved through quota systems, leadership development programs, and targeted initiatives that increase women's participation in decision-making bodies and senior leadership roles. Additionally, challenging stereotypes and addressing deep-rooted gender biases is essential. Raising awareness through awareness campaigns and highlighting the achievements of women leaders in environmental management can help break down barriers and inspire future generations.

By addressing these challenges and embracing these strategies, India can unlock the full potential of its women leaders in environmental management, contributing to a more sustainable and inclusive future. Women's leadership in this field can lead to innovative approaches, diverse perspectives, and effective solutions to environmental challenges. It is imperative that stakeholders across government, organizations, and institutions collaborate to create an enabling environment that supports and empowers women leaders in environmental management.

India is a rapidly developing country with a diverse environmental landscape and a growing need for effective environmental management. In recent years, there has been an increasing recognition of the importance of women's involvement in leadership positions, including environmental management. This chapter explores the opportunities and challenges faced by Indian women leaders in the field of environmental management, highlighting the progress made and the areas that require further attention.

OPPORTUNITIES FOR INDIAN WOMEN LEADERS IN ENVIRONMENTAL MANAGEMENT

Increasing awareness and recognition: Increasing awareness and recognition of the contributions made by Indian women leaders in environmental management can present several valuable opportunities for their advancement and influence in this field. In recent years, there has been a growing recognition of the vital role women play in environmental management. This awareness has led to more opportunities for women to take on leadership positions in government agencies, non-governmental organizations, research institutions, and corporate sectors thereby make a significant impact. *Here are some ways in which greater awareness and recognition can benefit women leaders:*

Visibility and Role Modeling: Increased awareness and recognition of successful women leaders in environmental management can serve as powerful role models for aspiring women in the field. Seeing successful women in leadership positions can inspire young women to pursue careers in environmental management, thus increasing the talent pool in the sector.

Empowerment and Confidence: When the achievements and efforts of women leaders are acknowledged and celebrated, it boosts their confidence and sense of empowerment. This, in turn, can motivate them to take on more significant challenges and lead impactful environmental initiatives with conviction.

Breaking Stereotypes: Greater awareness and recognition of women leaders in environmental management challenge traditional gender stereotypes that might have confined women to certain roles in the past. This can help break down barriers and prejudices, paving the way for more gender-inclusive work environments.

Access to Opportunities: As awareness about women's potential and capabilities in environmental management grows, it may lead to an increased focus on providing equal opportunities for women to access leadership roles, decision-making positions, and participation in key projects.

Networking and Collaboration: Increased recognition can open doors for women leaders to build stronger networks and collaborations with like-minded individuals, organizations, and government agencies. Networking can lead to more opportunities for joint initiatives, funding, and support for their environmental projects.

Policy Influence: When women leaders gain recognition for their expertise and accomplishments in environmental management, they are more likely to be invited to participate in policy discussions and decision-making processes. This can help shape environmental policies and strategies with a more gender-inclusive and holistic approach.

Increased Funding and Support: Greater awareness of women leaders' contributions can attract attention from potential donors, investors, and philanthropists, leading to increased funding and support for their environmental initiatives.

Strengthening Women's Networks: As more women leaders are recognized, it can lead to the strengthening of women's networks in the environmental sector. These networks can provide mentorship, support, and opportunities for collaboration and collective advocacy.

Global Visibility: Recognition of women leaders in environmental management can also attract attention at the international level, providing them with opportunities to represent India in global environmental forums, conferences, and partnerships.

To capitalize on these opportunities, it is crucial to promote and celebrate the achievements of Indian women leaders in environmental management through various platforms, media channels, and awards. Additionally, fostering an environment that appreciates diversity and actively supports women's participation and advancement in the sector can further enhance the impact of women leaders in driving sustainable environmental practices in India.

Government policies and initiatives: The Indian government has implemented various policies, initiatives and platforms to promote women's participation in environmental management to make a meaningful impact. For example, the National Policy for Women's Empowerment aims to create an enabling environment for women's leadership and decision-making across all sectors, including environmental management. Here are some ways in which government policies and initiatives can create opportunities for women in this field:

Gender-inclusive policies: When governments actively promote gender equality through policies and initiatives, it encourages the representation of women in various sectors, including environmental management. By implementing gender-inclusive policies, the government can ensure that women have equal access to opportunities, training, and leadership roles in the environmental sector.

Reservation and quotas: Some governments may introduce reservation or quota systems that mandate a certain percentage of leadership positions to be filled by women in environmental organizations, committees, and decision-making bodies. This approach helps to increase women's representation and influence in shaping environmental policies.

Skill development and training programs: Government-supported skill development and training programs can equip women leaders with the necessary knowledge and expertise in environmental management. These initiatives can enhance their qualifications and make them more competitive for leadership roles.

Financial support and funding: Governments can allocate funds and grants specifically dedicated to women-led environmental projects. Financial support can enable women to initiate and implement sustainable initiatives, research, and conservation programs effectively.

Recognition and awards: Governments can establish awards and recognition programs to celebrate the achievements of women leaders in environmental management. This recognition not only motivates women to continue their efforts but also sets examples for others to follow.

Women-focused NGOs and government agencies: Governments can establish or support NGOs and agencies dedicated to promoting women's participation in environmental management. These organizations can offer mentoring, networking, and capacity-building opportunities to aspiring women leaders.

Public awareness campaigns: Governments can run awareness campaigns highlighting the importance of women's participation in environmental conservation and sustainable practices. Such campaigns can challenge traditional gender roles and foster societal support for women leaders in this field.

Research and data collection: Governments can invest in research and data collection initiatives that specifically focus on women's contributions to environmental management. This data can be used to advocate for better representation and support for women in the sector.

International collaborations: Government-led international collaborations and partnerships can provide Indian women leaders with opportunities to participate in global environmental initiatives, conferences, and dialogues, thus expanding their networks and exposure.

By leveraging these policies and initiatives, Indian women leaders in environmental management can find a more conducive environment to thrive and contribute significantly to addressing environmental challenges. It is essential for governments to be committed to implementing and sustaining these measures, ensuring a more inclusive and sustainable future for India.

Education and skill development: The availability of quality education and skill development programs has improved significantly in India. This has empowered women to acquire the knowledge and skills needed for leadership roles in environmental management. Institutions and organizations are also offering specialized courses and training programs focused on environmental sustainability and conservation. Education and skill development play a crucial role in creating opportunities for Indian women leaders in environmental management. By focusing on these areas, women can enhance their knowledge, expertise, and leadership capabilities, enabling them to make a significant impact in the environmental sector. Here are some ways in which education and skill development provide opportunities for women leaders:

Enhancing Technical Knowledge: Education equips women with a strong foundation of technical knowledge in environmental sciences, ecology, sustainability, and related fields. This knowledge is essential for effective decision-making and problem-solving in environmental management.

Specialization in Environmental Disciplines: Skill development programs allow women to specialize in specific environmental disciplines, such as climate change, biodiversity conservation, renewable energy, water management, waste management, and more. Specialization enhances their expertise and credibility in these areas.

Leadership and Management Skills: Education and training programs focused on leadership and management equip women with the skills needed to lead teams, handle complex projects, and navigate challenging situations in the environmental sector.

Field Experience and Exposure: Skill development initiatives often include fieldwork and practical experiences, exposing women to real-world environmental challenges. This exposure helps them develop a deeper understanding of ground-level issues and the ability to implement practical solutions.

Research and Innovation: Education fosters a research-oriented mindset, enabling women to contribute to innovative solutions and advancements in environmental science and management practices.

Networking Opportunities: Educational institutions and skill development programs provide valuable networking opportunities. Building connections with peers, professors, and professionals can lead to collaboration, mentorship, and exposure to potential career opportunities.

Access to Job Opportunities: A well-educated and skilled workforce is more likely to be sought after by employers in the environmental sector. Educational qualifications and skill sets open doors to various job opportunities and leadership positions.

Empowerment and Confidence: Education and skill development empower women by boosting their self-confidence and self-belief. Women leaders who are equipped with knowledge and skills are more likely to overcome challenges and achieve their goals.

Entrepreneurship and Start-ups: Education and skill development can inspire women to start their environmental businesses or non-profit organizations, creating new avenues for sustainable projects and innovations.

International Collaborations: Through educational programs, women can participate in exchange programs, workshops, and conferences, leading to international collaborations and exposure to global best practices in environmental management.

To fully leverage these opportunities, it is essential to ensure equal access to education and skill development programs for women across all socio-economic backgrounds. Governments, non-governmental organizations, educational institutions, and private sectors can work together to provide scholarships, grants, and training opportunities specifically targeted towards women in environmental management. Creating a supportive and inclusive environment for women to pursue education and skill development will contribute to building a diverse and competent leadership base in the Indian environmental sector.

Networking and mentorship opportunities: Networking platforms and mentorship programs are emerging to connect aspiring women leaders with experienced professionals in the field of environmental management. These opportunities provide valuable guidance, support, and exposure to enhance leadership skills and navigate career paths effectively.

Networking and mentorship opportunities are powerful tools that can significantly benefit Indian women leaders in environmental management. These opportunities can open doors, provide support, and foster personal and professional growth. Here are some ways networking and mentorship can create opportunities for women leaders in this field:

Building Supportive Networks: Networking events, conferences, and professional gatherings provide women with the chance to connect with like-minded individuals, peers, and industry experts. These supportive networks can offer encouragement, advice, and a sense of community, helping women overcome challenges and stay motivated.

Collaboration and Partnerships: Networking can lead to collaborations and partnerships with other organizations, government agencies, and stakeholders. By working together, women leaders can pool resources, knowledge, and expertise to implement more substantial and impactful environmental projects.

Career Advancement: Strong professional networks can open doors to new job opportunities, promotions, and leadership roles in the environmental sector. Recommendations and referrals from within the network can enhance a woman's career prospects.

Learning and Knowledge Sharing: Networking allows women leaders to learn from experienced professionals, exchange ideas, and stay updated with the latest developments in environmental management. Knowledge sharing enhances their skills and competence in the field.

Access to Funding and Resources: A well-established network can provide access to potential donors, investors, and funding opportunities for environmental initiatives. It can also facilitate access to research facilities, equipment, and resources required for various projects.

Mentorship and Guidance: Mentorship programs connect women leaders with seasoned professionals who can offer valuable guidance, advice, and career support. Mentors can help navigate challenges, provide insights, and offer encouragement.

Exposure to Diverse Perspectives: Networking exposes women leaders to diverse perspectives and approaches to environmental management. This exposure broadens their understanding and encourages innovative thinking.

Overcoming Gender Barriers: A supportive network can help women leaders overcome gender barriers and biases that may exist in the environmental sector. It reinforces their confidence and empowers them to challenge stereotypes.

Advocacy and Influence: Women leaders who build strong networks can use their collective influence to advocate for gender equality and the importance of women's contributions in environmental management. This advocacy can lead to policy changes and a more inclusive work environment.

Global Engagement: Networking opportunities can extend beyond national borders, offering women leaders the chance to participate in international conferences, seminars, and collaborations. This global engagement can enhance their visibility and impact.

To make the most of networking and mentorship opportunities, it is essential to actively seek out and participate in relevant events, join professional associations and organizations, and engage in online communities and forums. Governments, NGOs, and private organizations can also play a role by establishing formal mentorship programs and hosting networking events specifically targeted towards supporting and promoting women leaders in environmental management. By nurturing strong networks and mentorship relationships, Indian women leaders can gain valuable resources and support that can propel their careers and contributions in the environmental sector.

Global partnerships and collaborations: India actively participates in global environmental initiatives and collaborations. This involvement opens doors for Indian women leaders to engage in international forums, exchange knowledge, and contribute to global environmental management agendas.

Global partnerships and collaborations present significant opportunities for Indian women leaders in environmental management to broaden their impact, access resources, and contribute to global sustainability efforts. Here are some ways in which global partnerships can create opportunities for women leaders in this field:

Knowledge Exchange and Learning: Global partnerships provide Indian women leaders with opportunities to engage with environmental experts and professionals from around the world. This exchange of knowledge and ideas can enhance their understanding of global environmental challenges and best practices.

Access to International Funding: Collaborating with international organizations and donors can increase the chances of securing funding for environmental projects. Global partnerships open doors to funding opportunities that may not be readily available within India.

Exposure to Cutting-Edge Technologies: Working with international partners exposes women leaders to the latest technologies and innovations in environmental management. This exposure can inspire the adoption of advanced techniques and solutions in their projects.

Scaling Impact: Collaborating on a global scale allows women leaders to tackle larger and more complex environmental issues. Partnering with organizations operating in different countries enables the scaling of successful initiatives to reach a broader audience.

Access to Global Markets: Women leaders involved in sustainable practices or eco-friendly products can access international markets through partnerships with global organizations. This opens up opportunities for growth and expansion of their initiatives.

Policy Advocacy: Global partnerships give Indian women leaders a platform to advocate for sustainable policies and practices at international forums and conferences. This advocacy can influence global agendas and policies related to environmental management.

Leveraging Diverse Perspectives: Collaborating with partners from different countries and cultures exposes women leaders to diverse perspectives and approaches to environmental challenges. This can lead to more comprehensive and innovative solutions.

Strengthening Leadership Skills: Engaging in global partnerships can strengthen women leaders' leadership skills, as they navigate cross-cultural interactions, negotiate partnerships, and manage diverse teams.

Enhancing Visibility and Recognition: Participation in global collaborations raises the visibility and recognition of Indian women leaders in environmental management. This recognition can attract more opportunities and support for their initiatives.

Building Resilience and Adaptability: Global collaborations often involve working in complex and dynamic environments. This experience can enhance the adaptability and resilience of women leaders, enabling them to better navigate challenges in their field.

To harness these opportunities effectively, Indian women leaders in environmental management can explore partnerships with international non-profit organizations, research institutions, governmental agencies, and multinational corporations. Collaborating with like-minded organizations that share similar environmental goals can create mutually beneficial relationships. Additionally, leveraging technology and digital platforms can facilitate global collaborations and knowledge sharing, even without physical travel.

It is essential for governments and organizations to actively support and promote global partnerships for women leaders in the environmental sector. This includes providing funding, creating networking opportunities, and facilitating cross-border collaborations through policy support and initiatives. With the right support and resources, Indian women leaders can significantly contribute to global efforts towards sustainability and environmental conservation.

Representation and diversity: As more organizations and governments prioritize gender diversity, there is a push for increased representation of women in leadership positions. This encourages the promotion and recognition of capable Indian women in the environmental sector. Representation and diversity offer crucial opportunities for Indian women leaders in environmental management to make a lasting impact and drive positive change. Here are some ways in which representation and diversity create opportunities for women in this field:

Empowerment and Inclusivity: Increased representation of women leaders in environmental management empowers them to actively participate in decision-making processes and shape policies that promote sustainability and conservation. It also fosters an inclusive work environment where diverse perspectives are valued.

Role Models and Inspiration: When young women see successful women leaders in the environmental sector, it inspires them to pursue careers in this field. Representation helps break gender stereotypes and encourages more women to enter and excel in environmental management.

Diverse Solutions and Approaches: Diversity in leadership brings a range of perspectives, experiences, and problem-solving approaches. Indian women leaders can offer unique insights and innovative solutions to environmental challenges, enhancing the overall effectiveness of conservation efforts.

Addressing Gender-Specific Issues: Women leaders can bring attention to and address gender-specific environmental issues, such as the impact of climate change on women, access to natural resources, and women's roles in sustainable development.

Access to Funding and Resources: Diverse leadership teams are more likely to attract funding and support from various sources, including those interested in promoting gender equality and diversity in environmental initiatives.

Improved Communication and Engagement: Women leaders often possess strong communication skills, enabling them to effectively engage with diverse stakeholders, including local communities and policymakers. This can lead to better collaboration and support for environmental projects.

Reducing the Gender Gap: Increased representation of women in leadership positions helps reduce the gender gap in the environmental sector, promoting equity and equal opportunities for women to advance in their careers.

Building Resilience and Adaptability: Diversity in leadership fosters a more resilient and adaptable sector. Women leaders, with their unique perspectives and experiences, can contribute to better preparedness for addressing environmental challenges.

Promoting Sustainable Practices: Women leaders often demonstrate a strong commitment to sustainability and the well-being of communities. Their leadership can drive the adoption of eco-friendly practices and social responsibility within organizations and projects.

Global Recognition and Collaboration: Diverse leadership teams are more likely to be recognized and sought after in international collaborations and partnerships. This recognition opens doors to global platforms for Indian women leaders to share their expertise and contribute to global sustainability efforts.

To fully realize the opportunities of representation and diversity, organizations and governments need to implement policies that promote gender equality in the environmental sector. This includes providing equal access to education, training, and leadership development programs for women, ensuring fair hiring practices, and fostering supportive work environments.

Additionally, creating mentorship programs and women-focused initiatives can provide the necessary support for aspiring women leaders to excel in environmental management. Embracing diversity and representation as core values within the environmental sector will not only benefit women leaders but also contribute to more comprehensive and sustainable approaches to environmental challenges in India.

CHALLENGES FOR INDIAN WOMEN LEADERS IN ENVIRONMENTAL MANAGEMENT

Many women has seen horrifying incidents of gender discrimination as a young female boss. She continues to endure more harassment than is appropriate, from being mistaken for a secretary to having her work continuously questioned by her more junior male coworkers. However, she focuses on conflicts she can win, as another wise woman once stated. “The gender conflicts that women typically experience at work and those that female entrepreneurs will experience are different. Because you are creating the organisation and the culture, the scenario is quite different for a woman entrepreneur. You must go out and market your story while you are developing your own brand. The hardest aspect, where discrimination enters, is that.

“Most people find it unsettling to be seated across from a woman who is on an equal footing. They are not at blame. An entire generation of men have only ever interacted with women in the roles of mother or wife, never in the workplace. They are unable to relate to women on an equal basis. They solely have knowledge of caring for or safeguarding women. Instead of trying to ignore the issue, it is critical to acknowledge it. “However, admitting it does not excuse this behaviour. She decided to let my work do the talking. To defuse these circumstances, women must concentrate on their strengths.

Figure 1. Key challenges

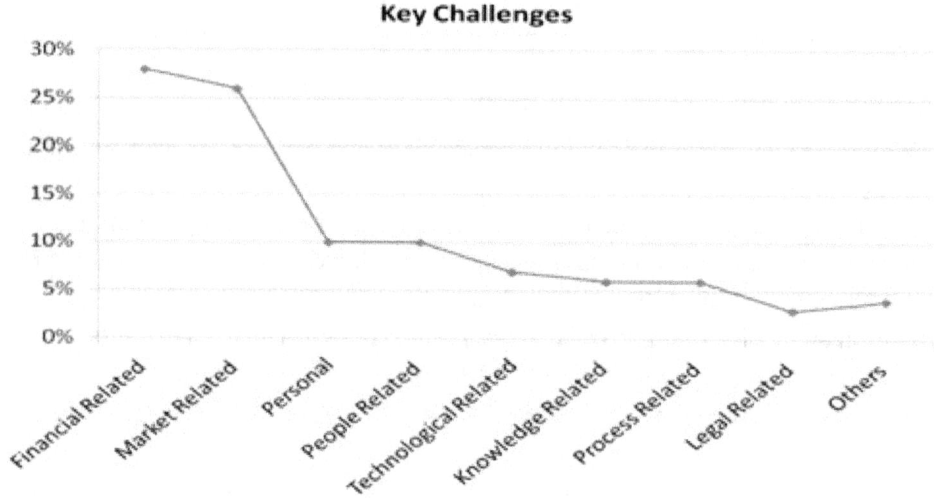

Gender biases and stereotypes: Deep-rooted gender biases and stereotypes still persist in society, which can hinder the progress of women in leadership positions. Women leaders often face challenges in breaking through traditional gender roles and stereotypes associated with leadership and decision-making.

Lack of representation: Despite progress, women remain underrepresented in senior leadership roles within the environmental management sector. Limited representation can result in fewer role models and mentors, making it difficult for women to navigate their careers and overcome barriers.

Work-life balance: Balancing professional responsibilities with family and societal expectations can be challenging for women leaders. The long working hours and demanding nature of environmental management roles may require significant dedication and sacrifice, affecting work-life balance.

Access to resources and funding: Limited access to resources, including financial support and research facilities, can pose challenges for women leaders in environmental management. Lack of adequate funding and resources can hinder their ability to implement projects and initiatives effectively.

Cultural and societal constraints: Cultural norms and societal expectations can act as barriers to women's leadership in environmental management. Traditional gender roles and cultural biases may limit the opportunities available to women and impact their decision-making authority.

Figure 2. Why women don't apply for additional financing

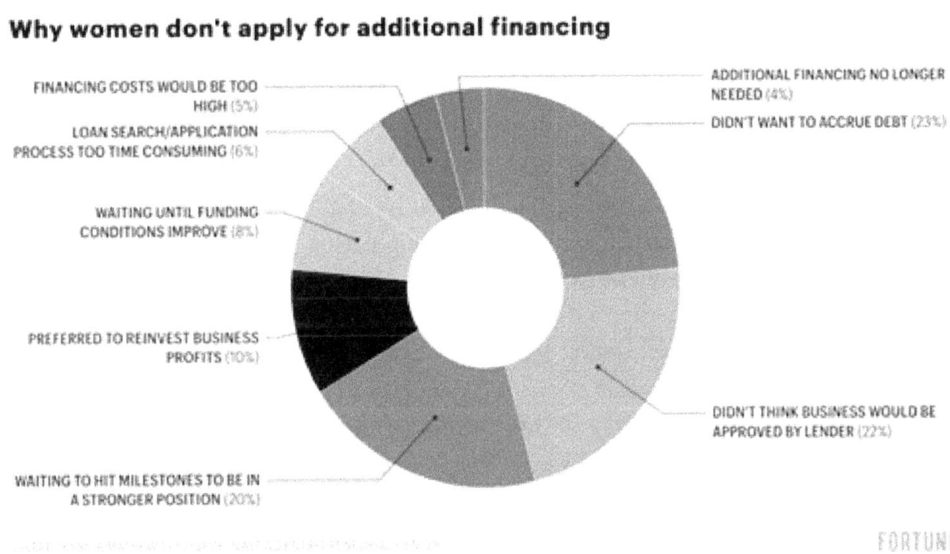

ROLE OF WOMEN IN SUSTAINABILITY AND ENVIRONMENT

The role of women in sustainability and environmental leadership is of paramount importance. Women play a crucial role in addressing environmental challenges and achieving sustainable development for several reasons:

Diverse perspectives: Women bring unique perspectives and experiences to the table, which can lead to more comprehensive and innovative solutions. Including women's voices in decision-making processes ensures a more holistic approach to tackling environmental issues.

Community engagement: Women often have strong connections to their communities and are actively involved in grassroots initiatives. They play vital roles in environmental conservation efforts at the local level, promoting sustainable practices within families and communities.

Conservation and stewardship: Women have historically played significant roles in environmental conservation and stewardship. They have been instrumental in preserving biodiversity, managing natural resources, and passing down traditional knowledge about sustainable practices.

Education and empowerment: When women are educated and empowered, they become powerful agents of change. Investing in women's education and leadership abilities can lead to more effective environmental policies and projects.

Sustainable development: Women's involvement in decision-making processes contributes to more balanced and equitable development. Sustainable development is not just about environmental conservation but also about ensuring social and economic well-being, where women's perspectives are vital.

Climate resilience: Women, especially in developing countries, are disproportionately affected by climate change impacts. Empowering women to take leadership roles in climate resilience and adaptation efforts can lead to more effective responses to climate-related challenges.

Role models and inspiration: Women leaders in sustainability and environmental fields serve as role models and inspiration for the next generation. Their presence encourages young girls and women to pursue careers in these areas and fosters a more diverse and inclusive workforce.

Collaboration and cooperation: Women often excel in building relationships and fostering collaboration. In environmental work, cooperation between diverse stakeholders is essential, and women's leadership can facilitate this process.

Population and resource management: Women's involvement in family planning initiatives and resource management can have a significant impact on population growth and resource consumption patterns, contributing to more sustainable development.

Global impact: Women make up approximately half of the world's population. Harnessing their full potential in sustainability and environmental leadership can lead to more substantial and far-reaching global impact.

In summary, the role of women in sustainability and environmental leadership is critical for achieving a more sustainable, equitable, and resilient future. Empowering and supporting women's participation in these fields is not only a matter of gender equality but also essential for the well-being of our planet and all its inhabitants.

STRATEGIES TO IMPROVE THE ROLE OF WOMEN IN SUSTAINABILITY AND ENVIRONMENT LEADERSHIP

Improving the role of women in sustainability and environmental leadership is crucial for achieving a more balanced and effective approach to addressing environmental challenges. Here are some strategies to promote and empower women in this field:

Promote education and training: Encourage and support girls and women to pursue education and training in fields related to sustainability, environmental science, and leadership. Providing scholarships, mentorship programs, and workshops can help break down barriers and create more opportunities.

Increase visibility and representation: Promote the visibility of women leaders and experts in sustainability and environmental sectors through media, conferences, and public events. This will serve as an inspiration for other women and challenge stereotypes about gender roles in these fields.

Create supportive networks: Establish networks and communities that facilitate collaboration, knowledge-sharing, and mentorship among women in sustainability and environmental leadership. These networks can provide valuable support and guidance throughout their careers.

Encourage women in decision-making roles: Advocate for the inclusion of women in decision-making bodies, committees, and boards related to environmental policy and sustainability initiatives. Diverse perspectives can lead to more comprehensive and effective solutions.

Address gender bias and stereotypes: Raise awareness about unconscious bias and gender stereotypes that may hinder women's advancement in sustainability and environmental leadership. Implement policies and training to eliminate such biases within organizations and institutions.

Support work-life balance: Create policies that support work-life balance, such as flexible working hours and parental leave. This will help women maintain their careers while managing family responsibilities.

Invest in women-led projects: Encourage funding and investment in projects and businesses led by women that focus on sustainable practices and environmental conservation. This can help break down financial barriers and foster innovative ideas.

Promote women's voices in media and storytelling: Support and encourage women to tell their stories and experiences in sustainability and environmental leadership through various media platforms. These stories can inspire others and showcase the impact of women's contributions in this field.

Integrate gender perspective in environmental projects: Consider the specific needs and experiences of women in the design and implementation of environmental projects. Addressing gender-specific challenges can lead to more inclusive and sustainable outcomes.

Advocate for policy changes: Support policies that promote gender equality and women's rights, as these factors are essential for empowering women to take on leadership roles in sustainability and environmental sectors.

By implementing these strategies, we can create a more inclusive and diverse leadership landscape, which will contribute to more effective and sustainable solutions for our planet's challenges.

CONCLUSION

Indian women leaders in environmental management have immense potential to contribute to sustainable development and environmental conservation. The opportunities available through increased awareness, government initiatives, education, networking, and global collaborations are paving the way for their participation. However, to fully harness this potential, it is crucial to address the challenges of gender biases, limited representation, work-life balance, resource access, and cultural constraints. By promoting gender equality, providing mentorship, creating inclusive policies, and breaking down societal barriers, India can unlock the full potential of its women leaders in environmental management, fostering a more sustainable and inclusive future.

ADDITIONAL READING

Ambec, S., & Lanoie, P. (2008). Does it pay to be green? A systematic overview. *The Academy of Management Perspectives*, *22*(4), 45–62. doi:10.5465/amp.2008.35590353

Banerjee, S. B., Iyer, E. S., & Kashyap, R. K. (2003). Corporate environmentalism: Antecedents and influence of industry type. *Journal of Marketing*, *67*(2), 106–122. doi:10.1509/jmkg.67.2.106.18604

Bass, B. M. (1985). *Leadership and performance beyond expectations*. Free Press.

Dubey, R., Gunasekaran, A., & Ali, S. S. (2015). Exploring the relationship between leadership, operational practices, institutional pressures and environmental performance: A framework for green supply chain. *International Journal of Production Economics*, *160*, 120–132. doi:10.1016/j.ijpe.2014.10.001

Saraite-Sariene, L., Galán-Valdivieso, F., Alonso-Cañadas, J., & García-Tabuyo, M. (2022), The role of women's leadership in environmental NGOs' online accountability. *Online Information Review, 46*(4).

Chapter 12
Women Leaders in Environmental Management

Sreevas T. V. T. K.
Presidency College, India

Prasad Vasantrao Kulkarni
Presidency College, India

ABSTRACT

This book chapter explores the significant role of women leaders in environmental management and their impact on sustainable development. Historically, women have been underrepresented in leadership positions in the environmental sector, but in recent years, their contributions and achievements have gained recognition worldwide. The chapter begins by discussing the barriers and challenges that women have faced in accessing leadership roles in environmental management. It examines social, cultural, and institutional factors that have limited women's participation and advancement in this field. Despite these obstacles, the chapter highlights the success stories of pioneering women leaders who have defied societal norms and shattered glass ceilings to make a difference in environmental conservation and sustainability. The chapter then delves into the unique perspectives and approaches that women bring to environmental management.

INTRODUCTION

The underrepresentation of women in leadership positions within the environmental sector has been a longstanding matter of concern. Despite their significant contributions to environmental management and sustainability, women have faced systemic barriers and discrimination (Sembiring & Nitivattananon, 2010) hindering their access to leadership roles and decision-making processes. This chapter aims to shed light on the historical challenges that women leaders have encountered in the environmental field and the transformative impact they have had in fostering a greener and more equitable future.

DOI: 10.4018/978-1-6684-5986-7.ch012

Historical Challenges in Environmental Leadership

Throughout history, the environmental sector, much like many other fields, has been predominantly male dominated. Women faced considerable challenges and biases that restricted their access to leadership positions within environmental organizations, government agencies, and policymaking bodies. The prevailing gender norms and societal expectations relegated women to supporting roles, limiting their opportunities to take on leadership roles and influence environmental policies.

Gender Bias and Discrimination

Gender bias and discrimination have been pervasive in the environmental sector, perpetuating the underrepresentation of women in leadership roles. Stereotypes about women's abilities and expertise in scientific and technical fields, such as ecology and conservation, hindered their career advancement and recognition (Crandall et al., 2021). Additionally, women have often been overlooked for leadership positions, with decision-makers favouring male candidates, further exacerbating the gender gap in environmental management.

Limited Access to Education and Resources

Historically, women faced limited access to education and professional development opportunities, particularly in the scientific and environmental fields. Unequal access to resources and mentorship programs hindered their career progression, making it challenging for women to break into leadership roles and excel in environmental management (Barkhuizen et al., 2022).

The Glass Ceiling Effect

The glass ceiling effect, a phenomenon where women encounter invisible barriers that prevent them from advancing to top leadership positions, has been prevalent in the environmental sector. Despite their expertise and dedication, women have often found themselves stuck in middle management roles, unable to ascend to top leadership positions, where decisions and policies are formulated.

Social Expectations and Family Responsibilities

Traditional social expectations often placed the burden of caregiving and family responsibilities primarily on women. Balancing family life with demanding leadership roles in the environmental sector posed additional challenges for women, dissuading some from pursuing leadership positions or advancing their careers (Bismark et al. 2015).

Women's Contributions and Resilience

Despite the barriers they faced, women have made remarkable contributions to environmental management and sustainability. Many women demonstrated exceptional leadership skills, pioneering groundbreaking research, community-based conservation efforts, and advocacy for climate action (Shaw, 2006). Their resilience and determination in the face of adversity have been instrumental in driving positive change and shaping the environmental agenda.

The Transformative Impact of Women Leaders

The inclusion of women in leadership positions has proven to be transformative for the environmental sector. Women leaders often bring diverse perspectives, collaborative approaches, and a keen focus on social and environmental justice. Their ability to connect with communities, consider the needs of vulnerable populations, and promote inclusive policies has led to more holistic and effective environmental solutions.

Overcoming Historical Challenges

Efforts to address the historical underrepresentation of women in environmental leadership have gained momentum in recent years. Advocacy for gender equality, diversity, and inclusion in the environmental sector has led to the formulation of policies and initiatives aimed at breaking down barriers and promoting women's empowerment.

Mentorship and Capacity-Building Programs

Mentorship and capacity-building programs have emerged to support aspiring women leaders in the environmental field. These initiatives provide training, networking opportunities, and mentorship to help women develop leadership skills and overcome career obstacles.

The Way Forward: A Greener and Equitable Future

The progress made in advancing gender equality in the environmental sector is an encouraging step toward fostering a greener and more equitable future. By promoting the inclusion of women in decision-making processes, leadership roles, and research endeavours, the environmental sector can harness the full potential of its diverse talent pool to address complex environmental challenges effectively (Robinson & Dechant, 1997).

The historical concern of underrepresentation of women in environmental leadership is a critical issue that must be addressed to achieve sustainable development goals. Women's contributions to environmental management have been invaluable, demonstrating their transformative impact on promoting a greener and more equitable future (Sharma et al., 2021). Overcoming historical challenges and promoting gender equality in the environmental sector will ensure that women continue to play a pivotal role in shaping the future of environmental conservation and sustainability. By fostering an inclusive and supportive environment for women leaders, we can collectively work toward a more sustainable and resilient world for all.

The underrepresentation of women in environmental leadership is a manifestation of deeply entrenched gender norms and societal expectations. Historically, women have been relegated to roles primarily associated with caregiving and domestic responsibilities, resulting in their exclusion from positions of influence in environmental management. Moreover, cultural biases and stereotypes have perpetuated the belief that men are inherently better suited for leadership roles in the environmental field, marginalizing the potential contributions of women (McKinnon & O'Connell, 2020). This unequal gender representation has led to a lack of diversity in perspectives, decision-making, and problem-solving strategies, hindering the holistic approaches required addressing complex environmental challenges effectively.

Institutional barriers also contribute to the limited presence of women in environmental leadership roles. Gender imbalances within environmental organizations and institutions have created environments where women find themselves underrepresented or excluded from leadership positions (McKinnon & O'Connell, 2020). Biases in hiring and promotion practices further exacerbate the gender gap in the sector, perpetuating a cycle of male-dominated leadership structures. The lack of mentorship and support networks can also deter aspiring women leaders from advancing their careers, as they may lack access to guidance and role models to navigate the challenges they encounter.

The pursuit of leadership in the environmental field often requires dedication to fieldwork, research, and networking, which can present unique challenges for women striving to balance career aspirations with family responsibilities. The lack of family-friendly policies and flexible work arrangements in some environmental organizations may further impede the professional advancement of women leaders, making it difficult to reconcile personal and professional commitments. These challenges can result in a loss of valuable talent and expertise, hampering the potential contributions of women to environmental management.

In male-dominated industries within the environmental sector, women leaders may encounter stereotypes questioning their technical expertise and competence. This implicit bias can undermine their confidence and diminish opportunities for advancement. Additionally, women leaders may face resistance or scepticism from male colleagues and subordinates, making it challenging for them to assert their authority and ideas. This dynamic can impede effective collaboration and teamwork, hindering the full potential of women leaders to drive positive environmental change.

Moreover, the intersectionality of gender with other dimensions of identity, such as race, ethnicity, and socioeconomic status, adds further complexity to the challenges faced by women in environmental leadership. Women from marginalized communities may experience compounded discrimination, limiting their access to resources, mentorship, and networking opportunities. Recognizing this intersectionality is crucial for developing inclusive policies and practices that address the unique challenges and needs of diverse groups of women in the environmental sector.

In light of these barriers and challenges, it is imperative to address the underrepresentation of women in environmental leadership and actively promote gender equality in the field. By fostering an inclusive and supportive environment that empowers women to lead, we can harness the full potential of their talent, expertise, and unique perspectives in environmental management. This chapter explores strategies for promoting gender equality, initiatives to support women leaders, and case studies of successful women environmentalists who have broken barriers and made a lasting impact on environmental conservation and sustainability. Embracing diversity and gender parity in leadership is not only a matter of equity but also a strategic imperative to build a more sustainable, resilient, and inclusive future for all.

Barriers and Challenges for Women in Environmental Leadership

Throughout history, women seeking leadership roles in the environmental sector have encountered numerous barriers and challenges that have hindered their progress and representation. One significant obstacle has been the persistence of gender norms and societal expectations that dictate traditional gender roles. These norms have often relegated women to the domestic sphere, leading to limited opportunities for them to be recognized as leaders in environmental management. Furthermore, cultural biases and stereotypes have perpetuated the perception that men are more suited for decision-making positions in the environmental field, marginalizing the potential contributions of women.

Institutional barriers also play a critical role in limiting women's access to leadership roles. The lack of gender diversity in environmental organizations and institutions has created an environment where women often find themselves underrepresented or excluded from leadership positions. Biases in hiring and promotion practices can further perpetuate the gender gap in the environmental sector. Additionally, the absence of mentorship and support networks can hinder the professional development of women aspiring to lead in the field, as they may not have access to role models or guidance to navigate the challenges they encounter.

Women in environmental leadership also face challenges related to work-life balance, especially when their roles require fieldwork or extensive travel. The demands of leadership positions, coupled with societal expectations around caregiving responsibilities, can create barriers for women who seek to advance in their careers while managing family commitments. Moreover, the lack of family-friendly policies and flexible work arrangements in some environmental organizations can add additional strain on women leaders, making it difficult for them to balance their professional and personal lives.

In male-dominated industries within the environmental sector, women may confront stereotypes that question their technical expertise and competence. This implicit bias can undermine their confidence and diminish opportunities for advancement. Additionally, women leaders may find themselves facing resistance or scepticism from male colleagues and subordinates, making it challenging for them to assert their authority and ideas.

The intersectionality of gender with other dimensions of identity, such as race, ethnicity, and socio-economic status, further compounds the challenges for women in environmental leadership. Women from marginalized communities may face compounded discrimination, limiting their access to resources, mentorship, and networking opportunities. This intersectional perspective highlights the need for inclusive policies and practices that address the unique challenges faced by diverse groups of women in environmental leadership.

In conclusion, women in environmental leadership have encountered numerous barriers and challenges, ranging from gender norms and cultural biases to institutional obstacles and work-life balance issues. Recognizing and addressing these challenges is vital to fostering an inclusive and diverse environmental sector. By breaking down barriers, promoting gender equality, and providing support and mentorship, we can create an environment that empowers women to lead and make a significant impact in environmental management. Embracing the talent and expertise of women leaders will contribute to more holistic, innovative, and sustainable solutions to the pressing environmental challenges we face.

Inspiring Success Stories of Women Environmental Leaders

Despite encountering substantial challenges, numerous women have demonstrated unwavering determination and resilience, surmounting barriers to emerge as trailblazers in the realm of environmental leadership. This section of the chapter celebrates the remarkable achievements of pioneering women who fearlessly defied societal norms and shattered the preconceived limitations placed upon them. These exceptional women leaders not only made significant contributions to environmental conservation and sustainability but also paved the way for future generations of women to follow in their footsteps. By shining a spotlight on their indelible impact, we underscore the utmost importance of recognizing and promoting women's invaluable leadership in environmental management.

One such inspiring success story is that of Dr. Jane Goodall, a renowned primatologist and conservationist, who devoted her life to studying chimpanzees in the wild. Dr. Goodall's groundbreaking research on chimpanzee behavior and social interactions challenged the prevailing scientific norms and reshaped our understanding of these remarkable creatures. Her tireless efforts in raising awareness about the threats faced by chimpanzees and their habitats have had a profound impact on global conservation initiatives.

Another trailblazing woman leader is Wangari Maathai, a Kenyan environmentalist and political activist. She became the first African woman to win the Nobel Peace Prize in 2004 for her pioneering Green Belt Movement, which focused on empowering women through tree planting and environmental conservation efforts. Maathai's vision of combining environmental sustainability with gender empowerment has inspired countless individuals worldwide to take action in tackling pressing environmental issues.

In the realm of renewable energy, Dr. Vandana Shiva, an Indian scholar, environmental activist, and ecofeminist, has been a guiding light in advocating for sustainable agricultural practices and the protection of biodiversity. Dr. Shiva's tireless activism against genetically modified organisms (GMOs) and her emphasis on the role of women as custodians of traditional agricultural knowledge have been transformative in shaping sustainable agricultural practices globally.

Additionally, Marina Silva, a Brazilian environmentalist and former Minister of Environment, has left an enduring legacy in advocating for the conservation of the Amazon rainforest and sustainable development. Her efforts to combat deforestation and promote environmentally responsible policies have gained international recognition and underscored the importance of indigenous knowledge and community engagement in environmental management.

Furthermore, Wangari Maathai, a Kenyan environmentalist and political activist, became the first African woman to win the Nobel Peace Prize in 2004 for her pioneering Green Belt Movement, which focused on empowering women through tree planting and environmental conservation efforts. Maathai's vision of combining environmental sustainability with gender empowerment has inspired countless individuals worldwide to take action in tackling pressing environmental issues.

These inspiring success stories exemplify the courage, dedication, and vision of women leaders who have made significant strides in environmental management. Their journeys have not only defied societal norms but have also paved the way for a more gender-inclusive and environmentally conscious future. By recognizing and celebrating the achievements of these pioneering women, we send a powerful message of encouragement and inspiration to women worldwide, affirming that their voices, contributions, and leadership are indispensable in shaping a sustainable and equitable world. These women leaders have demonstrated that gender should never be a barrier to environmental excellence and that by promoting gender equality, we unlock the full potential of talent and expertise necessary to tackle the most pressing environmental challenges of our time.

Women's Unique Perspectives in Environmental Management

Gender diversity in environmental leadership serves as a catalyst for embracing a wide spectrum of unique perspectives and approaches that have transformative implications for sustainable environmental management. Women leaders, often drawing from their distinct experiences and sensibilities, tend to prioritize not only the ecological aspect but also the social well-being of communities affected by environmental policies and initiatives. This section delves into the valuable contributions of women in decision-making processes within the environmental sector, shedding light on the profound impact of their perspectives on fostering long-term sustainability and community engagement.

One critical aspect of women's unique perspectives in environmental management is their emphasis on social well-being. Women leaders frequently recognize the interconnectedness between the environment and human communities, acknowledging that environmental policies and conservation efforts have significant implications for the quality of life and livelihoods of vulnerable populations. Their inclusive approach involves engaging with local communities and understanding their needs and aspirations, ensuring that environmental initiatives align with social equity and justice. By considering the social dimension of environmental issues, women leaders forge solutions that are more inclusive and sustainable, with a focus on improving the overall well-being of communities.

Moreover, women leaders tend to display a profound commitment to environmental stewardship that prioritizes long-term sustainability. Their approach involves embracing a holistic and forward-thinking outlook, which goes beyond short-term gains and considers the broader impact of decisions on future generations. This emphasis on sustainability is underpinned by an inherent understanding of the interconnectedness between ecological, economic, and social systems. By incorporating a long-term perspective, women leaders foster strategies that address the root causes of environmental challenges and offer enduring solutions to pressing issues, such as climate change, biodiversity loss, and resource depletion.

Community engagement stands as another hallmark of women's unique perspectives in environmental management. Women leaders often recognize the significance of engaging with local communities and stakeholders as essential partners in environmental decision-making processes. By involving diverse voices and building collaborative relationships, they gain a deeper understanding of the on-ground realities and the complex interactions between human activities and ecosystems. Community-driven approaches foster ownership and support for environmental initiatives, enhancing the effectiveness and sustainability of projects in the long run. Through their inclusive approach to community engagement, women leaders cultivate a sense of environmental responsibility and empowerment among local populations, fostering a shared commitment to conservation and sustainable development.

Research and case studies further illuminate the added value that women bring to environmental decision-making processes. Studies have demonstrated that diverse leadership teams, including women, exhibit higher levels of creativity, problem-solving abilities, and innovation. By drawing from a diverse range of perspectives and expertise, women leaders contribute to more robust and comprehensive environmental solutions. Case studies highlighting successful environmental initiatives led by women underscore the tangible impact of their unique approaches, demonstrating the transformative potential of gender diversity in leadership roles.

In conclusion, women's unique perspectives in environmental management offer an invaluable contribution to sustainable development and conservation efforts. By prioritizing social well-being, community engagement, and long-term sustainability, women leaders foster inclusive and innovative approaches to address environmental challenges. Embracing gender diversity in decision-making processes within the environmental sector ensures that policies and initiatives reflect the needs and aspirations of diverse communities while forging pathways to a greener and more sustainable future. By recognizing and supporting the distinct contributions of women in environmental leadership, we pave the way for a more resilient and equitable world, where environmental stewardship is harmoniously interwoven with social well-being and sustainable development.

CASE STUDIES OF WOMEN LEADERS' IMPACT IN ENVIRONMENTAL INITIATIVES

This section presents compelling case studies of women leaders who have spearheaded initiatives related to climate change mitigation, biodiversity conservation, renewable energy, and environmental justice. By examining their efforts and achievements, we illustrate how women's leadership contributes to positive environmental outcomes.

Case Study One: Christiana Figueres—Climate Change Mitigation

Introduction to the Case

Climate change, a defining challenge of our time, requires global cooperation and visionary leadership to combat its far-reaching impacts. Among the prominent figures leading the charge to address this crisis is Christiana Figueres, a Costa Rican diplomat, and environmental advocate. As the Executive Secretary of the United Nations Framework Convention on Climate Change (UNFCCC) from 2010 to 2016, Figueres played a pivotal role in mobilizing international efforts to negotiate and adopt the historic Paris Agreement in 2015. This landmark accord marked a turning point in the fight against global warming, uniting nearly 200 nations in a collective commitment to limit temperature rise and build climate resilience. This case explores the leadership and contributions of Christiana Figueres in the pursuit of a sustainable and resilient future for the planet.

Early Life and Background

Christiana Figueres was born in San José, Costa Rica, in 1956, into a family deeply committed to environmental preservation and public service. Her father, José Figueres Ferrer, was a former President of Costa Rica and played a significant role in advancing the country's democratic and environmental policies. This familial influence shaped Figueres' passion for environmental issues and sustainability from an early age.

Leading the UNFCCC: A Pivotal Appointment

In May 2010, Figueres was appointed as the Executive Secretary of the UNFCCC, a crucial international role in climate change negotiations. She assumed this position during a critical period when international cooperation and commitment to climate action were urgently needed. Her extensive background in environmental and climate policy, along with her diplomatic expertise, made her an ideal choice to lead the UNFCCC.

Building Consensus and Momentum

Figueres recognized that tackling climate change required more than technical solutions; it demanded a paradigm shift in how nations and stakeholders approached the issue. She sought to build bridges and create a positive atmosphere for dialogue between developed and developing countries, acknowledging their varying responsibilities and capacities to address climate change.

Under Figueres' guidance, the UNFCCC initiated a series of preparatory talks in the run-up to the Paris Climate Conference (COP21) in 2015. These talks provided a platform for nations to voice their concerns and outline their intended contributions to combat climate change, known as Intended Nationally Determined Contributions (INDCs).

Figueres understood the importance of involving diverse stakeholders in climate negotiations. She engaged with businesses, cities, civil society organizations, and academia to foster a collaborative approach to address climate challenges. This inclusive strategy not only broadened the base of support but also demonstrated the significance of collective action in achieving climate goals.

The Paris Agreement: A Turning Point

The pinnacle of Figueres' leadership came at the COP21 in Paris in December 2015. The summit aimed to achieve a legally binding and universal agreement on climate change, with the participation of all nations.

Figueres' adept diplomatic skills and relentless determination were instrumental in guiding the complex negotiations toward a successful outcome. She ensured that all voices were heard, and contentious issues were addressed with transparency and respect. Her empathetic and compassionate approach to the process helped bridge gaps between parties with differing priorities.

The Paris Agreement's key objective was to limit global warming to well below 2 degrees Celsius above pre-industrial levels and to pursue efforts to limit the temperature increase to 1.5 degrees Celsius. This ambitious goal recognized the urgency of the climate crisis and the need for immediate and collective action to prevent catastrophic consequences.

Securing Global Commitments

The success of the Paris Agreement hinged on securing commitments from nearly 200 countries. Figueres' leadership was instrumental in fostering an environment of trust and cooperation, encouraging nations to step up and submit their INDCs. By the time of the agreement's adoption, an overwhelming majority of countries had pledged to take significant steps to reduce greenhouse gas emissions and contribute to the global effort.

Figueres' unwavering dedication and persuasive advocacy were evident in her interactions with world leaders, encouraging them to set aside their differences and prioritize the greater good. Her efforts helped create a momentum for change that extended beyond the COP21 summit.

Legacy and Impact

The Paris Agreement's adoption marked a historic moment in international efforts to combat climate change. Christiana Figueres' visionary leadership and strategic approach were vital in achieving this significant milestone. Her ability to bring together governments, businesses, and civil society under a shared vision demonstrated the potential for collaborative leadership in tackling global challenges.

Figueres' legacy extends beyond the Paris Agreement. She inspired a new generation of climate activists and leaders, emphasizing the importance of hope, optimism, and perseverance in the face of daunting challenges. Her commitment to inclusive and transparent climate diplomacy set a precedent for future climate negotiations.

Conclusion

Christiana Figueres' tenure as the Executive Secretary of the UNFCCC was defined by her instrumental role in steering the world towards the adoption of the Paris Agreement. Her diplomatic finesse, strategic vision, and ability to unite diverse stakeholders were central to the success of this historic accord. The Paris Agreement stands as a testament to the power of collaborative leadership and the urgent need for collective action in combatting climate change. Figueres' legacy continues to inspire and serve as a guiding light in the ongoing global efforts to build a sustainable and resilient future for generations to come.

Case Study Two: Dr. Sylvia Earle—Biodiversity Conservation

Introduction to the Case

Dr. Sylvia Earle, a pioneering marine biologist and oceanographer, has spent her life exploring the mysteries of the world's oceans and advocating for their conservation. As the first female chief scientist of the U.S. National Oceanic and Atmospheric Administration (NOAA), she played a crucial role in advancing marine biodiversity conservation initiatives. One of her most significant accomplishments was the establishment of marine protected areas (MPAs) that serve as sanctuaries for diverse marine species, preserving delicate ecosystems from destructive human activities. This case study delves into the life and work of Dr. Sylvia Earle, highlighting her dedication to ocean exploration and her pivotal contributions to safeguarding the marine environment.

Early Life and Passion for the Oceans

Born on August 30, 1935, in Gibbstown, New Jersey, Sylvia Alice Earle developed an early fascination with nature and the sea. Spending her childhood on the shores of the Gulf of Mexico, she was captivated by the ocean's wonders and developed a deep love for marine life. This early exposure fueled her determination to become a marine scientist and advocate for the conservation of our planet's most vital resource.

Trailblazing Career and Ocean Exploration

Dr. Earle's academic journey began at Florida State University, where she earned a Bachelor of Science degree in 1955. She continued her education at Duke University, earning her Master's and Ph.D. degrees in phycology (the study of algae) in 1956 and 1966, respectively. Throughout her academic pursuits, she displayed a keen interest in understanding marine ecosystems and their intrinsic value to the planet's health.

Over the years, Earle participated in numerous oceanographic expeditions, breaking gender barriers in the male-dominated field of marine science. Her work involved deep-sea exploration, leading groundbreaking research on marine algae, and pioneering the use of underwater habitats. Earle's notable accomplishments earned her the nickname "Her Deepness" and established her as a prominent figure in oceanography.

Becoming Chief Scientist of NOAA

In 1990, Sylvia Earle was appointed as the Chief Scientist of the U.S. National Oceanic and Atmospheric Administration, a milestone achievement for both her career and the recognition of women in science. In this leadership role, she focused on raising awareness about the ocean's vulnerability to human-induced threats, such as overfishing, pollution, and climate change.

As the Chief Scientist, Earle advocated for a holistic approach to marine conservation, emphasizing the need to protect entire ecosystems rather than isolated species. She highlighted the interconnectedness of marine life and the dependence of human societies on healthy oceans.

Spearheading Marine Protected Areas

One of Dr. Earle's most influential contributions to marine conservation was her tireless effort to establish Marine Protected Areas (MPAs) around the world. MPAs are designated regions of the ocean where human activities are regulated or restricted to conserve marine biodiversity and safeguard fragile ecosystems.

Empowering Ocean Advocacy and Education

Throughout her career, Dr. Earle recognized the importance of public engagement and education in fostering ocean conservation. She co-founded the Deep Ocean Exploration and Research (DOER) program to facilitate undersea research and raise awareness about the need to protect marine environments. Additionally, she worked as an explorer-in-residence for the National Geographic Society, using media platforms to communicate the importance of ocean conservation to broader audiences.

The Impact of Sylvia Earle's Work

Dr. Sylvia Earle's unwavering dedication to ocean exploration and conservation has left a profound impact on marine science and environmental advocacy. Her pioneering efforts in establishing marine protected areas have significantly contributed to preserving some of the ocean's most pristine and vulnerable ecosystems.

Her work has also inspired a new generation of marine scientists and advocates, encouraging them to follow in her footsteps and continue the fight for ocean conservation. By breaking gender barriers in the field of oceanography, Earle has become a role model for women in science, proving that passion, determination, and expertise can create positive change.

Conclusion

Dr. Sylvia Earle's extraordinary career as a marine biologist and oceanographer has been marked by groundbreaking discoveries, environmental stewardship, and a relentless commitment to ocean conservation. As the first female chief scientist of NOAA, she advanced marine biodiversity conservation efforts and spearheaded the establishment of marine protected areas, preserving vital habitats and species.

Earle's advocacy and passion for the seas have helped raise awareness about the urgent need to protect our ocean's delicate ecosystems from human-induced threats. Her remarkable legacy continues to inspire future generations to explore, understand, and protect the vast wonders of the marine world, ensuring a sustainable and resilient future for our planet.

Case Study Three: Dr. Wangari Maathai—Environmental Justice

Introduction to the Case

Dr. Wangari Maathai, a Kenyan environmentalist and social activist, dedicated her life to addressing environmental degradation while advocating for gender equality. As the founder of the Green Belt Movement, she initiated a groundbreaking grassroots movement that empowered rural women in Kenya to plant trees, combat deforestation, and desertification, all while improving their livelihoods. Maathai's work exemplified the powerful intersection of environmental conservation and social justice, showcasing how environmental initiatives can lead to transformative change in communities. This case study explores the life and legacy of Dr. Wangari Maathai and the profound impact of the Green Belt Movement in Kenya and beyond.

Early Life and Educational Journey

Wangari Muta Maathai was born on April 1, 1940, in the rural village of Ihithe, Nyeri District, in colonial Kenya. Growing up in a farming community, she developed a deep connection with nature and witnessed firsthand the impacts of deforestation and environmental degradation on local communities.

In the early 1960s, Maathai received a scholarship to study biology in the United States. She pursued her higher education at Mount St. Scholastica College in Atchison, Kansas, and later earned a Bachelor of Science degree in biology from the University of Pittsburgh. She furthered her studies at the University of Giessen in Germany, where she obtained a master's degree in biological sciences. In 1971, she became the first East African woman to earn a doctorate, completing her studies at the University of Nairobi.

The Birth of the Green Belt Movement

Returning to Kenya in the late 1960s, Dr. Wangari Maathai was deeply concerned about the accelerating deforestation, soil erosion, and the deteriorating living conditions of rural communities. She noticed the increasing scarcity of firewood and clean drinking water, which disproportionately affected women who were responsible for gathering these resources.

Inspired by her background in biology and ecology, Maathai sought to address both environmental and social challenges simultaneously. In 1977, she founded the Green Belt Movement, an initiative that would come to be recognized globally for its innovative approach to conservation and women's empowerment.

Empowering Women Through Tree Planting

The Green Belt Movement's core mission was to combat deforestation, desertification, and soil erosion by empowering rural communities, particularly women, to plant trees. Maathai recognized that women were key stakeholders in environmental conservation, as they were deeply connected to the land and natural resources. Additionally, she understood that empowering women would lead to positive social and economic outcomes for communities.

Under the Green Belt Movement's guidance, women were encouraged to plant trees and care for them until they matured. In return, they received seedlings, technical training, and a small monetary incentive. The program not only aimed at restoring degraded landscapes but also offered opportunities for income generation through the sale of tree products, such as fruits, nuts, and firewood.

Overcoming Challenges and Expanding Impact

Despite facing initial resistance from traditional male-dominated societies, Maathai and her team persevered. They worked to dispel misconceptions about women's roles in the community and emphasized the benefits of tree planting and environmental conservation.

As the Green Belt Movement gained momentum, it expanded its focus beyond tree planting. Maathai recognized the need to address the root causes of environmental degradation, which often stemmed from political and economic issues. The movement began advocating for good governance, human rights, and democratic principles to promote sustainable development and protect the rights of marginalized communities.

Scaling Up and Global Recognition

Throughout the 1980s and 1990s, the Green Belt Movement continued to grow, spreading to other parts of Kenya and inspiring similar initiatives in neighbouring countries. By the turn of the millennium, millions of trees had been planted, and thousands of women had become actively involved in environmental conservation and community development.

In 2004, Dr. Wangari Maathai received global recognition for her tireless efforts when she was awarded the Nobel Peace Prize. The prestigious accolade not only celebrated her environmental leadership but also highlighted the critical link between environmental sustainability and peace-building.

The Green Belt Movement's Impact

The Green Belt Movement's impact extended far beyond tree planting. It revolutionized the way environmental initiatives integrated social justice and women's empowerment. The movement demonstrated that environmental conservation is not solely about protecting nature but also about fostering sustainable livelihoods and empowering marginalized communities.

Environmental Impact

Through tree planting and sustainable land management, the Green Belt Movement helped restore degraded landscapes, combat deforestation, and preserve biodiversity. The increased tree cover improved soil quality, reduced erosion, and mitigated the impacts of climate change. As a result, water sources were replenished, and local ecosystems were revitalized.

Women's Empowerment

The Green Belt Movement transformed the lives of countless women in rural Kenya. By participating in the initiative, women gained new skills, increased their income, and gained a sense of agency over their lives. As tree planting became a recognized and respected endeavor, women's roles within their communities evolved, leading to greater recognition and participation in decision-making processes.

Social and Political Impact

Beyond environmental and economic benefits, the Green Belt Movement played a crucial role in advocating for human rights and democratic governance. Maathai and her team were vocal in their opposition to land grabs and corrupt practices that threatened local communities and the environment. They called for social justice, good governance, and accountability from political leaders.

Legacy and Continued Impact

Dr. Wangari Maathai's legacy lives on through the Green Belt Movement and the countless lives it has touched. Her innovative approach to environmental conservation and women's empowerment has inspired similar initiatives worldwide. The movement's principles of environmental stewardship, social justice, and gender equality continue to shape conservation efforts and community development initiatives globally.

Conclusion

Dr. Wangari Maathai's visionary leadership and dedication to environmental conservation and social justice have left an indelible mark on Kenya and the world. Through the Green Belt Movement, she demonstrated that environmental initiatives can be a powerful vehicle for empowering communities and advancing gender equality. Her holistic approach to conservation, which integrated environmental restoration with women's empowerment and social change, serves as a shining example of the transformative potential of environmental justice movements. As her legacy continues to inspire generations, the lessons from the Green Belt Movement remind us of the importance of combining environmental protection with social progress in the pursuit of a sustainable and equitable future for all.

Case Study Four: Rachel Kyte—Renewable Energy

Introduction to Case

Rachel Kyte, a renowned advocate for renewable energy and sustainable development, has been at the forefront of the global fight against climate change. Her roles as a former World Bank Vice President and Special Envoy for Climate Change have enabled her to champion clean energy projects and drive the transition to low-carbon economies. Kyte's relentless efforts have focused on promoting access to clean and affordable energy in developing countries, mobilizing financial resources for renewable energy initiatives, and advocating for climate-resilient economies. This case study delves into the life and work of Rachel Kyte, highlighting her significant contributions to the renewable energy movement and sustainable development worldwide.

Early Career and Path to Sustainable Development

Rachel Kyte's journey into the realm of sustainable development and renewable energy began with her academic pursuits and early career experiences. Born in 1960 in the United Kingdom, Kyte earned a

Bachelor's degree in history and politics from the University of London and later obtained a Master's degree in international relations from the School of Advanced International Studies at Johns Hopkins University.

Kyte's passion for social justice and environmental conservation led her to work with various international organizations. She held roles with Oxfam, the International Development Law Organization, and the United Nations Development Programme (UNDP), where she focused on issues of poverty reduction and climate resilience.

World Bank Tenure and Renewable Energy Advocacy

In 2011, Rachel Kyte joined the World Bank as the Vice President for Sustainable Development, where she played a pivotal role in advancing the institution's climate and sustainability agenda. One of her primary focuses was promoting renewable energy initiatives in developing countries, aiming to increase energy access and reduce greenhouse gas emissions.

Under Kyte's leadership, the World Bank spearheaded numerous renewable energy projects worldwide, particularly in regions heavily dependent on fossil fuels. These projects aimed to accelerate the deployment of clean energy sources such as solar and wind power, offering affordable and sustainable alternatives to traditional energy sources.

Kyte's vision extended beyond individual projects; she sought to facilitate systemic changes in energy markets by promoting favorable policies and regulatory frameworks for renewable energy integration. Her efforts helped attract private sector investments into clean energy projects, catalyzing a shift towards sustainable and climate-resilient economies.

Advocating for Energy Access and Poverty Alleviation

Rachel Kyte recognized that access to clean and affordable energy was not only vital for combating climate change but also critical for poverty alleviation and social development. She emphasized the need for equitable energy access, particularly in underserved rural areas, where energy poverty hindered economic opportunities and human development.

Through various initiatives, Kyte worked to bring electricity to remote and marginalized communities, empowering them to improve their living conditions and livelihoods. Her work highlighted the importance of sustainable energy access in achieving the United Nations' Sustainable Development Goals (SDGs).

Mobilizing Financial Resources for Renewable Energy

A significant aspect of Kyte's impact as a renewable energy advocate was her ability to mobilize financial resources for sustainable development projects. She worked diligently to raise awareness among governments, investors, and international institutions about the potential and benefits of investing in clean energy solutions.

Through partnerships and collaborations with financial institutions and governments, Kyte facilitated the allocation of funding for renewable energy initiatives in developing countries. Her efforts aimed to create an enabling environment for investments in clean energy technologies, reducing the financial barriers associated with transitioning to renewable energy sources.

Special Envoy for Climate Change and Global Advocacy

In 2016, Rachel Kyte assumed the role of Special Envoy for Climate Change at the World Bank. In this capacity, she became a leading voice in global climate negotiations and advocacy efforts. Kyte passionately championed climate action, calling for ambitious and decisive measures to address the escalating threats of climate change.

As a prominent advocate for renewable energy and sustainability, Kyte used her position to influence international policies and commitments on climate change mitigation and adaptation. She played a crucial role in shaping the global discourse on renewable energy, highlighting its significance in achieving climate goals and advancing the Paris Agreement objectives.

Promoting Climate-Resilient Economies

Kyte's work extended beyond renewable energy advocacy to encompass building climate-resilient economies. She stressed the importance of integrating climate considerations into economic planning and decision-making processes. By promoting resilience to climate-related risks, Kyte aimed to minimize the vulnerability of communities and countries to the impacts of extreme weather events and environmental changes.

Through her initiatives, Kyte encouraged countries to embrace green and sustainable practices, fostering economic growth while reducing carbon emissions and resource depletion. She emphasized the potential for sustainable development to create jobs, spur innovation, and improve overall well-being.

Rachel Kyte's Legacy and Continued Impact

Rachel Kyte's dedication to renewable energy and sustainable development has left a legacy on the global stage. Her visionary leadership and tireless advocacy have been instrumental in accelerating the transition to clean energy and promoting climate resilience worldwide.

Kyte's work has inspired governments, businesses, and civil society to take more significant steps towards renewable energy adoption and climate action. Her emphasis on integrating environmental considerations with economic planning has influenced policy frameworks and development strategies at national and international levels.

CONCLUSION

Rachel Kyte's journey from a passion for social justice to a global leader in renewable energy and sustainable development exemplifies the power of individual commitment to creating positive change. Through her roles at the World Bank and as Special Envoy for Climate Change, Kyte has been a driving force in promoting renewable energy projects, mobilizing financial resources, and advocating for climate-resilient economies. Her relentless efforts have not only advanced the global renewable energy movement but also demonstrated the potential of renewable energy to drive social and economic transformation. As the world faces the urgent challenges of climate change, Rachel Kyte's legacy continues to inspire and guide efforts to create a sustainable, low-carbon, and resilient future for all.

The case studies of Christiana Figueres, Dr. Sylvia Earle, Dr. Wangari Maathai, and Rachel Kyte highlight the transformative impact of women leaders in spearheading initiatives related to climate change mitigation, biodiversity conservation, renewable energy, and environmental justice. These exceptional women have demonstrated visionary leadership and unwavering commitment to positive environmental outcomes. Their efforts have not only shaped international policies and agreements but also inspired action at the grassroots level, leading to tangible and sustainable environmental solutions. By recognizing and supporting women leaders in the environmental sector, we can harness the full potential of their talent and expertise in driving positive change for a greener and more sustainable future. These case studies underscore the imperative of gender diversity in environmental leadership and its significance in addressing the pressing environmental challenges of our time.

STRATEGIES FOR PROMOTING GENDER EQUALITY IN ENVIRONMENTAL LEADERSHIP

To address the underrepresentation of women in environmental leadership, we discuss practical strategies for promoting gender equality. Mentorship programs, networking opportunities, and policy changes are essential tools in empowering women to take on leadership roles in the environmental sector.

Mentorship Programs. Mentorship programs play a crucial role in supporting aspiring women leaders in the environmental sector. By pairing experienced women leaders with emerging talents, these programs provide valuable guidance, advice, and networking opportunities. Mentors can offer insights into navigating career challenges, overcoming barriers, and developing essential leadership skills. Mentorship fosters a supportive environment where women can build confidence, receive constructive feedback, and envision a path to leadership. These programs not only empower women but also create a culture of support and inclusivity within the environmental sector.

Networking Opportunities. Networking is instrumental in advancing careers and accessing leadership positions. To promote gender equality in environmental leadership, organizations can facilitate networking opportunities that bring together women leaders, professionals, and stakeholders. Conferences, workshops, and webinars tailored to women's leadership and environmental issues allow women to exchange ideas, collaborate, and build professional connections. By creating a platform for knowledge-sharing and collaboration, networking opportunities empower women to showcase their expertise, gain visibility, and explore potential leadership roles.

Policy Changes. Institutional and organizational policies can significantly impact gender representation in leadership positions. Implementing policies that promote gender equality and inclusivity is essential. Organizations can establish targets and benchmarks for women's representation in leadership roles, encouraging progress toward gender-balanced decision-making bodies. Transparent and unbiased recruitment and promotion practices should be adopted to ensure equal opportunities for women. Flexibility in work arrangements, including remote work options and family-friendly policies, can address work-life balance challenges faced by women in leadership roles. Furthermore, policies that address the gender pay gap and provide equal compensation for equal work support women's advancement in the environmental sector.

Leadership Development Programs. Investing in leadership development programs tailored to women's needs can nurture the skills and capacities required for effective environmental leadership. These programs should focus on enhancing communication, negotiation, and strategic thinking skills, equipping women leaders with the tools to navigate complex challenges and drive impactful change. By providing opportunities for skill-building and professional growth, leadership development programs help bridge the gender gap in environmental leadership.

Diverse Representation on Decision-Making Panels. To promote gender equality in environmental leadership, organizations and institutions should ensure diverse representation on decision-making panels and boards. By actively seeking women's participation and perspectives, organizations can benefit from a wider range of insights and approaches in addressing environmental challenges. Encouraging gender-balanced participation in committees, conferences, and policy-making bodies fosters more inclusive and effective decision-making processes.

Challenging Gender Stereotypes and Bias. Creating an inclusive environment that challenges gender stereotypes and biases is essential for promoting gender equality in environmental leadership. Education and awareness campaigns can raise consciousness about unconscious biases and their impact on women's advancement. By fostering a culture of respect and equal opportunities, organizations can ensure that women are judged on their merits and contributions rather than gender-based assumptions.

In conclusion, promoting gender equality in environmental leadership is imperative for addressing the underrepresentation of women in decision-making processes. By implementing mentorship programs, providing networking opportunities, enacting policy changes, investing in leadership development, ensuring diverse representation, and challenging gender biases, the environmental sector can unlock the full potential of women leaders. Empowering women to take on leadership roles contribute to a more equitable, innovative, and sustainable environmental sector, capable of tackling complex environmental challenges effectively. Gender equality in leadership not only fosters the career advancement of women but also leads to more holistic and inclusive solutions for a greener and more sustainable future.

REFERENCES

Barkhuizen, E. N., Masakane, G., & van der Sluis, L. (2022). In search of factors that hinder the career advancement of women to senior leadership positions. *SA Journal of Industrial Psychology*, *48*(1), 1–15. doi:10.4102ajip.v48i0.1986

Bismark, M., Morris, J., Thomas, L., Loh, E., Phelps, G., & Dickinson, H. (2015). Reasons and remedies for under-representation of women in medical leadership roles: A qualitative study from Australia. *BMJ Open*, *5*(11), e009384. doi:10.1136/bmjopen-2015-009384 PMID:26576814

Casad, B. J., Franks, J. E., Garasky, C. E., Kittleman, M. M., Roesler, A. C., Hall, D. Y., & Petzel, Z. W. (2021). Gender inequality in academia: Problems and solutions for women faculty in STEM. *Journal of Neuroscience Research*, *99*(1), 13–23. doi:10.1002/jnr.24631 PMID:33103281

Crandall, C., Baumann, J., Cooney, P., Croteau, A., Croxton, A., Flaherty-Walia, K., Kerns, J., Kraus, R., & Morales, N. (2021). How to be an ally to women in fisheries science. *Fisheries (Bethesda, Md.)*, *46*(3), 140–144. doi:10.1002/fsh.10565

McKinnon, M., & O'Connell, C. (2020). Perceptions of stereotypes applied to women who publicly communicate their STEM work. *Humanities & Social Sciences Communications*, *7*(1), 160. doi:10.105741599-020-00654-0

Robinson, G., & Dechant, K. (1997). Building a business case for diversity. *The Academy of Management Perspectives*, *11*(3), 21–31. doi:10.5465/ame.1997.9709231661

Sembiring, E., & Nitivattananon, V. (2010). Sustainable solid waste management toward an inclusive society: Integration of the informal sector. *Resources, Conservation and Recycling*, *54*(11), 802–809. doi:10.1016/j.resconrec.2009.12.010

Sharma, H. B., Vanapalli, K. R., Samal, B., Cheela, V. S., Dubey, B. K., & Bhattacharya, J. (2021). Circular economy approach in solid waste management system to achieve UN-SDGs: Solutions for post-COVID recovery. *The Science of the Total Environment*, *800*, 149605. doi:10.1016/j.scitotenv.2021.149605 PMID:34426367

Shaw, R. (2006). Community-based climate change adaptation in Vietnam: inter-linkages of environment, disaster, and human security. *Multiple dimensions of global environmental change*, 521-547. Research Gate.

Chapter 13
The Role of Women and Gender-Diverse Leaders in Building a Sustainable World

Joycia Thorat
University of KwaZulu-Natal, South Africa

ABSTRACT

Women and gender-diverse leaders play vital roles in society, ranging from maintaining households to protecting the environment. They are not homogenous but contribute significantly towards building a sustainable world. Being historically at the forefront of social and environmental movements, they have been advocating for change and are an integral part in promoting sustainability. This chapter explores the contributions of women and gender-diverse leaders in building a sustainable world. Women's unique perspective and skills will be highlighted.

INTRODUCTION

In our polarized world, leaders have a crucial role in promoting social justice. To create a better world for all, diverse leaders of different genders are needed in various sectors such as civil society, religion, culture, politics, and business. Different genders bring unique values to leadership. Despite progress in science, technology, and finance, we have not achieved an inclusive, equal, fair, and peaceful world for all citizens through the promotion of inclusive and diverse leadership. A global leadership that encompasses females, males, transgender individuals, and LGBTQIA+ individuals with equal participation could be the solution to address social injustices like inequality, poverty, conflict, communal violence, and climate change. In the context of achieving the UNSDG Agenda, this is crucial.

Leadership and gender have been subjects of constant exploration in social science research aimed at establishing a just and sustainable world order. Scholars have debated and explored this topic from various perspectives, yielding diverse results. Ecumenical human rights activist Bunch argues that while women have been leading families and communities for years, they have been excluded from decision-making positions that impact their lives. She emphasizes the need to empower women and diverse gender

DOI: 10.4018/978-1-6684-5986-7.ch013

leaders to create a more sustainable world. Historian Gordon notes that women social movement leaders are gifted, hardworking individuals whose self-recognition often takes a back seat to the visibility of their cause or followers in the movement. Agoy, a member of the World Council of Churches central committee, highlights that many women want to save the world but often experience burnout at a young age, while men become leaders later in life, benefiting from the work women have done. The historical achievements of women, although evident, have often been underplayed or recorded remotely in historical sources. Burnham emphasizes the importance of nurturing women leaders for social transformation.

Gender equality and women's empowerment are fundamental pillars for achieving sustainable development. However, the involvement of women and gender-diverse individuals in community leadership roles has historically been limited. Women have been champions of sustainable development in various ways. They have spearheaded initiatives to address climate change, promote renewable energy, advocate for environmental conservation, and promote sustainable agriculture practices. Women's leadership in these areas has led to innovative solutions and positive impacts on local communities and the planet. Gender-diverse community leaders, including those who identify as non-binary or transgender, also play an essential role in building a sustainable world. Their presence ensures that a wide range of perspectives are considered in decision-making processes, leading to more inclusive and equitable outcomes. By embracing gender diversity, communities can tap into a wealth of knowledge, creativity, and problem-solving abilities. Moreover, women and gender-diverse leaders bring a unique lens to sustainability efforts. They often prioritize social equity, health, and well-being, recognizing the interconnectedness of environmental and social issues. By addressing systemic inequalities and empowering marginalized communities, they foster a more just and sustainable society. Building a sustainable world requires collaboration, cooperation, and collective action. Women and gender-diverse community leaders have a vital role to play in rallying communities, fostering partnerships, and mobilizing resources. They can engage in advocacy, education, and awareness campaigns, inspiring others to embrace sustainable practices and lifestyles. Recognizing and supporting the leadership of women and gender-diverse individuals is key to achieving the United Nations' Sustainable Development Goals. It is essential to provide equal access to education, training, and resources, ensuring that women and gender-diverse leaders have the tools they need to drive change effectively.

Women have proven to be effective agents of change, advocating for social, economic, and environmental justice within their communities. The achievements and contributions of prominent women and gender-diverse community leaders have made significant strides in sustainable development. Examples include Wangari Maathai, the founder of the Green Belt Movement, who empowered women through environmental conservation efforts, and Christiana Figueres, who played a pivotal role in the adoption of the Paris Agreement as the Executive Secretary of the United Nations Framework Convention on Climate Change. The role of women and gender-diverse community leaders in building a sustainable world is indispensable. Their contributions span various domains, including environmental conservation, social equity, and community empowerment. By embracing gender diversity and empowering these leaders, we can foster a more inclusive, equitable, and sustainable future for all. To build a sustainable world, it is essential to empower and support women and gender-diverse individuals in leadership roles. This section explores the importance of representation and the benefits of diverse perspectives in decision-making processes. Studies have shown that gender diversity in leadership positions leads to better outcomes, increased innovation, and more inclusive policies.

Overall, the contributions of women and gender-diverse leaders are essential in shaping a sustainable world. Recognizing their contributions, empowering them, and providing equal opportunities for leadership are crucial steps toward creating a more inclusive and equitable society.

Understanding the Sustainable Development Goals (SDGs)

The Sustainable Development Goals (SDGs) are a collection of seventeen goals and 169 targets aimed at eradicating poverty, addressing inequality and injustice, and combating climate change by 2030. They are also known as the Global Goals and were built upon the foundation of the Millennium Development Goals (MDGs). The SDGs were adopted by the United Nations General Assembly on September 25, 2015, and were formulated through extensive consultations with over 100 countries and millions of citizen inputs, including online platforms. The SDGs were adopted as a call to action for people worldwide to address five critical areas: people, planet, prosperity, peace, and partnership. While progress has been made, a mid-term assessment reveals that only about 12% of countries are on track to meet the targets. Approximately half of the countries are off track or have regressed since 2015. This highlights the need for accelerated efforts to achieve the goals. The SDGs provide a comprehensive framework for addressing the world's most pressing challenges. The involvement of diverse stakeholders, including women and gender-diverse leaders, is vital for achieving these goals and creating a sustainable and equitable future for all.

Understanding Leadership

Leadership is defined as organizing and influencing individuals as well as groups of people to achieve a common goal. Another definition of leadership is it is an effort to influence and the power to induce compliance (Wren, 1995). Leadership is a process where a person can influence others and provide cohesive and coherent direction to accomplish a mission and task. Implicit political participation and representation could also be seen as leadership. Leadership building is the conscientiousness of people to realize their potential. (Thorat & Shekhar, 2015). Preconditions of effective leadership are easy to access to the structure of opportunities and resources and ability to exercise agency. Leaders articulate the needs and priorities related to rights, entitlements, and basic services, provision of their constituency, participate in social planning, and policy planning and act upon them. Also, how leadership emerges and unfolds through day-to-day experiences such as material and social conditions are thought to constitute leadership (Caroll et. al., 2008).

Though there are many definitions of leadership among scholars there is no agreement around the definition of leadership (CIPD, 2012). One of the terms often exchanged and used for a leader is the manager. The distinction is managers are people who do things right and leaders are people who do the right thing" (Bennis & Nanus, 1985). Managers are trans actors and leaders are transformers (Burns, 1978).

Gender Differences

Gender differences are socially defined differences between men, women, and gender nonbinary. The biological differences between different genders are basically sex differences. Gender differences are also regarded as stemming from different adoptive environment men, women non –binary people encounter. These different traits and strategies emerge to ensure survival and men to maximize reproductive success as per studies. Aspects of social life are generated from individual personal lives to the societal institutions. Sex differs on 3 counts only i.e. anatomy, hormones and chromosomes. Rest is socially constructed and has become a reality (PRIA, 2006).

Gender has become central to all sociological discourse since its usage in 1970.The distinction between 'sex' and 'gender' was made in 1968 when American psychiatrist and psychoanalyst Robert Stiller wrote in his work about individuals whose biological sex was defined ambiguous or whose sense of themselves conflicted with their assigned sex. Today the term transsexual, third sex, transgender, gender nonbinary etc. are used. Gender being a dynamic concept the understanding has changed over time. Gender changes from culture to culture, time to time, place to place, social groups and family to family. It is socio -cultural differences made by society. It refers to masculine, feminine, transgender qualities, behavior patterns, roles, and responsibilities. Gender can be changed as it is not a natural one. But Sex is biological and with visible differences in procreative functions and organs. As a natural process sex remains constant to place, time, and culture.

Gender organizes our identities and concepts of self. It structures our social interaction, and it is one basis on which power and resources are allocated. Gender organizes social and personal institutions like marriage, family, and parenthood as well as public institutions like education, religion, work, sports, legal, media, politics, etc. Gender is about culture and power; gender identity is an issue for/ men as it is for/ women. Gender denotes a hierarchical division between men and women which has roots in both institutions and social practices (Bhasin, 2002).

The paper highlights gender diversity from a non-binary point of view and includes all genders and their contribution towards a sustainable world.

The Objective of the Study

1. Examine the contributions and impact of women and gender-diverse community leaders in promoting sustainability across various sectors and domains.
2. Investigate the specific leadership roles, activities, and strategies employed by women and gender-diverse community leaders in driving sustainable development initiatives.
3. Explore the challenges and barriers faced by women and gender-diverse community leaders in their efforts to build a sustainable world and identify strategies to overcome them.
4. Assess the effectiveness and outcomes of gender diversity community leadership in achieving sustainable development goals at the local, regional, and global levels.
5. Provide evidence-based recommendations for policymakers, organizations, and communities to enhance the participation and influence of women and gender-diverse community leaders in sustainability efforts.
6. Contribute to the existing literature on gender and leadership in the context of sustainable development by generating empirical data and insights that can inform future research and practices.
7. Foster a greater understanding and appreciation of the importance of women and gender-diverse community leaders in addressing pressing sustainability challenges and creating a more equitable and inclusive world.

METHODOLOGY

The following research paper outlines the methodology employed to investigate the role of women and gender-diverse community leaders in building a sustainable world. The study aims to gather data and analyze the contributions and impact of these leaders in promoting sustainability. The review includes peer-reviewed journal articles, reports, policy documents, and relevant publications from reputable sources.

A systematic search was conducted using electronic databases such as PubMed, Google Scholar, and Scopus. The search incorporates keywords and phrases related to women & healthcare, accessibility, developing countries, barriers, and initiatives. The search also includes additional sources identified from the reference lists of relevant articles.

Inclusion and Exclusion Criteria

Articles and publications were included if they meet the following criteria:

- Focus on women & healthcare accessibility in developing countries.
- Published in English.
- Empirical studies, systematic reviews, policy documents, and reports.
- Pertinent to barriers, challenges, and initiatives related to accessibility.

Data Extraction and Analysis

The selected articles are critically reviewed, and relevant data was extracted using a standardized data extraction form. The data extracted include study characteristics, key findings, methodologies used, and recommendations. Thematic analysis was employed to identify common themes and initiatives related to the role of women and gender-diverse community leaders in promoting sustainability and building a more inclusive and equitable world.

Quality Assessment

The quality and relevance of the selected studies are evaluated using appropriate appraisal tools. The quality assessment will consider factors such as research design, sample size, methodology, data analysis, and validity of findings. This step helped to ensure the inclusion of robust and reliable studies in the review.

Data Synthesis

The findings from the selected studies were synthesized thematically to identify key challenges and barriers faced by women and gender-diverse community leaders in their efforts to build a sustainable world and identify strategies to overcome them. The synthesis also highlights the various initiatives, strategies, and interventions employed. The synthesized information will be organized and presented in a logical and coherent manner. The synthesized findings were discussed considering the research objectives and existing literature. The discussion identifies gaps, patterns, and emerging themes in the literature, and provides insights into the implications for policy, practice, and future research. Finally, a comprehensive conclusion was drawn, summarizing the key findings and highlighting the importance of addressing the Role of Women and Gender Diversity Community Leaders in Building a Sustainable World.

RESULTS AND DISCUSSION

The role of women and gender-diverse community leaders in building a sustainable world has gained significant attention in recent years. Gender equality and inclusivity are crucial components of sustainable development efforts, as recognized by international organizations like the United Nations. This literature review aims to provide an overview of existing research and scholarship on the contributions and impact of women and gender-diverse community leaders in fostering sustainability across various sectors.

"Women's Leadership in Sustainable Development: The Case of Water Management" by Raijeli et al. (2018) explores the role of women leaders in water management and their contributions to sustainable development. It highlights the importance of women's participation in decision-making, community engagement, and the implementation of sustainable water resource management practices.

"Gender Diversity on Corporate Boards and Environmental Performance: Evidence from Canadian Firms" by Flammer and Hong (2019) examines the relationship between gender diversity on corporate boards and environmental performance in Canadian firms. The findings suggest that gender-diverse boards are associated with improved environmental performance, indicating the positive impact of gender diversity in corporate sustainability efforts.

"Women's Leadership in Local Environmental Governance: Evidence from Community Forests in Nepal" by Upreti and Nepal (2019) investigates the role of women leaders in local environmental governance. It highlights the contributions of women in promoting sustainable forest management, biodiversity conservation, and community resilience through their active participation and decision-making.

"Gender and Climate Finance: Double Mainstreaming for Sustainable Development" by Harcourt (2019) explores the intersection of gender and climate finance in the context of sustainable development. It examines the role of gender-responsive climate finance mechanisms in empowering women and supporting gender equality, ultimately contributing to sustainable and inclusive development.

"Gender Mainstreaming in Climate Change Adaptation and Disaster Risk Reduction: Evidence from Case Studies" by Tilche et al. (2019) examines the role of women and gender-diverse community leaders in climate change adaptation and disaster risk reduction. It presents case studies from different regions, highlighting how gender mainstreaming strategies led by community leaders contribute to building resilience, enhancing adaptive capacity, and ensuring the sustainability of vulnerable communities in the face of climate-related challenges.

"Women as Agents of Change for Sustainable Development: Lessons from Rural India" by Mishra and Chopra (2017) explores the transformative role of women community leaders in sustainable development. It investigates their efforts in areas such as agriculture, natural resource management, and livelihood diversification. The findings emphasize the importance of empowering women as change agents to achieve sustainable development goals at the grassroots level.

"Gender and Natural Resource Management: Livelihoods, Mobility, and Interventions" by Upton and Rocheleau (2016) examines the intersection of gender, natural resource management, and sustainable livelihoods. It analyzes case studies from diverse geographical contexts and investigates how women and gender-diverse leaders contribute to sustainable resource use, biodiversity conservation, and community resilience. The study underscores the need for gender-responsive interventions that recognize and enhance the roles of women and gender diversity leaders in natural resource management.

"Empowering Women through Community Forest Management: Evidence from Nepal" by Kashwan (2019) explores how women and gender diversity leaders play a crucial role in promoting sustainable forest governance. It examines the impact of gender-inclusive policies and community-based decision-making processes on forest conservation, income generation, and social equity. The findings highlight the transformative potential of empowering women and gender-diverse individuals in community forestry for achieving sustainable development.

"Gender and Climate Change: Impacts, Responses, and Adaptation" by Roy et al. (2018) examines the gendered dimensions of climate change impacts, responses, and adaptation strategies. It explores the role of women and gender-diverse community leaders in addressing climate change vulnerabilities, promoting sustainable agricultural practices, and fostering climate resilience in vulnerable communities. The study emphasizes the need for gender-sensitive approaches to climate change mitigation and adaptation to build a sustainable and inclusive world.

A Comparative Analysis" by Vogt and Cavaghan (2017) examines women's leadership for sustainability in different cultural contexts. It explores the diverse approaches and strategies employed by women leaders in driving sustainability initiatives, emphasizing the need for gender-inclusive leadership models to address complex sustainability challenges.

"Women's Leadership in Water Governance for Sustainable Development" by Koppen et al. (2020) examines the role of women and gender-diverse leaders in water governance for sustainable development. It explores how women's involvement in decision-making processes related to water management contributes to more inclusive, equitable, and sustainable outcomes. The findings highlight the importance of recognizing and supporting the leadership of women and gender-diverse individuals in addressing water-related challenges and achieving sustainable water resource management.

"Gender Diversity in Entrepreneurship and Innovation: Perspectives on Sustainable Development" by Thai and Naudin (2019) investigates the impact of gender diversity in promoting sustainable development. It explores the contributions of women and gender-diverse leaders in fostering creativity, innovation, and sustainable business practices. The study emphasizes the need for inclusive entrepreneurial ecosystems that value diverse perspectives and experiences to drive sustainable economic growth.

"Participatory Forest Management and Gender: Examining the Role of Women in Building Sustainable Forests" by Mwangi and Bhagwat (2019) examines the involvement of women and gender-diverse leaders in participatory forest management and its impact on sustainable forest governance. It analyzes case studies from different regions and highlights the diverse roles played by women in forest conservation, livelihood enhancement, and community resilience. The findings underscore the need for gender-inclusive approaches in forest management to achieve sustainability objectives.

"Empowering Indigenous Women for Sustainable Development: Insights from Indigenous Communities" by Maza et al. (2020) explores the empowerment of indigenous women as community leaders in achieving sustainable development goals. It examines the roles and contributions of indigenous women in cultural preservation, environmental conservation, and sustainable livelihood practices. The study emphasizes the importance of recognizing indigenous knowledge and promoting gender equity to build sustainable and resilient indigenous communities.

"The Role of Women in Community-Based Tourism for Sustainable Development" by Jacob and McClure (2019) investigates the role of women and gender-diverse leaders in community-based tourism initiatives and their contributions to sustainable development. It examines case studies from different countries and highlights how women's participation in tourism decision-making and entrepreneurship fosters economic empowerment, cultural preservation, and environmental sustainability. The findings underscore the potential of community-based tourism to empower women and contribute to sustainable development.

These studies contribute to the literature on the role of women and gender-diverse community leaders in building a sustainable world. They shed light on the diverse contributions and challenges faced by these leaders in various aspects of sustainable development, providing valuable insights for policymaking, interventions, and future research endeavors. They provide valuable insights into the diverse contexts and domains where women and gender-diverse leaders make significant contributions to sustainability, emphasizing the need for gender equality, inclusivity, and empowerment for achieving sustainable development goals.

1. *Women Leaders and Sustainable Development-* Numerous studies have highlighted the positive correlation between women's leadership and sustainable development outcomes. Women leaders often prioritize social and environmental issues, leading to more equitable and environmentally conscious policies and practices (O'Brien & Williams, 2016). Research has shown that women's empowerment and leadership contribute to poverty reduction, improved health, and education outcomes, and enhanced environmental stewardship (Kabeer, 2016). Women's leadership in community-based natural resource management has been associated with more sustainable resource use, conservation efforts, and increased resilience to climate change (Agrawal & Gibson, 1999; Rocheleau et al., 2013).

    ◦ *Women in the boardroom and their impact on governance and performance.* This study examines the relationship between gender diversity on corporate boards and firm performance. The findings suggest that firms with a higher representation of women on their boards tend to exhibit better corporate governance practices and financial performance. The study highlights the positive role women play in shaping board dynamics and decision-making processes, ultimately contributing to corporate sustainability.
    ◦ *The bottom line: Corporate performance and women's representation on boards.* Catalyst's report explores the link between gender diversity on corporate boards and financial performance across a global sample of companies. The findings indicate that companies with a higher representation of women on their boards tend to outperform those with fewer women in terms of return on equity, return on sales, and return on invested capital. The study underscores the importance of gender diversity in driving corporate sustainability and overall business success.
    ◦ *The impact of gender diversity on the performance of business teams: Evidence from a field experiment.* This study investigates the impact of gender diversity within business teams on performance outcomes. The research findings suggest that teams with a balanced representation of men and women exhibit higher levels of creativity, problem-solving abilities, and overall performance. The study emphasizes the valuable contributions women can make to corporate sustainability by fostering inclusive team dynamics and enhancing collective decision-making processes.
    ◦ *Corporate social responsibility and women in corporate leadership.* Examining the relationship between corporate social responsibility (CSR) and gender diversity, this study finds that firms with a higher presence of women in leadership positions tend to engage in more CSR

activities. The research suggests that women leaders are more likely to prioritize social and environmental sustainability, leading to enhanced corporate sustainability outcomes. The study highlights the positive role of women in driving CSR initiatives within organizations.

- *The role of women directors on corporate boards and corporate sustainability reporting quality.* This study examines the relationship between the presence of women directors on corporate boards and the quality of sustainability reporting. The findings indicate that companies with a higher representation of women on their boards tend to disclose more comprehensive and transparent sustainability information. The study emphasizes the role of women in promoting corporate sustainability through enhanced reporting practices, which contributes to better stakeholder engagement and accountability.

2. *Community Leadership and Sustainable Community Development-* Women and gender diverse community leaders play a critical role in driving sustainable community development initiatives. Their involvement in grassroots organizations, community networks, and local governance structures contributes to improved social cohesion, participatory decision-making, and the integration of diverse perspectives (Buckingham & Lomax, 2017).Women-led community initiatives often prioritize social justice, poverty alleviation, and environmental conservation, resulting in more inclusive and sustainable communities (Kusakabe & Pearson, 2012).

- *Gender and community leadership for sustainable development.* This study explores the role of women in community leadership for sustainable development. It highlights how women's leadership contributes to community development through their involvement in decision-making processes, fostering social cohesion, and promoting environmental sustainability. The research emphasizes the need to recognize and support women's leadership to achieve sustainable community development.
- *Gender and forest conservation: The impact of women's participation in community forest governance.* Examining the role of women in community forest governance, this study reveals that their participation positively influences forest conservation efforts and sustainable resource management. The research underscores the importance of recognizing and involving women in decision-making processes related to natural resource management for sustainable community development.
- *Role of women in sustainable rural development: A critical analysis.* This study critically analyzes the role of women in sustainable rural development. It highlights how women's active involvement in community leadership contributes to the well-being of rural communities, including poverty reduction, improved health, and education, and enhanced environmental sustainability. The research emphasizes the need for gender-responsive policies and interventions to empower women as change agents for sustainable community development.
- *Gender equality in agriculture and sustainable food security: A review of evidence.* The study examines the link between gender equality in agriculture and sustainable food security. It highlights the critical role of women in agricultural production, food processing, and marketing, emphasizing their contribution to sustainable agricultural practices and resilient food systems. The study emphasizes the need for gender-inclusive policies and investments to empower women in agricultural communities for sustainable community development.
- *Women's leadership and sustainable development: Evidence from Indian agriculture.* This study focuses on women's leadership in Indian agriculture and its impact on sustainable

development. It highlights how women's leadership in farming communities positively influences agricultural productivity, food security, and natural resource conservation. The research emphasizes the importance of recognizing women's agencies and providing them with access to resources and decision-making power for sustainable community development in agriculture.

3. *Policy and Advocacy-* Women and gender diversity community leaders are often at the forefront of policy advocacy for sustainable development. They contribute to the formulation and implementation of gender-responsive policies, environmental regulations, and social welfare programs (Datta et al., 2018). The inclusion of diverse voices in policy dialogue leads to more comprehensive and effective policies that address the complex interplay between gender, environment, and development (Arora-Jonsson, 2011).
4. *Representation in Leadership-* According to the Inter-Parliamentary Union's data from 2021, women's representation in national parliaments globally stood at 26.9%. This indicates that there is still a significant gender gap in political leadership. In the business sector, women remain underrepresented in leadership positions. The World Economic Forum's Global Gender Gap Report 2020 estimated that only 36% of senior managerial roles worldwide were held by women.
5. *Contribution to Sustainable Development-* Women's involvement in sustainable development initiatives is critical. The United Nations Development Programme (UNDP) states that gender equality is a key driver of sustainable development and has a positive impact on poverty reduction, health, education, and environmental sustainability. The UNDP's report on gender and climate change highlights that women often play a central role in natural resource management, biodiversity conservation, and climate change adaptation at the community level.
6. *Economic Empowerment-* Closing the gender gap in economic participation can lead to significant benefits. The McKinsey Global Institute estimated in 2015 that advancing gender equality could add $12 trillion to global GDP by 2025. Women's entrepreneurship is crucial for economic growth and poverty reduction. According to the Global Entrepreneurship Monitor (GEM) Women's Report 2020/2021, women entrepreneurs represent 28.8% of the total entrepreneurial activity worldwide.
7. *Sustainable Development Goals (SDGs)-* The SDGs recognize the importance of gender equality and the empowerment of women and girls across all sectors. Achieving gender equality is considered a standalone goal (SDG 5), but it also cuts across various other goals, including poverty eradication, education, health, and environmental sustainability. Data from the UN Women's Progress of the World's Women 2019-2020 report indicates that women's leadership and participation have positive effects on the implementation of the SDGs.

The literature reviewed demonstrates the significant role of women and gender diversity community leaders in building a sustainable world. Their contributions span various sectors, including governance, business, community development, and policy advocacy. Women's leadership and gender diversity in decision-making processes are associated with improved social, economic, and environmental outcomes. Empowering and supporting women and gender-diverse individuals in leadership positions is crucial for achieving sustainable development goals and creating a more equitable and inclusive society.

Women and Gender Diversity Leadership: Shaping a Sustainable World

"Women should be part of any agreement on climate change - not as an afterthought or because it's politically correct, but because it's the right thing to do. Our future as humanity depends on unleashing the full potential of all human beings, and the full capacity of women, to bring about change." – Thoraya Ahmed Obaid, UNFPA Executive Director.

The global community faces an imminent risk with the advent of climate change, posing a significant threat to all forms of life. The survival of communities, ecosystems, and fundamental human rights is at stake, demanding a deliberate and comprehensive response. Unfortunately, it is the marginalized and vulnerable populations who bear the brunt of climate change, while the actions of the privileged and powerful exacerbate the crisis. The prevailing growth-oriented capitalist development paradigm further compounds the environmental challenges we face. To address this urgent issue, we must establish an alternative paradigm of inclusive and sustainable development as a just solution to combat climate change.

Women and gender diversity leaders play a crucial role in our society, contributing to the well-being of households, the maintenance of communities, and the protection of nature. They form a diverse group, encompassing women from various backgrounds and identities. At the grassroots level, these leaders make invaluable contributions, often through their invisible and underappreciated care work. Families, societies, and countries thrive due to the tireless dedication of women. Moreover, women in leadership positions, including heads of state and institutions, offer an alternative model of leadership at the national and international levels. Their significant contributions shape the world we live in. However, women's leadership faces both opportunities and challenges.

The rapid changes unfolding in our world are marked by alarming levels of loneliness, depression, anger, hatred, riots, and even climate catastrophes and wars. These challenges are further compounded by excessive industrialization, dangerous artificial intelligence, and the wanton destruction of the environment, pushing us closer to a tipping point. This current reality is largely influenced by the patriarchal systems that govern our societies. Yet, an unexplored alternative world awaits us, one that can be led by women, whose leadership has the power to bring about transformative change. Numerous examples exist of women leaders in countries and global institutions. Despite these examples, women and gender –diverse leaders remain largely invisible, despite significant efforts to promote gender equality and Sustainable Development Goals. Whether it is environmental degradation, warfare, economic inequality, gender disparities, communal tensions, or increasing insensitivity to social issues like corruption, society is witnessing a distressing decline. Political monopolies and the rise of ethnic and religious dictatorships further exacerbate these challenges. Women bear the brunt of these collective failures, as they are the most affected by the side effects of male-dominated cultures. To create a better society, we must change the direction and conditions in which we operate, and women must lead the way, with their leadership duly recognized. A transformative shift toward creating a just and peaceful world can only occur when women assume active leadership roles in close collaboration with nature. Countless examples exist of women on the ground, working tirelessly to build a sustainable world. This paper will delve into their transformative contributions, shedding light on their efforts to bring about positive change at the grassroots level. By recognizing and amplifying the work of women community leaders, we can pave the way for a truly sustainable world.

The Significance of Diversity in Leadership: Empowering Women and Gender Diverse Leaders for a Transformative Future

"I measure the progress of a community by the degree of progress women have achieved. I strongly believe in the movements run by women, if they are truly taken into confidence, they may change the present picture of the society, which is very miserable." – Dr. B.R. Ambedkar

In a world striving for progress and inclusivity, it is essential to recognize the indispensable role of leaders from diverse gender and sexual identities in various sectors, including civil society, religion, culture, politics, business, and education. The unique values and aspirations that women and trans-women bring to leadership are crucial for not only creating an alternative world but also fostering a community lifestyle closely aligned with nature. However, within a socio-cultural sphere dominated by male leadership, women often lack role models to emulate. The scarcity of women leaders, trans leaders, and other marginalized gender identities is a result of the prevailing male domination and the absence of women in fields where male presence and values predominate. To drive significant change and create more equitable and inclusive structures within society and institutions, there is a need to prioritize leadership development. Investing in leadership development is an effective way to impact organizations positively and tap into the limitless potential of individuals by recruiting and nurturing them as leaders.

Leadership is not solely about occupying positions of power; it encompasses life-changing actions and the conscious effort to provide platforms for the development of progressive women's leadership roles. As the famous quote suggests, "Do not follow where the path may lead. Go instead where there is no path and leave a trail." This quote highlights the immense potential for women to explore and establish their leadership in fields where their presence has historically been limited. Cultivating the minds of women and empowering them to lead will have a transformative effect not only on their own lives but also on society. Unfortunately, many religious and social spaces still struggle to fully actualize women's leadership, despite advocating for equal partnership and engagement. The achievement of women's leadership remains inadequate in terms of both quality and quantity. Within the church and ecumenical religious bodies, there has been a failure to support and place women in high leadership roles, except in a few exceptional cases, during the past 70 years of post-independence. This lack of vision to include women and gender-diverse leaders has created a significant gap in building a well-balanced Christian society and recording the contribution of women to nation-building. To address this issue, it is crucial to develop strong, value-based leaders within churches and ecumenical institutions. By doing so, these institutions can actively contribute to nation-building and the establishment of a just world order. Furthermore, it is imperative for ecumenical bodies to contextualize and upgrade themselves to meet the demands of the current scenario and bridge the leadership crunch they are facing.

The importance of diversity in leadership cannot be overstated. Empowering women and other marginalized gender identities to assume leadership roles is not only a matter of equity and justice but also a catalyst for transformative change. By embracing and supporting women's leadership, we can build a more balanced, inclusive, and sustainable society that thrives on the contributions of all its members.

Contribution of Gender-Diverse Leaders in the Community for a Sustainable World

Women and gender-diverse leaders have been advocating for universal principles such as respect, unity, independence, justice, sustainability, and equality. Their contribution at the community level plays a crucial role in preventing conflicts and fostering collective efforts toward a sustainable and peaceful society that includes everyone. They work to counter social norms that perpetuate violent masculinities and political systems that restrict the full participation of women and marginalized groups. Women's unpaid care work should be recognized as a significant contribution to the Sustainable Development Goals (SDGs), as they ensure the smooth running of homes, support the well-being of family members, and contribute to poverty and hunger eradication. Additionally, their work as unorganized laborers help protect nature and the environment.

Women and gender-diverse leaders actively participate in decision-making processes at various levels. They attend meetings, engage in discussions, and contribute to village development associations and committees, such as Thanda Mukti, school committees, ICDS, and safe village committees in the context of India. Economic empowerment is a focus area for women community leaders, where they help deprived women access government schemes, establish self-help groups, and engage in entrepreneurship activities. Political participation is encouraged through attending Gram Sabha meetings, contesting elections, and advocating for women's issues in the community. They also work to raise awareness of gender-based violence and support women in legal cases.

The unpaid care work performed by women is a major contribution to the SDGs, as it supports poverty and hunger reduction, ensures the smooth functioning of households and societies, and fosters community well-being. However, deliberate policies are needed to address gender inequality and value the importance of care work. The current unequal distribution of resources within households places a heavier burden on women, limiting their educational and economic opportunities and hindering their participation in society. Recognizing and valuing care work through appropriate remuneration and support services can reduce women's workload and contribute to their empowerment. For women, poverty goes beyond financial poverty. They face intra-household inequalities in resource distribution and are burdened with caregiving responsibilities, household chores, and informal sector work with low wages and occupational risks. Gender inequality is reflected in limited access to basic facilities, professional opportunities, ownership rights, and political awareness. Vulnerability, measured by factors such as poverty, lack of education, hazardous working conditions, limited access to resources and knowledge, and political awareness, disproportionately affects women. Overall, the contribution of women's and gender-diverse leaders in the community is instrumental in preventing conflicts and fostering a sustainable and inclusive society. Their efforts build peace within communities and transcend borders.

Gender diverse leaders' efforts aim to prevent conflicts and foster a collective approach to building a sustainable and inclusive society, leaving no one behind. This is particularly important in a world where social norms that promote violent masculinities and political systems rooted in war hinder the full participation of women and marginalized groups. It is crucial to acknowledge the significant role of women's unpaid care work in achieving the Sustainable Development Goals (SDGs), as they tirelessly ensure the smooth functioning of households, support the well-being of family members, and contribute to the eradication of poverty and hunger. Additionally, their extensive contributions to household tasks, agriculture, and community development all contribute to the SDGs. Moreover, women and gender-diverse leaders actively engage in community work to promote peace, generate income, protect the climate, and ensure the well-being of vulnerable individuals.

In terms of decision-making, women, and gender-diverse leaders at the community level participate in various levels of decision-making processes. They are encouraged to step out of their homes and engage in meetings, actively contributing to discussions and actively involving themselves in Village Development Associations and other committees at the village level. They emphasize the well-being of women households, paying attention to single women, widows, destitute, and those facing poverty, ensuring they can access government entitlements tailored to women's needs. Political participation is also encouraged through facilitating attendance at Gram Sabha meetings, supporting women's candidacy in elections, organizing and leading Women's Gram Sabha gatherings, and advocating for essential services such as water, roads, streetlights, and government schemes like MGNREGA. Legal awareness campaigns addressing gender-based violence, child marriage, domestic violence, and provisions such as POCSO (Prevention of Child Sexual Abuse) and POSH (Prevention of Sexual Harassment at Workplace) are conducted, along with providing legal services to support women in their cases with the police and courts. These efforts empower women and gender-diverse groups through their leadership, enabling active participation in building a sustainable world.

Addressing gender inequality requires deliberate policies aimed at bridging the gap. The importance of care work, mostly undertaken by women without pay, must be acknowledged to address this inequality. Both paid and unpaid care workers, predominantly women, face pressure and challenges, particularly in marginalized communities. Care work is predominantly carried out by women who often have limited time for education, earning a decent living, or participating in decision-making processes. This situation traps them at the bottom of the economy. Tax revenue could play a role in lifting women's status by implementing campaigns that require the richest 1% to pay an additional tax on their wealth over the next ten years, generating funds needed to create jobs in sectors such as elderly, child, and sick care, education, and healthcare. Funding vital services and infrastructure can reduce women's workload. It is important to note that women and girls currently contribute 12.5 billion hours of unpaid care work each day, amounting to USD 10.8 trillion annually, which is more than three times the size of the global tech industry. Recognizing and compensating for care work is crucial, as it should be valued on par with other sectors. This information is sourced from the Oxfam report, Credit Suisse Research Inst. Global wealth data book 2019, and Forbes 2019 Billionaires list.

However, it is important to note during the 27th Conference of the Parties (CoP27), an intermediate review of the implementation of the gender action plan was adopted. This review incorporated joint dialogues that focused on advancing the leadership and highlighting the solutions of women from local communities and indigenous women in climate policy actions. Self-Help Groups (SHGs) and women's groups also contribute to livelihood security, as highlighted in the Economic Affairs publication. Additionally, the importance of women's participation and leadership has been recognized through various United Nations Security Council resolutions, including Resolution 1325 and subsequent resolutions such as 1820, 1888, 1889, 1960, 2016, 2022, 2242, 2467, and 2494.

Organized grassroots women's groups play a significant role in sustainable development, leading to peace and security in rural areas, as emphasized by Maria Bagaso. These groups, such as CASA and similar organizations, contribute immensely to bridge-building through their various interventions. Grassroots women's leadership also contributes to peace and security, as highlighted by Monika Jeackel. Women's contributions at the community level encompass community building, community economy, community knowledge, health, safety, housing and security of tenure, governance, women's leadership, and partnerships. Their work on the ground strengthens security through preventive measures.

The Contribution of Trans and LGBTQI Leaders in Implementing Sustainable Development Goals

Transgender and LGBTQI (lesbian, gay, bisexual, transgender, queer, and intersex) leaders have made significant contributions to the implementation of Sustainable Development Goals (SDGs) by bringing their unique perspectives, advocacy, and activism to the forefront. Their efforts have played a crucial role in advancing the goals of equality, inclusivity, and social justice, ultimately fostering sustainable development for all.

One of the key contributions of trans and LGBTQI leaders lies in their advocacy for inclusive policies. They have been at the forefront of pushing for legislation that protects the rights and well-being of their communities. Through their tireless efforts, they have helped secure legal protections against discrimination, promote gender equality, and ensure equal access to healthcare, education, and employment opportunities. By advocating for inclusive policies, they contribute to SDG 5 (Gender Equality) and SDG 10 (Reduced Inequalities), paving the way for a more equitable society.

Moreover, trans and LGBTQI leaders have been instrumental in fostering social acceptance and understanding. By openly sharing their personal stories and lived experiences, they challenge stereotypes, combat stigma, and promote empathy and acceptance within society. Through education, awareness campaigns, and community engagement, they strive to create safe spaces and promote dialogue that leads to greater understanding and acceptance of diverse sexual orientations and gender identities. These efforts align with SDG 16 (Peace, Justice, and Strong Institutions) by fostering inclusive societies that respect and uphold the rights of all individuals.

Another significant contribution of trans and LGBTQI leaders is their dedication to enhancing the health and well-being of their communities. They advocate for accessible and culturally competent healthcare services that address the specific needs of transgender and LGBTQI individuals. They work to promote comprehensive sexual and reproductive health rights, combat HIV/AIDS, and address mental health issues that disproportionately affect their communities. By advocating for better healthcare services and promoting well-being, they contribute to SDG 3 (Good Health and Well-being) and ensure that no one is left behind in the pursuit of a healthier society.

Trans and LGBTQI leaders also play a vital role in empowering marginalized communities within their ranks. They create safe spaces, support networks, and mentorship programs that uplift and empower individuals who face multiple forms of discrimination. By addressing the intersectional challenges faced by marginalized populations, they contribute to SDG 1 (No Poverty) and SDG 11 (Sustainable Cities and Communities). Their efforts aim to build resilient communities where everyone has equal opportunities to thrive.

Furthermore, trans and LGBTQI leaders actively engage in advocating for global LGBTQI rights and influencing international policies. They participate in international forums, such as the United Nations and regional human rights organizations, to highlight the importance of LGBTQI-inclusive policies and the recognition of LGBTQI rights. Their advocacy contributes to SDG 16 by promoting peaceful and inclusive societies at the global level and ensuring that the voices of marginalized communities are heard and respected. Hence, the contributions of trans and LGBTQI leaders in implementing Sustainable Development Goals are significant and far-reaching. Through their advocacy, they promote inclusive policies, foster social acceptance, enhance health and well-being, empower marginalized communities, and advocate for global LGBTQI rights. Their efforts align with the principles of sustainable development and contribute to creating a more equitable and inclusive world for all. It is essential to recognize, support, and amplify their voices and leadership to ensure that the SDGs are achieved and that no one is left behind in the pursuit of a sustainable future.

Way Forward

1. Recognition and Representation: Ensure the recognition and representation of women and gender-diverse leaders at all levels of decision-making processes. This includes government bodies, community organizations, and international forums. Create spaces that encourage and promote their active participation and leadership.
2. Capacity Building and Leadership Development: Invest in capacity-building programs and leadership development initiatives tailored specifically for women and gender-diverse leaders. Provide them with the necessary skills, knowledge, and resources to effectively advocate for sustainable development, address climate change challenges, and promote gender equality.
3. Supportive Policies and Legislation: Enact supportive policies and legislation that protect and promote the rights of women and gender-diverse individuals. This includes laws addressing gender-based violence, equal access to resources and opportunities, and the inclusion of gender perspectives in climate and sustainability policies.
4. Access to Resources and Funding: Ensure equal access to financial resources and funding opportunities for women and gender-diverse leaders. This can be achieved through targeted grants, scholarships, and financial support for sustainable projects and initiatives led by these leaders.
5. Collaboration and Networking: Foster collaboration and networking among women and gender-diverse leaders to create a supportive ecosystem. Facilitate platforms for knowledge exchange, sharing of best practices, and collaboration on joint projects. Encourage partnerships between community leaders, civil society organizations, businesses, and governments to work together towards sustainable development goals.
6. Data Collection and Research: Invest in gender-disaggregated data collection and research to understand the specific challenges faced by women and gender-diverse individuals in building a sustainable world. Use this data to inform policymaking, program development, and monitoring of progress toward gender equality and sustainable development.
7. Awareness and Advocacy: Raise awareness about the important role of women and gender-diverse leaders in building a sustainable world. Advocate for gender-responsive policies and practices, promoting gender equality and social inclusion in all spheres of society. Highlight success stories and positive impacts generated by women and gender-diverse leaders to inspire others and drive further change.
8. Empowering Local Communities: Support initiatives that empower local communities, particularly marginalized groups, by providing them with the necessary resources, education, and opportunities to actively participate in decision-making processes and sustainable development initiatives. Recognize and value the traditional knowledge and practices of indigenous communities, incorporating their wisdom into sustainable development strategies.

CONCLUSION

Women and gender-diverse leaders make numerous contributions to creating a sustainable world. The role of women and gender-diverse community leaders in building a sustainable world are of utmost importance in achieving social, economic, and environmental progress. Through their unique perspectives, experiences, and leadership qualities, women, and gender-diverse individuals contribute to the development and implementation of sustainable solutions at the grassroots level. Their involvement is crucial for promoting inclusive decision-making, advancing gender equality, and addressing the complex challenges we face in creating a sustainable future. Unfortunately, their contributions often go unrecognized and undervalued, with male leadership being predominantly acknowledged. Men's leadership, driven by profit and exploitation of the Earth, contributes to the deterioration of the world. Women and gender-diverse individuals, facing the consequences of such actions, work towards building a sustainable world for all. The pre-existing inequalities, gender roles, responsibilities, expectations, and unequal access to resources deepen the marginalization of certain groups, as highlighted by the UN Secretary-General. Addressing security issues related to local resource competition, livelihood insecurity, migration, volatile food prices, and climate change policies requires integrated policy agendas, coordinated efforts, increased financing, and gender inclusion at all levels. Women's prevention and preparedness work at the community level, including unpaid care contributions, enhance community resilience and security.

One of the key aspects of the role of women and gender-diverse community leaders is their contribution to social development and empowerment. Women leaders often prioritize the well-being of their communities, advocating for access to quality education, healthcare, and social services. They play a pivotal role in addressing gender-based inequalities, promoting women's rights, and empowering marginalized groups. By actively engaging with community members, these leaders foster dialogue, bridge divides, and promote social cohesion, thereby building more inclusive and resilient communities.

Moreover, women and gender-diverse community leaders make substantial contributions to economic development and poverty alleviation. They champion entrepreneurship, support income-generating activities, and enhance financial inclusion for women and marginalized groups. By promoting sustainable livelihoods and economic opportunities, these leaders contribute to poverty reduction, economic growth, and the overall well-being of their communities. Their efforts help break down barriers, challenge gender norms, and create pathways for economic empowerment, ultimately leading to more equitable and sustainable societies.

In the realm of environmental sustainability, women, and gender-diverse community leaders play a crucial role in conservation, natural resource management, and climate change adaptation. Their deep connection to the environment, traditional knowledge, and community-based approaches provides valuable insights and innovative solutions. Women are often at the forefront of climate resilience initiatives, advocating for sustainable agricultural practices, forest conservation, and clean energy alternatives. Their leadership fosters environmental stewardship, biodiversity conservation, and the preservation of ecosystems, thereby contributing to the long-term sustainability of our planet.

Furthermore, the active participation of women and gender diverse community leaders in policy-making and governance processes is vital for shaping sustainable development agendas. Their voices and perspectives bring diversity, inclusivity, and new ideas to decision-making spaces. By advocating for gender-responsive policies, these leaders contribute to the advancement of women's rights, gender equality, and social justice. Their efforts also promote accountability, transparency, and good governance practices, ensuring that sustainable development initiatives are effectively implemented at all levels.

However, despite their significant contributions, women and gender diverse community leaders continue to face various challenges and systemic barriers. Gender-based discrimination, limited access to resources, unequal power dynamics, and societal norms often hinder their full participation and influence. Addressing these challenges requires a collective effort, including supportive policies, capacity-building initiatives, and gender-responsive institutions that enable and amplify the voices of women and gender-diverse individuals.

It is interesting to note that the feminist movement not only liberated women, but also gave expression to their gender and voice. This helped in gaining equality and helped place them in leadership and management positions. The women's liberation movement was successful in establishing gender equality between men and women, also recognized as unique and distinct from males. and this holds true across all organizational settings, like business, politics, military, and religion (Hausman et al., 2010). Women's advancement is hampered by external discrimination, (Leyners & Heilman, 2006). Higher performance standard, limited access to career development measure in the organization, also promotion of women to precarious position adds to it. The gender gap in leadership is one of the central topics that need to be addressed. Also, the reason to understand the underrepresentation of women in leadership (Eagly & Carli, 2007; Iben et al., 2008).

In conclusion, the role of women and gender diversity community leaders in building a sustainable world is indispensable. Their leadership, resilience, and commitment to social, economic, and environmental sustainability contribute to the achievement of Sustainable Development Goals and the creation of more just and equitable societies. Recognizing and supporting their contributions is not only a matter of gender equality but also a fundamental requirement for achieving sustainable development and ensuring that no one is left behind. By harnessing the potential of women and gender-diverse leaders, we can pave the way for a more sustainable, inclusive, and prosperous future for all.

REFERENCES

Agarwal, B. (2010). *Gender and Green Governance: The Political Economy of Women's Presence Within and Beyond Community Forestry*. Oxford University Press. doi:10.1093/acprof:oso/9780199569687.001.0001

Brouwer, E., & Schwenke, K. (2019). Trans and non-binary leadership in the LGBTQI movement: Challenges, strategies, and opportunities. *Journal of Homosexuality*, *66*(6), 768–788.

Chakraborty, D., & Ghosh, D. (2020). The politics of transgender rights and inclusive development in India. *Asian Journal of Women's Studies*, *26*(2), 209–227.

Chant, S. (2016). *Women, Girls, and World Poverty: Empowerment, Equality, and the Sustainable Development Goals*. Routledge.

Flammer, C., & Hong, B. (2019). Gender Diversity on Corporate Boards and Environmental Performance: Evidence from Canadian Firms. *Strategic Management Journal*, *40*(3), 361–381.

Food and Agriculture Organization of the United Nations (FAO). (2019). *The State of Food and Agriculture 2019: Moving Forward on Food Loss and Waste Reduction*. https://www.fao.org/documents/card/en?details=ca6030en

Harcourt, W. (2019). Gender and Climate Finance: Double Mainstreaming for Sustainable Development. In A. Phillips (Ed.), *The Routledge Handbook of Gender and Development* (pp. 437–449). Routledge.

Hines, S., & Hungerford, C. (2019). The role of transgender leaders in advancing LGBTQ+ workplace inclusion. *Human Resource Development Review*, *18*(3), 272–298.

Inter-Parliamentary Union (IPU) and United Nations Development Program (UNDP). (2012). *Women in Parliament: Beyond Numbers*. http://archive.ipu.org/pdf/publications/SG12_en.pdf

International Union for Conservation of Nature (IUCN). (2016). *Gender and Climate Change: Overview Report*. https://www.iucn.org/resources/publications/gender-and-climate-change-overview-report

Jones, N., & Macvarish, J. (2018). Queering social reproduction: Achieving sustainability, equality, and justice for LGBTQI+ people. *Social Politics*, *25*(2), 160–184.

Kabeer, N. (2005). Gender equality and women's empowerment: A critical analysis of the third-millennium development goal. *Gender and Development*, *13*(1), 13–24. doi:10.1080/13552070512331332273

Kashwan, P. (2019). Empowering Women through Community Forest Management: Evidence from Nepal. *World Development*, *120*, 71–83.

Keleher, T. (2019). Transgender leadership and the politics of gender identity. *International Journal of Politics Culture and Society*, *32*(1), 95–112.

Kingston, S. (2018). From invisibility to influence: LGBTQ2 leadership and community engagement in Canada. *Canadian Journal of Nonprofit and Social Economy Research*, *9*(1), 54–67.

Leurs, K., & Ponzanesi, S. (2019). Migrant and LGBTQI+ youth: Performing digital diaspora and effective citizenship. *Feminist Media Studies*, *19*(7), 989–1006.

Mendoza, B. S., & Binford, L. (2020). Queer and trans-Latinx immigrants and the politics of vulnerability in the United States. *Gender & Society*, *34*(1), 79–104.

Millar, K. (2020). Transgender leadership development for educational change: Possibilities and pitfalls. *Journal of Homosexuality*, *67*(1), 23–40.

Mishra, A., & Chopra, K. (2017). Women as Agents of Change for Sustainable Development: Lessons from Rural India. *Journal of Rural Studies*, *53*, 141–152.

OECD Development Centre. (2017). *Social Institutions and Gender Index*. https://www.genderindex.org/

Palacios, R. C. (2018). Beyond the gender binary: Trans and gender non-conforming leadership in organizations. *Equality, Diversity and Inclusion*, *37*(7), 612–627.

Pfeffer, C. A., & Nugent, N. R. (2019). Transgender leaders in the workplace: Visibility and diversity climate. *Journal of Vocational Behavior*, *114*, 81–95.

Pitts, V. (2020). Expanding the gender binary in leadership: Insights from transgender executives. *The Journal of Leadership Studies*, *14*(2), 112–115.

Reck, J. M., & Lombardi, E. L. (2020). The role of transgender and gender nonconforming leaders in higher education. *New Directions for Student Services*, *172*, 37–46.

Roy, J., Islam, A., Mishra, A., & Ahmad, S. (2018). Gender and Climate Change: Impacts, Responses, and Adaptation. *Climate Policy*, *18*(7), 844–863.

Rugalema, G., & Owusu-Adjei, E. (2016). *Women's Empowerment through Climate-Smart Agriculture: Evidence from Africa*. https://www.unwomen.org/sites/default/files/Headquarters/Attachments/Sections/Library/Publications/2015/FPI%20BriefAgriculture%20LocalUSweb.pdf

Santos, C., & Sciascia, S. (2019). Transgender entrepreneurship and sustainable development. *Business & Society*, *58*(6), 1224–1252.

Schwenke, K., & Brouwer, E. (2018). Non-binary and transgender leadership: Success factors and challenges. In R. J. Thompson & J. Storberg-Walker (Eds.), *Leadership and Power in International Development* (pp. 185–201). Routledge.

Sorial, S., & White, F. A. (2018). Transgender employees in the workplace: A systematic review of the literature. *International Journal of Management Reviews*, *20*(4), 889–911.

Stone, D. (2019). *LGBTQ-inclusive governance: A global perspective*. Routledge.

Taga, R., McNamara, K., & Amoafoa-Smart, I. (2018). Women's Leadership in Sustainable Development: The Case of Water Management. *Journal of Cleaner Production*, *171*, 498–507.

Testa, R. J., & Simpson, B. (2019). Gender diversity and transgender inclusion in the workplace. *The Journal of Social Issues*, *75*(2), 585–610.

Tilche, A., Mysiak, J., Feldmann, U., Gressmann, H., Tramonti, G., & Zebisch, M. (2019). Gender Mainstreaming in Climate Change Adaptation and Disaster Risk Reduction: Evidence from Case Studies. *Sustainability*, *11*(17), 4682.

United Nations. (2011). *The Millennium Development Goals Report 2011*. https://www.un.org/millenniumgoals/pdf/(2011_E)%20MDG%20Report%202011_Book%20LR.pdf

United Nations. (2015). *Transforming Our World: The 2030 Agenda for Sustainable Development*. https://sdgs.un.org/2030agenda

United Nations Development Programme (UNDP). (2020). *Gender Equality Strategy 2018–2021*. https://www.undp.org/sites/g/files/zskgke326/files/2023-03/UNDP-Gender-Equality-Strategy-2018-2021-Annual-Report.pdf

United Nations Development Programme (UNDP). (2021). *Women's Empowerment and Sustainable Development*. https://www.undp.org/women-empowerment-and-sustainable-development

United Nations Environment Programme (UNEP). (2018). *Gender and Environment: Making the Connections*. https://www.unep.org/resources/gender-and-environment-making-connections

United Nations Framework Convention on Climate Change (UNFCCC). (2017). *Gender and Climate Change: A Framework for Action*. https://unfccc.int/gender

United Nations Population Fund (UNFPA). (2021). *Gender Equality*. https://www.unfpa.org/gender-equality

Upreti, B. R., & Nepal, S. (2019). Women's Leadership in Local Environmental Governance: Evidence from Community Forests in Nepal. *Gender and Development*, *27*(3), 409–425.

Upton, C., & Rocheleau, D. (2016). Gender and Natural Resource Management: Livelihoods, Mobility, and Interventions. *The International Journal of the Commons*, *10*(2), 865–886.

Valentine, S., & Funk, D. (2020). Examining the work experiences of transgender individuals: A systematic review. *Human Resource Management Review*, *30*(1), 100706.

Vogt, M., & Cavaghan, R. (2017). Women's Leadership for Sustainability: A Comparative Analysis. *Journal of Cleaner Production*, *140*(2), 1674–1686.

Witten, T. M., & Eyler, A. E. (2019). *Transgender leadership: Theory and practice for managing diversity*. Routledge.

Women, U. N. (2020). *Women's Leadership and Political Participation*. https://www.unwomen.org/en/what-we-do/leadership-and-political-participation

Women, U. N. (2020). *Gender Equality and Women's Empowerment: Women's Leadership and Political Participation*. https://www.unwomen.org/en/what-we-do/leadership-and-political-participation

World Bank. (2019). *Women, Business, and the Law 2019: A Decade of Reform*. https://openknowledge.worldbank.org/handle/10986/31752

World Bank Group. (2018). *Women, Business, and the Law 2018: Measuring Gender Equality in the Law*. https://thedocs.worldbank.org/en/doc/102741522965756861-0050022018/original/WBLKeyFindingsWebFINAL.pdf

Compilation of References

Abd Majid, N., & Jaaffar, A. H. (2023). The Effect of Women's Leadership on Carbon Disclosure by the Top 100 Global Energy Leaders. *Sustainability (Basel), 15*(11), 8491. Advance online publication. doi:10.3390u15118491

Abdulai, A. M. (2022). Could Eco-Feminists or Female Environmentalists Boost the Fight Against Climate Change. *African Journal of Climate Change and Resource Sustainability, 1*(1), 107–114. doi:10.37284/ajccrs.1.1.1039

Abdulrahman, M., Hussein, A., & Amoush, M. (2020). Female Characteristics and Their New Roles in Leadership. *Journal of Business and Management Sciences, 8*(2), 38–47. doi:10.12691/jbms-8-2-1

Adger, W. N. (2000). Social and ecological resilience: Are they related? *Progress in Human Geography, 24*(3), 347–364. doi:10.1191/030913200701540465

Adger, W. N., Barnett, J., Chapin, F. S. III, & Ellemor, H. (2011). This Must Be the Place: Under representation of Identity and Meaning in Climate Change Decision-Making. *Global Environmental Politics, 11*(2), 1–25. doi:10.1162/GLEP_a_00051

Adjei Mensah, C., Antwi, K. B., & Dauda, S. (2014). Female Traditional Leaders (Queen Mothers) and Community Planning and Development in Ghana. *Environmental Management and Sustainable Development, 3*(1), 205. doi:10.5296/emsd.v3i1.5142

Africa Progress Panel. (2015). *Africa progress report: power people planet - seizing energy and climate opportunities.* Africa Progress Panel.

African Development Group (2012). *Annual development effectiveness review.* African Development Group.

Agarwal, B. (2000). Conceptualizing environmental collective action: Why gender matters. *Cambridge Journal of Economics, 24*(3), 283–310. https://www.jstor.org/stable/23601169. doi:10.1093/cje/24.3.283

Agarwal, B. (2001). Participatory Exclusions, Community Forestry, and Gender: An Analysis for South Asia and a Conceptual Framework. *World Development, 29*(10), 1623–1648. doi:10.1016/S0305-750X(01)00066-3

Agarwal, B. (2010). *Gender and Green Governance: The Political Economy of Women's Presence Within and Beyond Community Forestry.* Oxford University Press. doi:10.1093/acprof:oso/9780199569687.001.0001

Agrawal, A., & Gibson, C. C. (1999). Enchantment and Disenchantment: The Role of Community in Natural Resource Conservation. *World Development, 27*(4), 629–649. doi:10.1016/S0305-750X(98)00161-2

Agrawala, S., Ota, T., Ahmed, A. U., Smith, J., & Van Aalst, M. (2003). *Development and climate change in Bangladesh: focus on coastal flooding and the Sundarbans.* OECD.

Ahern, J. (2011). From fail-safe to safe-to-fail: Sustainability and resilience in the new urban world. *Landscape and Urban Planning, 100*(4), 341–343. doi:10.1016/j.landurbplan.2011.02.021

Akinsemolu, A. A., & Olukoya, O. A. P. (2020). The vulnerability of women to climate change in coastal regions of Nigeria: A case of the Ilaje community in Ondo State. *Journal of Cleaner Production, 246.*

Akter, S., & Mallick, B. (2013). The poverty–vulnerability–resilience nexus: Evidence from Bangladesh. *Ecological Economics*, *96*, 114–124. doi:10.1016/j.ecolecon.2013.10.008

Akter, S., Rutsaert, P., Luis, J., Htwe, N. M., San, S. S., Rahajo, B., & Pustika, A. (2017). Women's empowerment and gender equity in Agriculture: A different perspective from Southeast Asia. *Food Policy*, *69*, 270–279. doi:10.1016/j.foodpol.2017.05.003

Aldrich, D. P. (2012). *Building resilience: Social capital in post-disaster recovery*. University of Chicago Press. doi:10.7208/chicago/9780226012896.001.0001

Ambec, S., & Lanoie, P. (2008). Does it pay to be green? A systematic overview. *The Academy of Management Perspectives*, *22*(4), 45–62. doi:10.5465/amp.2008.35590353

Amin, M. (2004). Introduction: Real-time Systems. In R. Zimmerman & T. Horan (Eds.), *Digital Infrastructures: Enabling civil and environmental systems through information technology* (pp. 116–140). Routledge.

Ani, P. A., & Correa, A. B. (2016). *Agricultural Extension Policies in the Philippines: Towards Enhancing the Delivery of Technological Services*. FFC. https://ap.fftc.org.tw/article/1092

Apolitical.co. (2022). *100 Most influential people in climate policy 2022/23*. Apolitical.co. https://apolitical.co/list/en/most-influential-climate-100-2022

Asian Development Bank. (2009). *Poverty in the Philippines: Causes, Constraints, and Opportunities*. ADB. https://www.adb.org/publications/poverty-philippines-causes-constraints-and-opportunities

Asteria, D., Herdiansyah, H., & Apriana, I. W. A. (2016). Women's Environmental Literacy as Social Capital in Environmental Management for Environmental Security of Urban Area. In *IOP Conference Series: Earth and Environmental Science*. IOP Publishing. 10.1088/1755-1315/30/1/012014

Atchison, A. L., & Down, I. (2019). The effects of women officeholders on environmental policy. *The Review of Policy Research*, *36*(6), 805–834. doi:10.1111/ropr.12346

Awal, M. A., Rashid, M. H. A., Islam, A. F. M. T., Imam, M. F. & Sarkar, M. A. R. (2016). Measuring Climate Change Vulnerability and its Adaptive Capacity: Policies and Planning for Bangladesh. *British Journal of Environment & Climate Change.*

Bahadur, A., & Pichon, F. (2017). *Analysis of resilience measurement frameworks and approaches*. Overseas Development Institute.

Balabantaray, S. R. (2023). *Women's Leadership and Sustainable Environmental Initiatives: A macroscopic investigation from Ecofeminism framework Women's Leadership and Sustainable Environmental Initiatives: A macroscopic investigation from Ecofeminism framework. August.* Research Gate.

Balabantaray, S. R. (2023). Women's Leadership and Sustainable Environmental Initiatives: A macroscopic investigation from Ecofeminism framework. *International Journal of Multidisciplinary Research and Growth Evaluation*, *4*(4), 1039–1046.

Banerjee, S. B., Iyer, E. S., & Kashyap, R. K. (2003). Corporate environmentalism: Antecedents and influence of industry type. *Journal of Marketing*, *67*(2), 106–122. doi:10.1509/jmkg.67.2.106.18604

Barkhuizen, E. N., Masakane, G., & van der Sluis, L. (2022). In search of factors that hinder the career advancement of women to senior leadership positions. *SA Journal of Industrial Psychology*, *48*(1), 1–15. doi:10.4102ajip.v48i0.1986

Barnes, M. L., Bodin, Ö., McClanahan, T. R., Kittinger, J. N., Hoey, A. S., Gaoue, O. G., & Graham, N. A. (2019). Social-ecological alignment and ecological conditions in coral reefs. *Nature Communications*, *10*(1), 1–10. doi:10.103841467-019-09994-1 PMID:31053708

Barrios, L. M., Prowse, A., & Vargas, V. R. (2020). Sustainable development and women's leadership: A participatory exploration of capabilities in Colombian Caribbean fisher communities. *Journal of Cleaner Production*, *264*, 121277. doi:10.1016/j.jclepro.2020.121277

Barth, N. (2020). 10 Woman Environmentalists You Should Know About. *Greenpop*. https://greenpop.org/10-woman-environmentalists-you-should-know-about/

Bass, B. M. (1985). *Leadership and performance beyond expectations*. Free Press.

Basu, M. (n.d.). Impact of Environmental Factors on Women Leadership. In M. Basu (Ed.), *Emerging Trends and Innovations in Modern Management* (pp. 365–376). Inspira Research Association.

Berkes, F. (1993). Traditional ecological knowledge in perspective. *Traditional ecological knowledge: Concepts and cases, 1*.

Berkes, F. (2001). *Managing small-scale fisheries: alternative directions and methods*. IDRC.

Bhakta, H., Raja, K., Shankar, V. R., & Prakash, V. (2020). *Since January 2020 Elsevier has created a COVID-19 resource centre with free information in English and Mandarin on the novel coronavirus COVID- 19. The COVID-19 resource centre is hosted on Elsevier Connect, the company's public news and information* . Research Gate.

Birindelli, G., Iamuzzi, A. P., & Savioli, M. (2019, November). The impact of women leaders on environmental performance: Evidence on gender diversity in banks. *Corporate Social Responsibility and Environmental Management, 26*(6), 1485-1499. doi:1002/csr.1762

Birindelli, G., Iannuzzi, A. P., & Savioli, M. (2019). The impact of women leaders on environmental performance: Evidence on gender diversity in banks. *Corporate Social Responsibility and Environmental Management*, *26*(6), 1485–1499. doi:10.1002/csr.1762

Bismark, M., Morris, J., Thomas, L., Loh, E., Phelps, G., & Dickinson, H. (2015). Reasons and remedies for under-representation of women in medical leadership roles: A qualitative study from Australia. *BMJ Open*, *5*(11), e009384. doi:10.1136/bmjopen-2015-009384 PMID:26576814

Bloor, M., Frankland, M., Thomas, M., & Stewart, K. (2000). Focus groups on social research. *Sage (Atlanta, Ga.)*.

Botella, C., Rueda, S., López-Iñesta, E., & Marzal, P. (2019). Gender diversity in STEM disciplines: A multiple factor problem. *Entropy (Basel, Switzerland)*, *21*(1), 1–17. doi:10.3390/e21010030 PMID:33266746

Botswana. (2009). *National energy policy for Botswana - Final Draft*. Botswana.

Brahma, S., Nwafor, C., & Boateng, A. (2021). Board gender diversity and firm performance: The UK evidence. *International Journal of Finance & Economics*, *26*(4), 5704–5719. doi:10.1002/ijfe.2089

Brouwer, E., & Schwenke, K. (2019). Trans and non-binary leadership in the LGBTQI movement: Challenges, strategies, and opportunities. *Journal of Homosexuality*, *66*(6), 768–788.

Bullough, A., Guelich, U., Manolova, T. S., & Schjoedt, L. (2022). Women's entrepreneurship and culture: Gender role expectations and identities, societal culture, and the entrepreneurial environment. *Small Business Economics*, *58*(2), 985–996. doi:10.100711187-020-00429-6

Burkhardt, K., Nguyen, P., & Poincelot, E. (2020). Agents of change: Women in top management and corporate environmental performance. *Corporate Social Responsibility and Environmental Management*, *27*(4), 1591–1604. doi:10.1002/csr.1907

Buss, D., Rutherford, B., Kumah, C., & Spear, M. (2021). Beyond the rituals of inclusion: The environment for women and resource governance in Africa's artisanal and small-scale mining sector. *Environmental Science & Policy*, *116*, 30–37. doi:10.1016/j.envsci.2020.10.019

Campanella, T. J. (2006). Urban resilience and the recovery of New Orleans. *Journal of the American Planning Association*, *72*(2), 141–146. doi:10.1080/01944360608976734

Carrington, D. (2020). Christiana Figueres on the climate emergency: 'This is the decade, and we are the generation. *The Guardian.* https://www.theguardian.com/environment/2020/feb/15/christiana-figueres-climate-emergency-this-is-the-decade-the-future-we-choose

Casad, B. J., Franks, J. E., Garasky, C. E., Kittleman, M. M., Roesler, A. C., Hall, D. Y., & Petzel, Z. W. (2021). Gender inequality in academia: Problems and solutions for women faculty in STEM. *Journal of Neuroscience Research*, *99*(1), 13–23. doi:10.1002/jnr.24631 PMID:33103281

Chakraborty, D., & Ghosh, D. (2020). The politics of transgender rights and inclusive development in India. *Asian Journal of Women's Studies*, *26*(2), 209–227.

Chant, S. (2016). *Women, Girls, and World Poverty: Empowerment, Equality, and the Sustainable Development Goals.* Routledge.

Choudhury, A. M., Neelormi, S., & Quadir, D. A. (2003). Socio-economic and physical perspectives of water related vulnerability to climate change: Results of field study in Bangladesh. *Science and Culture*, *23*, 225–238.

Chowdhury, Z. (2001). *Final Report on Network/ Political Analysis Study of the Shrimp Component of the Social Feasibility Studies for the Fourth Fisheries Project.* University of Dhaka and Bangladesh Centre for Advance Studies.

Christia, L. M., & Vijayanand, A. (2023). Women Leadership in Environmental Conservation in View of Laudato SI. *Journal of Pharmaceutical Negative Results*, *14*, 226–230.

Cinner, J. E., Adger, W. N., Allison, E. H., Barnes, M. L., Brown, K., Cohen, P. J., & Marshall, N. A. (2018). Building adaptive capacity to climate change in tropical coastal communities. *Nature Climate Change*, *8*(2), 117–123. doi:10.103841558-017-0065-x

Cinner, J. E., & Barnes, M. L. (2019). Social Dimensions of Resilience in Social-Ecological Systems. *One Earth*, *1*(1), 51–56. doi:10.1016/j.oneear.2019.08.003

Clark, R. (2019, October 15). What I learned about writing from reading GretaThunberg's speech to the UN. *Poynter.*

CNN (Cable News Network). (2023) "Christiane Amanpour: Chief international anchor." *Cable News Network.* https://edition.cnn.com/profiles/christiane-amanpour-profile

COG. UN Women & UNCDF (Council of County Governors, UN Women and the United Nations Capital Development Fund). (2022). The Impact of Devolution on Women and Girls in Kenya. *Assessment of the Impact of Devolution on Women and Girls Report.* Council of Governors (CoG) in partnership with UN Women and the United Nations Capital Development Fund. www.maarifa.cog.go.ke

Coser, L. A., & Rosenberg, B. (Eds.). (1965). Sociological theory: A book of readings (No. 301.082 C834 1964). Macmillan.

Crandall, C., Baumann, J., Cooney, P., Croteau, A., Croxton, A., Flaherty-Walia, K., Kerns, J., Kraus, R., & Morales, N. (2021). How to be an ally to women in fisheries science. *Fisheries (Bethesda, Md.)*, *46*(3), 140–144. doi:10.1002/fsh.10565

Cutter, S. L., Barnes, L., Berry, M., Burton, C., Evans, E., Tate, E., & Webb, J. (2008). A place-based model for understanding community resilience to natural disasters. *Global Environmental Change*, *18*(4), 598–606. doi:10.1016/j.gloenvcha.2008.07.013

Das, A., & Jain, R. (2021). The influence of women's leadership on community-based conservation efforts in India. *Conservation & Society*, *19*(2), 158–170.

Dasgupta, S., Kamal, F. A., Khan, Z. H., Choudhury, S., & Nishat, A. (2014a). *River Salinity and Climate Change: Evidence from Coastal Bangladesh* (Policy Research Working Paper No. 6817). Development Research Group, World Bank. https://documents.worldbank.org/en/publication/documents-reports/documentdetail/522091468209055387/river-salinity-and-climate-change-evidence-from-coastal-bangladesh

Dawkins, R. (2017). *Richard Dawkins: We need a new party – the European Party*. New Stateman. https://www.newstatesman.com/politics/2017/03/richard-dawkins-we-need-new-party-european-party

De Masi, S., Słomka-Gołębiowska, A., & Paci, A. (2022). Women on boards and corporate environmental performance in Italian companies: The importance of nomination background. *Business Ethics, the Environment & Responsibility*, *31*(4), 981–998. doi:10.1111/beer.12467

Dewi, K. H. (2017). Protecting Environment and Securing Power: Female Leaders' Political Initiative in the Age of Climate Change in Indonesia and Japan. *International Journal of Indonesian Studies*, *1*, 122–160.

Dilokwanich, S., Kaewsawang, S., Savatdipap, S., Iochawna, J., & Limganjanawat, S. (2015). Lessons Learnt of Thai Women Environmental Leaders. *Environment Asia*, *8*, 18–25.

DMB. (2010). *National Plan for Disaster Management 2010-2015*. Disaster Management Bureau, Bangladesh. https://ddm.portal.gov.bd/sites/default/files/files/ddm.portal.gov.bd/page/332124ba_20b1_42cc_816b_d97d7eb4f478/npdm_final.pdf

Dubey, R., Gunasekaran, A., & Ali, S. S. (2015). Exploring the relationship between leadership, operational practices, institutional pressures and environmental performance: A framework for green supply chain. *International Journal of Production Economics*, *160*, 120–132. doi:10.1016/j.ijpe.2014.10.001

Einarsen, S., Raknes, B. R. I., & Matthiesen, S. B. (1994). Bullying and harassment at work and their relationships to work environment quality: An exploratory study. *European Journal of Work and Organizational Psychology*, *4*(4), 381–401. doi:10.1080/13594329408410497

Elix, J., & Lambert, J. (2014). 16. Passionate defenders, accidental leaders: Women in the Australian environment movement. *Diversity in Leadership, 301*.

Elliott, R. (2018). The sociology of climate change as a sociology of loss. *European Journal of Sociology. Archives Européennes de Sociologie*, *59*(3), 301–337. doi:10.1017/S0003975618000152

Environment Agency Abu Dhabi. (2021). *H.E. Razan Khalifa Al Mubarak*. EAD. https://www.ead.gov.ae/en/About-Us/Our-Leadership/Leaders/Razan-Khalifa-Al-Mubarak

ESMAP (2011). *Improved cookstoves to conserve biomass resources*. ESMAP.

FAO. (2023). *The status of women in agrifood systems*. FAO.

Faroque, A. R., Quader, E., Gani, M. O., & Mortazavi, S. (2022). The Role of NGOs in CSR. *Comparative CSR and Sustainability, November*, 109–125.

Flammer, C., & Hong, B. (2019). Gender Diversity on Corporate Boards and Environmental Performance: Evidence from Canadian Firms. *Strategic Management Journal, 40*(3), 361–381.

Fleisher, B. M., & Chen, J. (1997). The coast-noncoast income gap, productivity, and regional economic policy in China. *Journal of Comparative Economics, 25*(2), 220–236. doi:10.1006/jcec.1997.1462

Flor, L. S., Friedman, J., Spencer, C. N., Cagney, J., Arrieta, A., Herbert, M. E., Stein, C., Mullany, E. C., Hon, J., Patwardhan, V., Barber, R. M., Collins, J. K., Hay, S. I., Lim, S. S., Lozano, R., Mokdad, A. H., Murray, C. J. L., Reiner, R. C. Jr, Sorensen, R. J. D., & Gakidou, E. (2022). Quantifying the effects of the COVID-19 pandemic on gender equality on health, social, and economic indicators: A comprehensive review of data from March, 2020, to September, 2021. *Lancet, 399*(10344), 2381–2397. doi:10.1016/S0140-6736(22)00008-3 PMID:35247311

Folke, C., Carpenter, S. R., Walker, B., Scheffer, M., Chapin, T., & Rockström, J. (2010). Resilience thinking integrates resilience, adaptability, and transformability. *Ecology and Society, 15*(4), 1–9. doi:10.5751/ES-03610-150420

Food and Agriculture Organization of the United Nations (FAO). (2019). *The State of Food and Agriculture 2019: Moving Forward on Food Loss and Waste Reduction.* https://www.fao.org/documents/card/en?details=ca6030en

Forster, M. E., Palmer, F., & Barnett, S. (2016). Karanga mai ra: Stories of Māori women as leaders. *Leadership, 12*(3), 324–345. doi:10.1177/1742715015608681

Frank, A. (2023). What is scientism, and why is it a mistake? *Bigthink.* https://bigthink.com/13-8/science-vs-scientism/

Fraser, E. D., Dougill, A. J., Mabee, W. E., Reed, M., & McAlpine, P. (2006). Bottom-up and top-down: Analysis of participatory processes for sustainability indicator identification as a pathway to community empowerment and sustainable environmental management. *Journal of Environmental Management, 78*(2), 114–127. doi:10.1016/j.jenvman.2005.04.009 PMID:16095806

Freudenberg, N., & Steinsapir, C. (2014). Not in our backyards: The grassroots environmental movement. In R. E. Dunlap & A. G. Mertig (Eds.), *American Environmentalism* (pp. 27–37). Taylor & Francis.

Friedmann, J. (2008). The uses of planning theory: A bibliographic essay. *Journal of Planning Education and Research, 28*(2), 247–257. doi:10.1177/0739456X08325220

Friends of Bonobo. (2023). *Our founder, Claudine André: A hero for bonobos*. Bonobos. bonobos.org/our-founders-story.

Frostrup, M. (2023). Mariella Frostrup. *MariellaOnline.* https://mariellafrostrup.co.uk/britains-novel-landscapes-with-mariella-frostrup-which-airs-on-thursdays-at-9pm-from-3-february-on-more4/

Galizzi, G., Meliou, E., & Steccolini, I. (2021). Theme: Experiences and challenges with gender budgeting and accounting. Moving towards gender-responsive forms of accountability?: Editorial: Institutionalizing gender budgeting—contemporary experiences and future challenges. *Public Money & Management, 41*(7), 499–501. doi:10.1080/09540962.2021.1971862

GFDR (Global Facility for Disaster Reduction and Recovery). (2016). *The Making of a Riskier Future: How Decisions Are Shaping Future Disaster Risk.* World Bank.

Ghosh, S., & Rao, D. (2019). Women as key drivers of sustainable practices: An empirical study of rural India. *Sustainability, 11*(3), 689.

Giddens, A., & Sutton, P. W. (2013). *Sociology* (8th ed.). Polity.

Glass, C., Cook, A., & Ingersoll, A. R. (2016). Do women leaders promote sustainability? Analyzing the effect of corporate governance composition on environmental performance. *Business Strategy and the Environment, 25*(7), 495–511. doi:10.1002/bse.1879

Glemarec, Y., Bayut-Renoux, F., & Waissbein, O. (2016). *Removing barriers to women entrepreneurs' engagement in decentralized sustainable energy solutions for the poor.* Research Gate.

GLOPP. (2008). *DFID's Sustainable Livelihoods Approach and its Framework.* GLOPP. http://www.glopp.ch/B7/en/multimedia/B7_1_pdf2.pdf

GoB. (2008a). *Cyclone Sidr in Bangladesh: Damage, Loss and Needs Assessment for Disaster Recovery and Reconstruction.* GFDRR. https://www.gfdrr.org/sites/default/files/2275_CycloneSidrinBangladeshExecutiveSummary.pdf

Goldstein, J. (2018). *Planetary Improvement: Cleantech Entrepreneurship and the contradictions of green Capitalism.* MIT Press.

Gomez, C., & White, Z. (2021). The Role of Women CEOs in Promoting Green Corporate Agenda. *Corporate Sustainability Review, 18*(8), 110–125.

Gore, T., Alestig, M., & Ratcliff, A. (2020). Confronting Carbon Inequality. *Oxfam Media Briefing, September.* https://oxfamilibrary.openrepository.com/bitstream/handle/10546/621052/mb-confronting-carbon-inequality-210920-en.pdf

Government Service Insurance System. (2021). *The Gender and Development (GAD) Program.* GSIS. https://www.gsis.gov.ph/gad/

Greenly. (2023). *The International Council for Local Environmental Initiatives.* Greenly. https://greenly.earth/en-us/blog/company-guide/the-international-council-for-local-environmental-initiatives-iclei

Gupta, D., Lele, S., & Sahu, G. (2020). Promoting a responsive state: The role of NGOs in decentralized forest governance in India. *Forest Policy and Economics, 111*(January), 102066. doi:10.1016/j.forpol.2019.102066

Gwagwa, N. (1991). Women in local government: Towards a future in South Africa. *Environment and Urbanization, 3*(1), 70–78. doi:10.1177/095624789100300107

Hallegatte, S., Bangalore, M., Bonzanigo, L., Fay, M., Kane, T., Narloch, U., & Vogt-Schilb, A. (2015). *Shock waves: managing the impacts of climate change on poverty.* The World Bank.

Hallegatte, S., Vogt-Schilb, A., Bangalore, M., & Rozenberg, J. (2016). *Unbreakable: building the resilience of the poor in the face of natural disasters.* World Bank Publications.

Hannam, K. (1999). Environmental management in India: Recent challenges to the Indian Forest Service. *Journal of Environmental Planning and Management, 42*(2), 221–233. doi:10.1080/09640569911226

Haque, M. A., Rahman, D., & Rahman, M. H. (2016). The importance of community-based approach to reduce sea level rise vulnerability and enhance resilience capacity in the coastal areas of Bangladesh: A review. *Journal of Sustainability Science and Management, 11*(2), 81–100.

Harari, Y. N. (2019). *21 Lessons for the 21st Century.* Vintage.

Harcourt, W. (2019). Gender and Climate Finance: Double Mainstreaming for Sustainable Development. In A. Phillips (Ed.), *The Routledge Handbook of Gender and Development* (pp. 437–449). Routledge.

Hassaniyan, A. (2021). The environmentalism of the subalterns: A case study of environmental activism in Eastern Kurdistan/Rojhelat. *Local Environment, 26*(8), 930–947. doi:10.1080/13549839.2021.1933927

Haugen, S. E. (2009). *Measures of labor underutilization from the current population survey* (Working Paper No. 424). Bureau of Labor Statistics. https://www.bls.gov/osmr/pdf/ec090020.pdf

Haynes, S. (2019, May 24). Students from 1,600 cities just walked out of school to protest climate change: It could be great Thunberg's biggest strike yet. *Time.*

He, G., Lu, Y., Mol, A. P., & Beckers, T. (2012). Changes and challenges: China's environmental management in transition. *Environmental Development, 3*, 25–38. doi:10.1016/j.envdev.2012.05.005

Highfield, W. E., Peacock, W. G., & Van Zandt, S. (2014). Mitigation planning: Why hazard exposure, structural vulnerability, and social vulnerability matter. *Journal of Planning Education and Research, 34*(3), 287–300. doi:10.1177/0739456X14531828

Hillman, R. (2020). *Are Women Leaders of Nature? -a quantitative study on the effect of women's political representation on environmental politics in European countries*. Gupea. https://gupea.ub.gu.se/handle/2077/63686

Hines, S., & Hungerford, C. (2019). The role of transgender leaders in advancing LGBTQ+ workplace inclusion. *Human Resource Development Review, 18*(3), 272–298.

Holling, C. S., Schindler, D. W., Walker, B. W., &Roughgarden, J. (1995). Biodiversity in the functioning of ecosystems: an ecological synthesis. *Biodiversity loss: economic and ecological issues*, 44-83.

Hossain, M. A., Reza, M. I., Rahman, S., & Kayes, I. (2011). Climate Change and its Impacts on the Livelihoods of the Vulnerable People in the Southwestern Coastal Zone in Bangladesh. *Climate Change Management*, 237–259

Howard, P. L. (2003). *Women and Plants: Gender Relations in Biodiversity Management and Conservation*. Zed Books.

Howell, B. (2022). The 7 Biggest Polluters by Industry in 2022, as Ranked in New Research. *EP Online*. https://eponline.com/articles/2022/10/14/the-7-biggest-polluters.aspx

Huq, N., Hugé, J., Boon, E., & Gain, A. K. (2015). Climate change impacts in agricultural communities in rural areas of coastal Bangladesh: A tale of many stories. *Sustainability (Basel), 7*(7), 8437–8460. doi:10.3390u7078437

Husby, T., de Groot, H. L., Hofkes, M. W., & Filatova, T. (2018). Flood protection and endogenous sorting of households: The role of credit constraints. *Mitigation and Adaptation Strategies for Global Change, 23*(2), 147–168. doi:10.100711027-015-9667-7 PMID:30093827

Ikram, M., Zhou, P., Shah, S. A. A., & Liu, G. Q. (2019). Do environmental management systems help improve corporate sustainable development? Evidence from manufacturing companies in Pakistan. *Journal of Cleaner Production, 226*, 628–641. doi:10.1016/j.jclepro.2019.03.265

International Consultation. (1993). *To Advance Women in Ecosystem Management and to Promote Cooperation, Co-ordination and Strategic Planning Among Multilateral, Regional and Bilateral Agencies and Organisations*. Washington D.C. https://digitallibrary.un.org/record/241092

International Fund for Agricultural Development. (2014). *2014 Annual Report Highlights*. IFAD. https://beta.ifad.org/documents/38714170/39625423/Highlights+AR+2014_E.pdf/74576162-ae01-48a8-8db9-acd53a250682

International Household Survey Network. (1991). *National Survey 1991*. IHSN. https://catalog.ihsn.org/index.php/catalog/4672/study-description

International Union for Conservation of Nature (IUCN). (2016). *Gender and Climate Change: Overview Report*. https://www.iucn.org/resources/publications/gender-and-climate-change-overview-report

Inter-Parliamentary Union (IPU) and United Nations Development Program (UNDP). (2012). *Women in Parliament: Beyond Numbers*. http://archive.ipu.org/pdf/publications/SG12_en.pdf

IPCC (Intergovernmental Panel on Climate Change). (2018, October 6). Summary for Policymakers. In: Masson-Delmotte, V., P. Zhai, H.-O.Pörtner, D. Roberts, J. Skea, P.R. Shukla, W. Moufouma-Okia, C. Péan, R. Pidcock, S. Connors, J.B.R. Matthews, Y. Chen, X. Zhou, M.I. Gomis, E. Lonnoy, T. Maycock, M. Tignor, and T. Waterfield (eds.) Global Warming of 1.5oC. An IPCC Special Report. Cambridge University Press.

IPPC. (2012). *Managing the Risks of Extreme Events and Disasters to Advance Climate Change Adaptation*. IPPC.

IRENA (International Renewable Energy Agency) (2013). *Renewable energy and jobs, annual review*. IRENA.

Islam, M. S. (2018*). Sustainable approaches to Drought Management: Exploratory Study on selected regions of North-west in Bangladesh* [Unpublished PhD Thesis, University of Dhaka].

Islands, M. (2016). *Consideration of reports submitted by state parties under Article 18 of the convention*. UN. https://digitallibrary.un.org/record/849531?ln=en

Jackson, C. (1993). Doing What Comes Naturally? Women and Environment in Development. *World Development*, *21*(12), 1947–1963. doi:10.1016/0305-750X(93)90068-K

Jackson, C. (1995). From conjugal contracts to environmental relations: Some thoughts on labour and technology. *IDS Bulletin*, *26*(1), 33–39. doi:10.1111/j.1759-5436.1995.mp26001005.x

Jagnani, M., & Mahadevan, M. (2021). *Women Leaders Improve Environmental Outcomes: Evidence from Crop Fires in India*. IZA. https://conference.iza.org/conference_files/Gender_2021/mahadevan_m31449.pdf

Jamwal, P., & Nambi, A. (2018). Wastewater treatment for carbon mitigation. *Current Science*, *114*(5), 1031.

Jane Goodall Institute USA. (2023). *About Jane*. Jane Goodall Institute. https://janegoodall.org/our-story/about-jane/

Jane Goodall Institute USA. (2023). *Our Story*. Jane Goodall Institute. https://janegoodall.org/our-story/about-jane/#conservationist

Jewitt, S. (2000). Mothering earth? Gender and environmental protection in the Jharkhand, India. *The Journal of Peasant Studies*, *27*(2), 94–131. doi:10.1080/03066150008438733

Johnson, L., & Lee, M. (2015). Eco-feminism and Environmental Advocacy. *Green Women Studies*, *8*(2), 30–47.

Jones, L. (2019). Resilience isn't the same for all: Comparing subjective and objective approaches to resilience measurement. *Wiley Interdisciplinary Reviews: Climate Change*, *10*(1), e552. doi:10.1002/wcc.552

Jones, L., & D' Errico, M. (2019). *Whose resilience matters? like-for-like comparisons of objective and subjective measures of resilience*. London School of Economics and Political Science, LSE Library.

Jones, L., & Tanner, T. (2017). 'Subjective resilience': Using perceptions to quantify household resilience to climate extremes and disasters. *Regional Environmental Change*, *17*(1), 229–243. doi:10.100710113-016-0995-2

Jones, M. S., & Solomon, J. (2019). Challenges and supports for women conservation leaders. *Conservation Science and Practice*, *1*(6), e36. doi:10.1111/csp2.36

Jones, N., & Macvarish, J. (2018). Queering social reproduction: Achieving sustainability, equality, and justice for LGBTQI+ people. *Social Politics*, *25*(2), 160–184.

Jongman, B., Winsemius, H. C., Aerts, J. C., De Perez, E. C., Van Aalst, M. K., Kron, W., & Ward, P. J. (2015). Declining vulnerability to river floods and the global benefits of adaptation. *Proceedings of the National Academy of Sciences of the United States of America, 112*(18), E2271–E2280. doi:10.1073/pnas.1414439112 PMID:25902499

Jouber, H. (2022). Women leaders and corporate social performance: Do critical mass, CEO managerial ability and corporate governance matter? *Management Decision, 60*(5), 1185–1217. doi:10.1108/MD-07-2020-0953

Justin, M. S. M., Aquino, P. G. Jr, & Le, D. H. (2020). Roles and Strategies of 20th and 21st Century Women Environmentalists. In S. A. Khan (Ed.), *Global Perspectives on Green Business Administration and Sustainable Supply Chain Management* (pp. 163–180). IGI Global. doi:10.4018/978-1-7998-2173-1.ch009

Kabeer, N. (2005). Gender equality and women's empowerment: A critical analysis of the third-millennium development goal. *Gender and Development, 13*(1), 13–24. doi:10.1080/13552070512331332273

Kartiki, K. (2011). Climate change and migration: A case study from rural Bangladesh. *Gender and Development, 19*(1), 23–38. doi:10.1080/13552074.2011.554017

Kashwan, P. (2019). Empowering Women through Community Forest Management: Evidence from Nepal. *World Development, 120*, 71–83.

Kates, R. W., Travis, W. R., & Wilbanks, T. J. (2012). Transformational adaptation when incremental adaptations to climate change are insufficient. *Proceedings of the National Academy of Sciences of the United States of America, 109*(19), 7156–7161. doi:10.1073/pnas.1115521109 PMID:22509036

Katherine. (2023). Guardians of the Planet 16 Women Environmentalists You Should Know. *Almighty Girl.* https://www.amightygirl.com/blog?p=11863

Keleher, T. (2019). Transgender leadership and the politics of gender identity. *International Journal of Politics Culture and Society, 32*(1), 95–112.

Khalil, G. M. (1992). Cyclones and storm surges in Bangladesh: Some mitigative measures. *Natural Hazards, 6*(1), 11–24. doi:10.1007/BF00162096

Khan, A. (2017). Women in Green Tech: Breaking Barriers. *Sustainable Tech Review, 10*(1), 12–25.

Khatibi, F. S., & Indira, M. (2011). Empowerment of women through self-help groups and environmental management: Experiences of NGOs in Karnataka State, India. *Journal of Human Ecology (Delhi, India), 34*(1), 29–40. doi:10.1080/09709274.2011.11906366

Khatibi, F. S., & Indira, M. (2017). Empowerment of Women through Self Help Groups and Environmental Management: Experiences of NGOs in Karnataka State, India. *Journal of Human Ecology (Delhi, India), 61*(3), 213–224.

Kijima, R., Yang-yoshihara, M., & Maekawa, M. S. (2021). Using-design-thinking-to-cultivate-the-next-generation-of-female-STEAM-thinkersInternational-Journal-of-STEM-Education.pdf. *International Journal of STEM Education, 6.* https://stemeducationjournal.springeropen.com/articles/10.1186/s40594-021-00271-6

Kilincarslan, E., Elmagrhi, M. H., & Li, Z. (2020) Impact of Governance Structure on Environmental Disclosures in the Middle East. *The International Journal of Business in Society.*

KimballA.PalmerD.MarquisC. (2012). The Impact of Women Top Managers and Directors on Corporate Environmental Performance. https://ssrn.com/abstract=2211826Electroniccopyavailableat:https://ssrn.com/abstract=2211826Electroniccopyavailableat:http://ssrn.com/abstract=2211826 doi:10.2139/ssrn.2211826

Kingston, S. (2018). From invisibility to influence: LGBTQ2 leadership and community engagement in Canada. *Canadian Journal of Nonprofit and Social Economy Research*, *9*(1), 54–67.

Koohi, E., Shobeiri, S. M., Koohi, E., & Meiboudi, H. (2016). Women's participation in Environmental Management and Development Promotion Culture. *International Journal of Resistive Economics*, *4*(1), 23–35.

Koohi, E., Shobeiri, S.M., & Koohi, E., & Meiboudi. (2016). Women's participation in environmental management and development Promotion Culture. *The Open Access Journal of Resistive Economics*, *4*(1), 23–35.

Kumar, A., & Patel, N. (2017). Gender and green governance in India. *Indian Journal of Gender Studies*, *24*(3), 345–367.

Kumari, S. (2019). Role of Women in Environmental Management. *International Research Journal of Management Sociology & Humanity*, *10*(1), 432–436. doi:10.32804/IRJMSH

Labaris, A. (2009). Women involvement in environmental protection and management: A case of Nasarawa state. *Journal of Sustainable Development in Africa*, *10*, 179–191.

Lall, S. V., & Deichmann, U. (2010). *Density and disasters: economics of urban hazard risk*. The World Bank. doi:10.1596/1813-9450-5161

Leach, M. (2007). Earth mother myths and other ecofeminist fables: How a strategic notion rose and fell. *Development and Change*, *38*(1), 67–85. doi:10.1111/j.1467-7660.2007.00403.x

Leurs, K., & Ponzanesi, S. (2019). Migrant and LGBTQI+ youth: Performing digital diaspora and effective citizenship. *Feminist Media Studies*, *19*(7), 989–1006.

Levin, M. (2021, November 24). *Why female leadership is crucial to tackling climate change and other crises*. Climate Change Champions. https://climatechampions.unfccc.int/why-female-leadership-is-crucial-to-tackling-climate-change-and-other-crises/

Li, J., Zhao, F., Chen, S., Jiang, W., Liu, T., & Shi, S. (2017). Gender diversity on boards and firms' environmental policy. *Business Strategy and the Environment*, *26*(3), 306–315. doi:10.1002/bse.1918

Liu, C. (2018). Are women greener? Corporate gender diversity and environmental violations. *Journal of Corporate Finance*, *52*, 118–142. doi:10.1016/j.jcorpfin.2018.08.004

Liu, Z., Cai, L., & Zhang, Y. (2022). Gender Diversity and Environmental Performance: New Evidence from China. *Sustainability (Basel)*, *14*(21), 13775. doi:10.3390u142113775

Loayza, N. V., Olaberria, E., Rigolini, J., & Christiaensen, L. (2012). Natural disasters and growth: Going beyond the averages. *World Development*, *40*(7), 1317–1336. doi:10.1016/j.worlddev.2012.03.002

Lohani, M., & Aburaida, L. (2017). Women empowerment: A key to sustainable development. *The Social ION*, *6*(2), 26–29. doi:10.5958/2456-7523.2017.00006.4

Lu, J., & Herremans, I. M. (2019). Board gender diversity and environmental performance: An industries perspective. *Business Strategy and the Environment*, *28*(7), 1449–1464. doi:10.1002/bse.2326

Luo, M., & Chen, L. (2018). Female Leadership: A Sustainable Future in Business. *Journal of Business Ethics*, *20*(6), 123–140.

MacGrath, M., Jones, N., & Burho, E. (2022). The world's 100 most powerful women. *Forbes*. https://www.forbes.com/lists/power-women/?sh=6c438baf5a95

Magill, R. S. (1993). Focus groups, program evaluation, and the poor. *Journal of Sociology and Social Welfare*, *20*(1), 103. doi:10.15453/0191-5096.2059

Magis, K. (2010). Community resilience: An indicator of social sustainability. *Society & Natural Resources*, *23*(5), 401–416. doi:10.1080/08941920903305674

Mahmood, S. M., & Ansary, B. S. (2013). Shrimp Fry Collection as Alternative Livelihood: A Case Study on the South-west Coastal Region of Bangladesh. *ASA University Review*, *7*(2), 139–149.

Malik, K., & Verma, R. (2018). Women-led environmental movements in India: Exploring the role of gender. *Environment and Natural Resources Research*, *8*(4), 65–76.

Marshall, G. (2012). *Economics for collaborative environmental management: renegotiating the commons*. Routledge. doi:10.4324/9781849771047

Mayunga, J. S. (2007). Understanding and applying the concept of community disaster resilience: a capital-based approach. *Summer academy for social vulnerability and resilience building, 1*(1), 1-16.

Mccormick, J. (2010). *International Regimes*. Research Gate.

McKinnon, M., & O'Connell, C. (2020). Perceptions of stereotypes applied to women who publicly communicate their STEM work. *Humanities & Social Sciences Communications*, *7*(1), 160. doi:10.105741599-020-00654-0

Mediawire. (2021, Sep 27). Women Leaders: The better half for building a better planet. *Economic Times*. https://economictimes.indiatimes.com/industry/csr/initiatives/women-leaders-the-better-half-for-building-a-better-planet/articleshow/86563194.cms?from=mdr

Mendoza, B. S., & Binford, L. (2020). Queer and trans-Latinx immigrants and the politics of vulnerability in the United States. *Gender & Society*, *34*(1), 79–104.

Mensah, I. K. (2020). Impact of Government Capacity and E-Government Performance on the Adoption of E-Government Services. *International Journal of Public Administration*, *43*(4), 303–311. doi:10.1080/01900692.2019.1628059

Merchant, C. (1981). Earthcare: Women and the environment. *Environment*, *23*(5), 6–40. doi:10.1080/00139157.1981.9933143

Mies, M., & Shiva, V. (1993). *Ecofeminism*. Fernwood Publications. doi:10.5040/9781350219786

Millar, K. (2020). Transgender leadership development for educational change: Possibilities and pitfalls. *Journal of Homosexuality*, *67*(1), 23–40.

Mishra, A., & Chopra, K. (2017). Women as Agents of Change for Sustainable Development: Lessons from Rural India. *Journal of Rural Studies*, *53*, 141–152.

Mishra, V. (2014). Handmade in India: The craft of sustainable fashion. *Fashion Theory*, *18*(4), 465–484.

Mitchell, B. (2005). Participatory partnerships: Engaging and empowering to enhance environmental management and quality of life? *Social Indicators Research*, *71*(1-3), 123–144. doi:10.100711205-004-8016-0

Moser, S. C., Melillo, J. M., Jacobs, K. L., Moss, R. H., & Buizer, J. L. (2016). Aspirations and common tensions: Larger lessons from the third US national climate assessment. *Climatic Change*, *135*(1), 187–201. doi:10.100710584-015-1530-z

Moss, M. L., & Townsend, A. M. (2004). Moving information in the twenty-first century city. In R. Hanley (Ed.), *Moving People, Goods, and Information in the 21st Century: The Cutting-Edge Infrastructures of Networked Cities* (pp. 69–84). Routledge.

Murty, T. S., & El-Sabh, M. I. (1992). Mitigating the effects of storm surges generated by tropical cyclones: A proposal. *Natural Hazards*, *6*(3), 251–273. doi:10.1007/BF00129511

Murty, T. S., & Neralla, V. R. (1992). On the recurvature of tropical cyclones and the storm surge problem in Bangladesh. *Natural Hazards*, *6*(3), 275–279. doi:10.1007/BF00129512

Mutahara, M., Warner, J. F., Wals, A. E. J., Khan, M. S. A., & Wester, P. (2017). Social learning for adaptive delta management: Tidal River Management in the Bangladesh Delta. *International Journal of Water Resources Development*, 1–21.

Nadeem, M., Bahadar, S., Gull, A. A., & Iqbal, U. (2020). Are women eco-friendly? Board gender diversity and environmental innovation. *Business Strategy and the Environment*, *29*(8), 3146–3161. doi:10.1002/bse.2563

Narain, S., & Ghosh, A. (2017). Why Should I Not Be Angry? State of India's Environment 2017. In *Figures*. Centre for Science and Environment.

Nasrin, F. (2012). Women, environment and environmental advocacy: Challenges for Bangladesh. *Asian journal of social sciences & humanities, 1*(3), 149-172.

Navdanya International. (2023). *Vandana Shiva*. Navdanya International. https://navdanyainternational.org/our-staff/vandana-shiva/

Ngwa, E. N. (1995, April). The role of women in environmental management: An overview of the Cameroonian situation. *GeoJournal*, *35*(4), 515–520. https://www.jstor.org/stable/41146448. doi:10.1007/BF00824366 PMID:12291368

Niu, Y., & Sims, C. (2022). Women Leaders in the Non-Governmental Organization (NGO) Sector. In J. Ke & G. N. McLean (Eds.), *Chinese Women in Leadership* (pp. 203–226). Springer International Publishing.

Nordstrom, Louise (2019, January 25). The Swedish teen holding world leaders accountable for climate change. *France 24*.

Norgaard, K., & York, R. (2005). Gender equality and state environmentalists. *Gender & Society*, *19*(4), 506–522. doi:10.1177/0891243204273612

Nugent, C., & Shandra, J. M. (2009). State environmental protection efforts, women's status, and world polity: A cross-national analysis. *Organization & Environment*, *22*(2), 208–229. doi:10.1177/1086026609338166

O'Reilly, S. (2019). The Influence of Gender on Environmental Policy Making. *Policy and Sustainability*, *9*(3), 15–29.

OECD Development Centre. (2017). *Social Institutions and Gender Index*. https://www.genderindex.org/

Olsson, L., Jerneck, A., Thoren, H., Persson, J., & O'Byrne, D. (2015). Why resilience is unappealing to social science: Theoretical and empirical investigations of the scientific use of resilience. *Science Advances*, *1*(4), e1400217. doi:10.1126ciadv.1400217 PMID:26601176

Ornelas, R. (2011). Water and the Indigenous Women's Leadership Project. *Journal of Research on Women and Gender*, 2, 111–134.

Östros, B. B. (2023, March 8). *Business case for women's leadership*. EIB. https://www.eib.org/en/stories/climate-women-profits

Palacios, R. C. (2018). Beyond the gender binary: Trans and gender non-conforming leadership in organizations. *Equality, Diversity and Inclusion*, *37*(7), 612–627.

Pandey, P., & Pandey, D. M. (2015). *Research Methodology: Tools and Techniques*. Bridge Center.

Patankar, A. (2015). *The exposure, vulnerability, and ability to adapt of poor households to recurrent floods in Mumbai*. World Bank. https://documents1.worldbank.org/curated/en/418931467991989631/pdf/WPS7481.pdf

Patel, A. (2016). From Grassroots to Global: Women Leading in Conservation. *International Journal of Conservation, 15*(4), 78–95.

Patel, Z. (1996). Women and Environmental Management. Agenda. *Empowering Women for Gender Equity, 29*, 66–71.

Paul, M. (2014) Isatou Ceesay. *One Plastic Bag*. http://oneplasticbag.com/the-women

Personal Communication. (2016). *Filipino Women in International Migration Situation Analysis, Policy Context and International Mechanisms*. UN. https://www.unwomen.org/sites/default/files/Headquarters/Attachments/Sections/Library/Publications/2016/Filipino-Women-International-Migration.pdf

Pfeffer, C. A., & Nugent, N. R. (2019). Transgender leaders in the workplace: Visibility and diversity climate. *Journal of Vocational Behavior, 114*, 81–95.

Philippine Commission on Women. (2009). *Annual Accomplishment Report*. PCW. https://pcw.gov.ph/assets/files/2020/05/2009_pcw_accomplishment_report.pdf

Philippine Commission on Women. (2016). *A Handbook on the Application of the Enhanced Gender Mainstreaming Evaluation Framework*. PCW. https://library.pcw.gov.ph/wp-content/uploads/2020/12/PCW-A-Handbook-on-the-Application-of-the-Enhanced-Gender-Mainstreaming-Evaluation-Framework-2016.pdf?fbclid=IwAR2ehD2vaFFqQxbjGy4MIEyst7P7PAK83lRWaxROyyLpgQ74maFTqCtl19E

Philippine Commission on Women. (2023). *Citizen's Charter*. https://pcw.gov.ph/citizens-charter/

Pierli, G., Murmura, F., & Palazzi, F. (2022). Women and Leadership: How Do Women Leaders Contribute to Companies' Sustainable Choices? *Frontiers in Sustainability, 3*(July), 1–10. doi:10.3389/frsus.2022.930116

Pinker, S. (2018). *Enlightenment Now: The Case for Reason, Science, Humanism, and Progress*. Penguin Books.

Piper, N. (2003). Bridging gender, migration and governance: Theoretical possibilities in the Asian context. *Asian and Pacific Migration Journal, 12*(1-2), 21–48. doi:10.1177/011719680301200102

Pitts, V. (2020). Expanding the gender binary in leadership: Insights from transgender executives. *The Journal of Leadership Studies, 14*(2), 112–115.

Pless, N. M., & Maak, T. (2004). Building an inclusive diversity culture: Principles, processes and practice. *Journal of Business Ethics, 54*(2), 129–147. doi:10.100710551-004-9465-8

Pohlmann, L. M. (2013). *Women leaders of environmental advocacy organizations: A qualitative study*. University of Southern Maine.

Prindeville, D. M. (2003). Latina and American Indian women leaders, political representation, and environmental policy. *Journal of Latino/Latin. American Studies (Lawrence, Kan.), 1*, 55–76.

Puhazhendhi, V., & Jayaraman, B. (1999). Increasing Women's Participation and Employment Generation among Rural Poor: An Approach through Informal Groups. *Indian Journal of Agricultural Economics. Indian Society of Agricultural Economics, 54*(3), 287–295. doi:10.22004/ag.econ.297680

Purushothaman, S. (1998). The Empowerment of Women in India: Grassroots Women's Networks and the State. *Sage (Atlanta, Ga.), 8*(1), 109–111. doi:10.1177/1018529119980107

Quadir, D. A., & Iqbal, A. (2008). *Tropical Cyclones: Impacts on Coastal Livelihoods. Investigation of the Coastal Inhabitants of Bangladesh*. IUCN.

Ramirez-Santos, A. G., Ravera, F., Rivera-Ferre, M. G., & Calvet-Nogués, M. (2023). Gendered traditional agroecological knowledge in agri-food systems: A systematic review. *Journal of Ethnobiology and Ethnomedicine*, *19*(1), 1–19. doi:10.118613002-023-00576-6 PMID:37024903

Ramstetter, L., & Habersack, F. (2020). Do women make a difference? Analysing environmental attitudes and actions of Members of the European Parliament. *Environmental Politics*, *29*(6), 1063–1084. doi:10.1080/09644016.2019.1609156

Read, R. (2021) Opinion: This year's Nobel Peace Prize should've gone to Greta Thunberg. *The Independent.*

Reck, J. M., & Lombardi, E. L. (2020). The role of transgender and gender nonconforming leaders in higher education. *New Directions for Student Services*, *172*, 37–46.

Resurrección, B. P. (2013). Persistent women and environment linkages in climate change and sustainable development agendas. *Women's Studies International Forum*, *40*, 33–43. doi:10.1016/j.wsif.2013.03.011

Ricci, F., Scafarto, V., & della Corte, G. (2023). Board Gender Diversity, Female Leaders and Corporate Environmental Responsibility: A Structured Literature Review. When the Crisis Becomes an Opportunity: *The Role of Women in the Post-Covid Organization,* 411-432.

Rico, M. N. (1998). Gender, the Environment and the Sustainability. In M. N. Rico (Ed.), *Gender, the Environment and the Sustaibility of Development* (pp. 15–20). United Nation.

Robert, B., & Brown, E. B. (2004).Covariance Structure Analysis of Health-Related Indicators in Elderly People Living at Home with a Focus on Subjective Health Perception, (1).

Robinson, G., & Dechant, K. (1997). Building a business case for diversity. *The Academy of Management Perspectives*, *11*(3), 21–31. doi:10.5465/ame.1997.9709231661

Roots & Shoots. (2023). *About Roots & Shoot*. Roots & Shoots. https://rootsandshoots.org/about-us/

Roy, A. (1999). The greater common good. *India International Centre Quarterly*, *26*(1), 25–36.

Roy, J., Islam, A., Mishra, A., & Ahmad, S. (2018). Gender and Climate Change: Impacts, Responses, and Adaptation. *Climate Policy*, *18*(7), 844–863.

Rugalema, G., & Owusu-Adjei, E. (2016). *Women's Empowerment through Climate-Smart Agriculture: Evidence from Africa*. https://www.unwomen.org/sites/default/files/Headquarters/Attachments/Sections/Library/Publications/2015/FPI%20BriefAgriculture%20LocalUSweb.pdf

Saikia, U. (2021). Role Of Women In Environmental Movement In India. *Ilkogretim Online - Elementary Education Online, 20*(6), 2473-2478. doi:10.17051/ilkonline.2021.06.228

Santos, C., & Sciascia, S. (2019). Transgender entrepreneurship and sustainable development. *Business & Society*, *58*(6), 1224–1252.

Saraçli, S., Yilmaz, V., & Arslan, T. (2014). The Effects of Mothers' Educational Levels on University Students' Environmental Protection Commitments and Environmental Behaviors. *Eurasian Journal of Educational Research*, *14*(55), 177–199. doi:10.14689/ejer.2014.55.11

Saraite-Sariene, L., Galán-Valdivieso, F., Alonso-Cañadas, J., & García-Tabuyo, M. (2022), The role of women's leadership in environmental NGOs' online accountability. *Online Information Review, 46*(4).

Schipper, E. L. F., & Langston, L. (2015). *A comparative overview of resilience measurement frameworks. Analyzing Indicators and Approaches.* ODI. https://odi.org/en/publications/a-comparative-overview-of-resilience-measurement-frameworks-analysing-indicators-and-approaches/

Schueman, L. J. (2023). Climate Hero: Marina Silva. *One Earth.* https://www.oneearth.org/climate-hero-marina-silva/

Schwenke, K., & Brouwer, E. (2018). Non-binary and transgender leadership: Success factors and challenges. In R. J. Thompson & J. Storberg-Walker (Eds.), *Leadership and Power in International Development* (pp. 185–201). Routledge.

Sembiring, E., & Nitivattananon, V. (2010). Sustainable solid waste management toward an inclusive society: Integration of the informal sector. *Resources, Conservation and Recycling, 54*(11), 802–809. doi:10.1016/j.resconrec.2009.12.010

Shambel, G. (2012). The gender dimension of climate change and women's role in environmental protection and natural resource management in South Wollo, Ethiopia. *Research on Humanities and Social Sciences, 2*, 32–40.

Sharma, H. B., Vanapalli, K. R., Samal, B., Cheela, V. S., Dubey, B. K., & Bhattacharya, J. (2021). Circular economy approach in solid waste management system to achieve UN-SDGs: Solutions for post-COVID recovery. *The Science of the Total Environment, 800*, 149605. doi:10.1016/j.scitotenv.2021.149605 PMID:34426367

Sharma, P. (2015). Women's role in environmental conservation in India. *Journal of Sustainable Development, 8*(2), 250–255.

Shashikala, S. (2007). *India: Promoting Urban Social Development through Self Help Groups in Karnataka. A Report to the Operations Evaluation Department for the Project Performance Evaluation Report of Karnataka Urban Infrastructure Development Project in India.* Asian Development Bank (ADB). https://www.adb.org/sites/default/files/evaluation-document/35856/files/karnataka-urban-infrastructure-developmentproject.pdf

Shattuck, A. (2022). One Person, One Tree: Wangari Maathai and Ecofeminism in the Greenbelt Movement. *Women Leading Change: Case Studies in Women, Gender, and Feminism, 6*(2), 21–35.

Shaw, R. (2006). Community-based climate change adaptation in Vietnam: inter-linkages of environment, disaster, and human security. *Multiple dimensions of global environmental change*, 521-547. Research Gate.

Shinbrot, X. A., Wilkins, K., Gretzel, U., & Bowser, G. (2019). Unlocking Women's Sustainability Leadership Potential: Perceptions of Contributions and Challenges for Women in Sustainable Development. *World Development, 119*, 120–132. doi:10.1016/j.worlddev.2019.03.009

Shiva, V. (1992). *Women's Indigenous Knowledge and Biodiversity Conservation.* India International Centre Quarterly, 205-2014.

Shiva, V. (2005). *Earth democracy: Justice, sustainability, and peace.* South End Press. doi:10.5040/9781350219755

Shrestha, R. K., L'Espoir Decosta, J. N. P., & Shrestha, R. (2023). Creating value networks through a learning society for Indigenous women tourism entrepreneurs. *International Journal of Contemporary Hospitality Management.* doi:10.1108/IJCHM-03-2022-0317

SIDA. (1996). Promoting Sustainable Livelihoods: A Report from the Task Force on Poverty. Stockholm: Swedish International Development Cooperation Agency (SIDA).

Siegle, L. (2021). Revealed: why hundreds of thousands of tonnes of recycling are going up in smoke. *The Guardian.* https://www.theguardian.com/environment/2021/mar/07/revealed-why-hundreds-of-thousands-of-tonnes-of-recycling-are-going-up-in-smoke

Siegle, L. (2022, February 25). Can we enjoy fast fashion without destroying the planet? *The Guardian.* https://www.theguardian.com/environment/2022/feb/25/rent-clothes-fashion-sustainable-climate

Sigita Strumskyte, S. R. (2022). *Women's leadership in environmental action.* Organisation for Economic Co-operation and Development.

Silesbarcena, I. (2019). Climate change and sustainable development. *Ecofeminism*. https://sites.manchester.ac.uk/global-social-challenges/2019/06/08/why-women-as-environmental-activists-will-bring-about-necessary-political-change-for-a-sustainable-future/

Silva, M. (2019). The Hollow Consensus on the Amazon. *Jordan Times*.

Smith, J. (2014). Feminine Perspectives on Green Leadership. *Journal of Environmental Leadership*, *12*(3), 45–60.

Solsvik, T. (2020, February 26). Climate activist Thunberg heads growing field of Nobel Prize candidates. *Reuters*.

Sorial, S., & White, F. A. (2018). Transgender employees in the workplace: A systematic review of the literature. *International Journal of Management Reviews*, *20*(4), 889–911.

Srivastava, N. (2020). Gender dynamics in forest management: A case study from Uttarakhand, India. *Forest Policy and Economics*, *112*, 102097.

Stone, D. (2019). *LGBTQ-inclusive governance: A global perspective*. Routledge.

Sultana, P., Thompson, P. M., & Ahmed, M. (2002). *Women-led fisheries management-a case study from Bangladesh*. World Fish Center. https://digitalarchive.worldfishcenter.org/

Taga, R., McNamara, K., & Amoafoa-Smart, I. (2018). Women's Leadership in Sustainable Development: The Case of Water Management. *Journal of Cleaner Production*, *171*, 498–507.

Testa, R. J., & Simpson, B. (2019). Gender diversity and transgender inclusion in the workplace. *The Journal of Social Issues*, *75*(2), 585–610.

The Elders. (2023). Gro Harlem Brundtland. *The Elders*. https://theelders.org/profile/gro-harlem-brundtland

The Green Belt Movement. (2023) Wangari Matthai. Biography. *The Green Belt Movement*. https://www.greenbeltmovement.org/wangari-maathai/biography

The Green Belt Movement. (2023). *Our History*. Green Belt Movement. http://www.greenbeltmovement.org/who-we-are/our-history

The Guardian. (2023). Lucy Siegle. *The Guardian*. https://www.theguardian.com/profile/lucysiegle

The Japan International Cooperation Agency. (2016). *Transforming Gender Relations in Disaster Risk Reduction*. https://www.jica.go.jp/Resource/jica-ri/publication/booksandreports/l75nbg00000697z9-att/JICA_Philippines_Sri_Lanka.pdf

The New York Times. (2020). Greta Thunberg joins climate march on her last day in Davos. *The New York Times*.

Tilche, A., Mysiak, J., Feldmann, U., Gressmann, H., Tramonti, G., & Zebisch, M. (2019). Gender Mainstreaming in Climate Change Adaptation and Disaster Risk Reduction: Evidence from Case Studies. *Sustainability*, *11*(17), 4682.

Torres, R. (2020). Empowering Women in the Era of Climate Change. *Climate Change and Gender*, *7*(5), 60–77.

U. N. Women Africa (2017). Gender, Energy and Policy - A Review of Energy Policies in East and Southern Africa. UN Women Africa.

UAE MOCCAE (United Arab Emirates Ministry of Climate Change and Environment). (2021). *The UAE State of Climate Report: A Review of the Arabian Gulf Region's Changing Climate & Its Impacts*. UAE MOCCAE. www.moccae.gov.ae

UN DESA (United Nations Department of Economic and Social Affairs). (2010). *The world's women 2010, trends and statistics*. UN.

UN Foundation. (n.d.). Gro Harlem Brundtland (Norway). UN Foundation. https://unfoundation.org/who-we-are/our-board/gro-harlem-brundtland-norway-vice-chair/

UNDP-UNEP PEI (United Nations Development Programme - UN Environment Programme Poverty-Environment Initiative) (2015). *Mainstreaming Environment and Climate Change for Poverty Reduction and Sustainable Development - A Handbook*. UN.

UNEP. (2020). *Gender Equality and the Environment: Policy and Strategy*. UNEP. https://wedocs.unep.org/bitstream/handle/20.500.11822/7655/-Gender_equality_and_the_environment_Policy_and_strategy-2015Gender_equality_and_the_environment_policy_and_strategy.pdf

UNESCAP (United Nations Economic and Social Commission for Asia and the Pacific). (2012). *Statistical Yearbook for Asia and the Pacific 2012*. UNESCAP. https://www.unescap.org/stat/data/syb2012/index.asp

UNFCCC. (2023, June 19,). Razan Al Mubarak. razanalmubarak.com. https://www.ead.gov.ae/en/About-Us/Our-Leadership/Leaders/Razan-Khalifa-Al-Mubarak

United Arab Emirates University. (2022, November 21). H.E. Mariam Bint Mohammed Saeed Hareb Almheiri. *Board of Trustees*. United Arab Emirates University. https://www.uaeu.ac.ae/en/about/board-of-trustees/mariam-almheiri.shtml

United Nations Department of Economic and Social Affairs. (n.d). *2030 Agenda for Sustainable Development: Sustainable Development Goals*. UN. https://social.desa.un.org/2030agenda-sdgs

United Nations Development Programme (UNDP). (2001). *Annual Report 2001*. UNDP. https://www.undp.org/publications/annual-report-2001

United Nations Development Programme (UNDP). (2020). *Gender Equality Strategy 2018–2021*. https://www.undp.org/sites/g/files/zskgke326/files/2023-03/UNDP-Gender-Equality-Strategy-2018-2021-Annual-Report.pdf

United Nations Development Programme (UNDP). (2021). *Women's Empowerment and Sustainable Development*. https://www.undp.org/women-empowerment-and-sustainable-development

United Nations Environment Programme (UNEP). (2018). *Gender and Environment: Making the Connections*. https://www.unep.org/resources/gender-and-environment-making-connections

United Nations Framework Convention on Climate Change (UNFCCC). (2017). *Gender and Climate Change: A Framework for Action*. https://unfccc.int/gender

United Nations Population Fund (UNFPA). (2021). *Gender Equality*. https://www.unfpa.org/gender-equality

United Nations. (2011). *The Millennium Development Goals Report 2011*. https://www.un.org/millenniumgoals/pdf/(2011_E)%20MDG%20Report%202011_Book%20LR.pdf

United Nations. (2015). *The Paris Agreement*. UN. https://unfccc.int/process-and-meetings/the-paris-agreement

United Nations. (2015). *Transforming Our World: The 2030 Agenda for Sustainable Development*. https://sdgs.un.org/2030agenda

United Nations. (2019). *Sustainable Development Goals Report 2019*. UN. https://digitallibrary.un.org/record/478154?ln=en

United Nations. (2023, March 10). Delegates Highlight Rural Women's, Girls' National, Economic Contributions, But Point to Gaps in Education, Food Security, as Commission Session Continues. https://press.un.org/en/2023/wom2225.doc.htm

United Nations. (n.d.). Rio 1992: *United Nations Conference on Environment & Development*. UN. https://www.un.org/en/conferences/environment/rio1992

Upreti, B. R., & Nepal, S. (2019). Women's Leadership in Local Environmental Governance: Evidence from Community Forests in Nepal. *Gender and Development*, *27*(3), 409–425.

Upton, C., & Rocheleau, D. (2016). Gender and Natural Resource Management: Livelihoods, Mobility, and Interventions. *The International Journal of the Commons*, *10*(2), 865–886.

USAIF-LEAF. (2014). *Women's Inclusion in REDD+ in The Philippines Lessons from Good Practices in Forest, and Other Natural Resources Management Sectors*. https://redd.unfccc.int/uploads/2234_1_philippines_gender_and_redd_report_sm.pdf

Valdez, I. K. M., Arevalo, M.P.N., Robredo, J.P.G., Gacad, S.L.S., Villaceran, M.A.J., Libang, G.R., Dela Paz, E.P., Rubin, K.S.T., & Eala, M.A.B. (2022). Violence against women in the Philippines: Barriers to seeking support. *The Lancet Regional Health - Western Pacific, 23*, 1-2.

Valentine, S., & Funk, D. (2020). Examining the work experiences of transgender individuals: A systematic review. *Human Resource Management Review*, *30*(1), 100706.

Villarin, T., & Ramos, S. (2009). *Transformative Power: Case Studies on Political Participation of Community Women Leaders* (1st ed.). Institute of Politics and Governance and Friedrich Ebert Stiftung.

Vogt, M., & Cavaghan, R. (2017). Women's Leadership for Sustainability: A Comparative Analysis. *Journal of Cleaner Production*, *140*(2), 1674–1686.

Wallace-Wells, D. (2019, September 17). It's Greta's world. *Intelligencer.*

Warren, M. A., Donaldson, S. I., Lee, J. Y., & Donaldson, S. I. (2019). Reinvigorating Research on Gender in the Workplace Using a Positive Work and Organizations Perspective. *International Journal of Management Reviews*, *21*(4), 498–518. doi:10.1111/ijmr.12206

Watts, J. (2019, April 23). The Greta Thunberg effect: At last MPs focus on climate change. *The Guardian.*

Weber, E. U. (2010). What shapes perceptions of climate change? *Wiley Interdisciplinary Reviews: Climate Change*, *1*(3), 332–342. [REMOVED HYPERLINK FIELD] doi:10.1002/wcc.41

WEF (World Economic Forum). (2023). *Hindou Oumarou Ibrahim*. World Economic Forum. https://www.weforum.org/people/hindou-oumarou-ibrahim

Weinstein, J. (2010). *The Problem of Rural Education in the Philippines*. Josh Weinstein. https://joshweinstein.wordpress.com/2010/03/02/the-problem-of-education-in-the-philippines/

WeinsteinJ. (2010). *The Problem of Rural Education in the Philippines*. https://joshweinstein.wordpress.com/2010/03/02/the-problem-of-education-in-the-philippines/

Wei, S., Sial, M. S., Zhou, W., Badulescu, A., & Badulescu, D. (2021). Improving the environmental footprint through employees: A case of female leaders from the perspective of CSR. *International Journal of Environmental Research and Public Health*, *18*(24), 13082. doi:10.3390/ijerph182413082 PMID:34948692

Wilkinson, C. (2012). Social-ecological resilience: Insights and issues for planning theory. *Planning Theory*, *11*(2), 148–169. doi:10.1177/1473095211426274

Winsemius, H. C., Jongman, B., Veldkamp, T. I., Hallegatte, S., Bangalore, M., & Ward, P. J. (2018). *Disaster risk, climate change, and poverty: assessing the global exposure of poor people to floods and droughts*. LSE. https://eprints.lse.ac.uk/87536/1/Bangalore_Disaster%20risk_2018.pdf

Witten, T. M., & Eyler, A. E. (2019). *Transgender leadership: Theory and practice for managing diversity*. Routledge.

WMO (World Meteorological Organization). (2022). WMO's state of the climate in Europe Report for 2022 urges immediate action on Europe's climate crisis. *State of Climate in Europe 2022.* WMO. https://community.wmo.int/en/news/wmos-state-climate-europe-report-2022-urges-immediate-action-europes-climate-crisis

WOCAN. (2021). *RWL Project Philippines Report.* https://www.wocan.org/resource/rwl-project-philippines-report/

Women Deliver. (2021). The link between climate change and sexual and reproductive health and rights: An evidence review. *Women Deliver.*

Women Political Leaders. (2020). *The G7 Reykjavík Index for Leadership.* Kantar.

Women U. N. (2015). The Beijing Declaration and Platform for Action Turns 20. New York: United Nations Entity for Gender Equality and the Empowerment of Women. UN Women.

Women U. N. (2022). *Impact of devolution on women and girls in Kenya.* UN. https://africa.unwomen.org/sites/default/files/2022-12/IMPACT%20OF%20DEVOLUTION11%20eVersion.pdf

Women U. N. (2022). *Impact of devolution on women and girls in Kenya.* UN.https://africa.unwomen.org/sites/default/files/2022-12/IMPACT%20OF%20DEVOLUTION11%20eVersion.pdf

Women U. N. (2022, March 11). *Five young women on the forefront of climate action across Europe and Central Asia.* Europe and Central Asia. https://eca.unwomen.org/en/stories/news/2022/03/five-young-women-on-the-forefront-of-climate-action-across-europe-and-central-asia

Women U. N. (2023). EmPower: Women for Climate-Resilient Societies Programme Brief. UN. https://asiapacific.unwomen.org/en/digital-library/publications/2023/06/empower-women-for-climate-resilient-societies-programme-brief#:~:text=From%202018%20to%202022%2C%20UN,across%20Asia%20and%20the%20Pacific

Women Watch U. N. (2009). *Fact Sheet: Women, Gender Equality and Climate Change.* UN. https://www.un.org/womenwatch/feature/climate_change/

Women, U. N. (2014). *United Nations Framework Convention on Climate Change.* https://www.unwomen.org/en/how-we-work/intergovernmental-support/climate-change-and-the-environment/united-nations-framework-convention-on-climate-change

Women, U. N. (2020). *Gender Equality and Women's Empowerment: Women's Leadership and Political Participation.* https://www.unwomen.org/en/what-we-do/leadership-and-political-participation

Women, U. N. (2020). *Women's Leadership and Political Participation.* https://www.unwomen.org/en/what-we-do/leadership-and-political-participation

World Bank and Australian Aid. (2014). *Where are we during flooding?' A qualitative assessment of Poverty and Social Impacts of Flooding in selected neighborhoods of HCMC.* World Bank. https://documents1.worldbank.org/curated/en/669141603288540994/pdf/People-in-Harms-Way-Flood-Exposure-and-Poverty-in-189-Countries.pdf

World Bank Group. (2016). *Impact of Climate Change and Aquatic Salinization on Fish Habitats and Poor Communities in Southwest Coastal Bangladesh and Bangladesh Sundarbans.* World Bank. https://elibrary.worldbank.org/doi/abs/10.1596/1813-9450-7593

World Bank Group. (2018). *Women, Business, and the Law 2018: Measuring Gender Equality in the Law.* https://thedocs.worldbank.org/en/doc/102741522965756861-0050022018/original/WBLKeyFindingsWebFINAL.pdf

World Bank. (2019). *Women, Business, and the Law 2019: A Decade of Reform.* https://openknowledge.worldbank.org/handle/10986/31752

Wuyep, S., Dung, V., Buhari, A., Madaki, D., & Bitrus, B. (2014). Women participation in environmental protection and management: Lessons from plateau state, Nigeria. *American Journal of Environmental Protection*, *2*, 32–36. doi:10.12691/env-2-2-1

Yachina, N. P., Khuziakhmetov, A. N., & Gabdrakhmanova, R. G. (2018). Formation and development of the regional system of continuous environmental education of a teacher. *Ekoloji*, *27*, 1315–1322.

Yarram, S. R., & Adapa, S. (2021). Board gender diversity and corporate social responsibility: Is there a case for critical mass? *Journal of Cleaner Production*, *278*, 123319. doi:10.1016/j.jclepro.2020.123319

Zimmerman, R., & Horan, T. A. (Eds.). (2004). *Digital Infrastructures: Enabling civil and environmental systems through information technology*. Psychology Press. doi:10.4324/9780203357125_chapter_ONE

Related References

To continue our tradition of advancing academic research, we have compiled a list of recommended IGI Global readings. These references will provide additional information and guidance to further enrich your knowledge and assist you with your own research and future publications.

Ajiboye, O. E., & Yusuff, O. S. (2017). Foreign Land Acquisition: Food Security and Food Chains – The Nigerian Experience. In I. Management Association (Ed.), Natural Resources Management: Concepts, Methodologies, Tools, and Applications (pp. 1524-1545). Hershey, PA: IGI Global. https://doi.org/ doi:10.4018/978-1-5225-0803-8.ch072

Alapiki, H. E., & Amadi, L. A. (2018). Sustainable Food Consumption in the Neoliberal Order: Challenges and Policy Implications. In A. Obayelu (Ed.), *Food Systems Sustainability and Environmental Policies in Modern Economies* (pp. 90–123). Hershey, PA: IGI Global. doi:10.4018/978-1-5225-3631-4.ch005

Altaş, A. (2018). Geographical Information System Applications Utilized in Museums in Turkey Within the Scope of the Cultural Heritage Tourism: A Case Study of Mobile Application of Müze Asist. In S. Chaudhuri & N. Ray (Eds.), *GIS Applications in the Tourism and Hospitality Industry* (pp. 42–60). Hershey, PA: IGI Global. doi:10.4018/978-1-5225-5088-4.ch002

Andreea, I. R. (2018). Beyond Macroeconomics of Food and Nutrition Security. *International Journal of Sustainable Economies Management*, *7*(1), 13–22. doi:10.4018/IJSEM.2018010102

Anwar, J. (2017). Reproductive and Mental Health during Natural Disaster: Implications and Issues for Women in Developing Nations – A Case Example. In I. Management Association (Ed.), Gaming and Technology Addiction: Breakthroughs in Research and Practice (pp. 446-472). Hershey, PA: IGI Global. https://doi.org/ doi:10.4018/978-1-5225-0778-9.ch021

Awadh, H., Aksissou, M., Benhardouze, W., Darasi, F., & Snaiki, J. (2018). Socioeconomic Status of Artisanal Fishers in the West Part of Moroccan Mediterranean. *International Journal of Social Ecology and Sustainable Development*, *9*(1), 40–52. doi:10.4018/IJSESD.2018010104

Aye, G. C., & Haruna, R. F. (2018). Effect of Climate Change on Crop Productivity and Prices in Benue State, Nigeria: Implications for Food Security. In V. Erokhin (Ed.), *Establishing Food Security and Alternatives to International Trade in Emerging Economies* (pp. 244–268). Hershey, PA: IGI Global. doi:10.4018/978-1-5225-2733-6.ch012

Azizan, S. A., & Suki, N. M. (2017). Consumers' Intentions to Purchase Organic Food Products. In T. Esakki (Ed.), *Green Marketing and Environmental Responsibility in Modern Corporations* (pp. 68–82). Hershey, PA: IGI Global. doi:10.4018/978-1-5225-2331-4.ch005

Barakabitze, A. A., Fue, K. G., Kitindi, E. J., & Sanga, C. A. (2017). Developing a Framework for Next Generation Integrated Agro Food-Advisory Systems in Developing Countries. In I. Management Association (Ed.), Agri-Food Supply Chain Management: Breakthroughs in Research and Practice (pp. 47-67). Hershey, PA: IGI Global. doi:10.4018/978-1-5225-1629-3.ch004

Beachcroft-Shaw, H., & Ellis, D. (2018). Using Successful Cases to Promote Environmental Sustainability: A Social Marketing Approach. In I. Management Association (Ed.), Sustainable Development: Concepts, Methodologies, Tools, and Applications (pp. 936-953). Hershey, PA: IGI Global. doi:10.4018/978-1-5225-3817-2.ch042

Behnassi, M., Kahime, K., Boussaa, S., Boumezzough, A., & Messouli, M. (2017). Infectious Diseases and Climate Vulnerability in Morocco: Governance and Adaptation Options. In I. Management Association (Ed.), Public Health and Welfare: Concepts, Methodologies, Tools, and Applications (pp. 91-109). Hershey, PA: IGI Global. https://doi.org/ doi:10.4018/978-1-5225-1674-3.ch005

Bekele, F., & Bekele, I. (2017). Social and Environmental Impacts on Agricultural Development. In W. Ganpat, R. Dyer, & W. Isaac (Eds.), *Agricultural Development and Food Security in Developing Nations* (pp. 21–56). Hershey, PA: IGI Global. doi:10.4018/978-1-5225-0942-4.ch002

Benaouda, A., & García-Peñalvo, F. J. (2018). Towards an Intelligent System for the Territorial Planning: Agricultural Case. In F. García-Peñalvo (Ed.), *Global Implications of Emerging Technology Trends* (pp. 158–178). Hershey, PA: IGI Global. doi:10.4018/978-1-5225-4944-4.ch010

Bhaskar, A., Rao, G. B., & Vencatesan, J. (2017). Characterization and Management Concerns of Water Resources around Pallikaranai Marsh, South Chennai. In P. Rao & Y. Patil (Eds.), *Reconsidering the Impact of Climate Change on Global Water Supply, Use, and Management* (pp. 102–121). Hershey, PA: IGI Global. doi:10.4018/978-1-5225-1046-8.ch007

Bhyan, P., Shrivastava, B., & Kumar, N. (2022). Requisite Sustainable Development Contemplating Buildings: Economic and Environmental Sustainability. In A. Hussain, K. Tiwari, & A. Gupta (Eds.), *Addressing Environmental Challenges Through Spatial Planning* (pp. 269–288). IGI Global. https://doi.org/10.4018/978-1-7998-8331-9.ch014

Bogataj, D., & Drobne, D. (2017). Control of Perishable Goods in Cold Logistic Chains by Bionanosensors. In S. Joo (Ed.), *Applying Nanotechnology for Environmental Sustainability* (pp. 376–402). Hershey, PA: IGI Global. doi:10.4018/978-1-5225-0585-3.ch016

Bogataj, D., & Drobne, D. (2017). Control of Perishable Goods in Cold Logistic Chains by Bionanosensors. In I. Management Association (Ed.), Materials Science and Engineering: Concepts, Methodologies, Tools, and Applications (pp. 471-497). Hershey, PA: IGI Global. https://doi.org/ doi:10.4018/978-1-5225-1798-6.ch019

Bogueva, D., & Marinova, D. (2018). What Is More Important: Perception of Masculinity or Personal Health and the Environment? In D. Bogueva, D. Marinova, & T. Raphaely (Eds.), *Handbook of Research on Social Marketing and Its Influence on Animal Origin Food Product Consumption* (pp. 148–162). Hershey, PA: IGI Global. doi:10.4018/978-1-5225-4757-0.ch010

Bouzid, M. (2017). Waterborne Diseases and Climate Change: Impact and Implications. In M. Bouzid (Ed.), *Examining the Role of Environmental Change on Emerging Infectious Diseases and Pandemics* (pp. 89–108). Hershey, PA: IGI Global. doi:10.4018/978-1-5225-0553-2.ch004

Bowles, D. C. (2017). Climate Change-Associated Conflict and Infectious Disease. In M. Bouzid (Ed.), *Examining the Role of Environmental Change on Emerging Infectious Diseases and Pandemics* (pp. 68–88). Hershey, PA: IGI Global. doi:10.4018/978-1-5225-0553-2.ch003

Buck, J. J., & Lowry, R. K. (2017). Oceanographic Data Management: Quills and Free Text to the Digital Age and "Big Data". In P. Diviacco, A. Leadbetter, & H. Glaves (Eds.), *Oceanographic and Marine Cross-Domain Data Management for Sustainable Development* (pp. 1–22). Hershey, PA: IGI Global. doi:10.4018/978-1-5225-0700-0.ch001

Buse, C. G. (2017). Are Climate Change Adaptation Policies a Game Changer?: A Case Study of Perspectives from Public Health Officials in Ontario, Canada. In M. Bouzid (Ed.), *Examining the Role of Environmental Change on Emerging Infectious Diseases and Pandemics* (pp. 230–257). Hershey, PA: IGI Global. doi:10.4018/978-1-5225-0553-2.ch010

Calderon, F. A., Giolo, E. G., Frau, C. D., Rengel, M. G., Rodriguez, H., Tornello, M., ... Gallucci, R. (2018). Seismic Microzonation and Site Effects Detection Through Microtremors Measures: A Review. In N. Ceryan (Ed.), *Handbook of Research on Trends and Digital Advances in Engineering Geology* (pp. 326–349). Hershey, PA: IGI Global. doi:10.4018/978-1-5225-2709-1.ch009

Carfì, D., Donato, A., & Panuccio, D. (2018). A Game Theory Coopetitive Perspective for Sustainability of Global Feeding: Agreements Among Vegan and Non-Vegan Food Firms. In I. Management Association (Ed.), Game Theory: Breakthroughs in Research and Practice (pp. 71-104). Hershey, PA: IGI Global. https://doi.org/ doi:10.4018/978-1-5225-2594-3.ch004

Castagnolo, V. (2018). Analyzing, Classifying, Safeguarding: Drawing for the Borgo Murattiano Neighbourhood of Bari. In G. Carlone, N. Martinelli, & F. Rotondo (Eds.), *Designing Grid Cities for Optimized Urban Development and Planning* (pp. 93–108). Hershey, PA: IGI Global. doi:10.4018/978-1-5225-3613-0.ch006

Chekima, B. (2018). The Dilemma of Purchase Intention: A Conceptual Framework for Understanding Actual Consumption of Organic Food. *International Journal of Sustainable Economies Management*, 7(2), 1–13. doi:10.4018/IJSEM.2018040101

Chen, Y. (2017). Sustainable Supply Chains and International Soft Landings: A Case of Wetland Entrepreneurship. In B. Christiansen & F. Kasarcı (Eds.), *Corporate Espionage, Geopolitics, and Diplomacy Issues in International Business* (pp. 232–247). Hershey, PA: IGI Global. doi:10.4018/978-1-5225-1031-4.ch013

Çıtak, L., Akel, V., & Ersoy, E. (2018). Investors' Reactions to the Announcement of New Constituents of BIST Sustainability Index: An Analysis by Event Study and Mean-Median Tests. In M. Risso & S. Testarmata (Eds.), *Value Sharing for Sustainable and Inclusive Development* (pp. 270–289). Hershey, PA: IGI Global. doi:10.4018/978-1-5225-3147-0.ch012

D'Aleo, V., D'Aleo, F., & Bonanno, R. (2018). New Food Industries Toward a New Level of Sustainable Supply: Success Stories, Business Models, and Strategies. In V. Erokhin (Ed.), *Establishing Food Security and Alternatives to International Trade in Emerging Economies* (pp. 74–97). Hershey, PA: IGI Global. doi:10.4018/978-1-5225-2733-6.ch004

Dagevos, H., & Reinders, M. J. (2018). Flexitarianism and Social Marketing: Reflections on Eating Meat in Moderation. In D. Bogueva, D. Marinova, & T. Raphaely (Eds.), *Handbook of Research on Social Marketing and Its Influence on Animal Origin Food Product Consumption* (pp. 105–120). Hershey, PA: IGI Global. doi:10.4018/978-1-5225-4757-0.ch007

Danisman, G. O. (2022). What Drives Eco-Design Innovations in European SMEs? In U. Akkucuk (Ed.), *Disruptive Technologies and Eco-Innovation for Sustainable Development* (pp. 191–206). IGI Global. https://doi.org/10.4018/978-1-7998-8900-7.ch011

Deenapanray, P. N., & Ramma, I. (2017). Adaptations to Climate Change and Climate Variability in the Agriculture Sector in Mauritius: Lessons from a Technical Needs Assessment. In I. Management Association (Ed.), Natural Resources Management: Concepts, Methodologies, Tools, and Applications (pp. 655-680). Hershey, PA: IGI Global. https://doi.org/ doi:10.4018/978-1-5225-0803-8.ch030

Deenapanray, P. N., & Ramma, I. (2017). Adaptations to Climate Change and Climate Variability in the Agriculture Sector in Mauritius: Lessons from a Technical Needs Assessment. In I. Management Association (Ed.), Natural Resources Management: Concepts, Methodologies, Tools, and Applications (pp. 655-680). Hershey, PA: IGI Global. https://doi.org/ doi:10.4018/978-1-5225-0803-8.ch030

Deshpande, S., Basu, S. K., Li, X., & Chen, X. (2017). Smart, Innovative and Intelligent Technologies Used in Drug Designing. In I. Management Association (Ed.), Pharmaceutical Sciences: Breakthroughs in Research and Practice (pp. 1175-1191). Hershey, PA: IGI Global. https://doi.org/ doi:10.4018/978-1-5225-1762-7.ch045

Dlamini, P. N. (2017). Use of Information and Communication Technologies Tools to Capture, Store, and Disseminate Indigenous Knowledge: A Literature Review. In P. Ngulube (Ed.), *Handbook of Research on Theoretical Perspectives on Indigenous Knowledge Systems in Developing Countries* (pp. 225–247). Hershey, PA: IGI Global. doi:10.4018/978-1-5225-0833-5.ch010

Dolejsova, M., & Kera, D. (2017). The Fermentation GutHub Project and the Internet of Microbes. In S. Konomi & G. Roussos (Eds.), *Enriching Urban Spaces with Ambient Computing, the Internet of Things, and Smart City Design* (pp. 25–46). Hershey, PA: IGI Global. doi:10.4018/978-1-5225-0827-4.ch002

Dolunay, O. (2018). A Paradigm Shift: Empowering Farmers to Eliminate the Waste in the Form of Fresh Water and Energy Through the Implementation of 4R+T. In I. Management Association (Ed.), Sustainable Development: Concepts, Methodologies, Tools, and Applications (pp. 882-892). Hershey, PA: IGI Global. https://doi.org/ doi:10.4018/978-1-5225-3817-2.ch039

Dube, P., Heijman, W. J., Ihle, R., & Ochieng, J. (2018). The Potential of Traditional Leafy Vegetables for Improving Food Security in Africa. In V. Erokhin (Ed.), *Establishing Food Security and Alternatives to International Trade in Emerging Economies* (pp. 220–243). Hershey, PA: IGI Global. doi:10.4018/978-1-5225-2733-6.ch011

Duruji, M. M., & Urenma, D. F. (2017). The Environmentalism and Politics of Climate Change: A Study of the Process of Global Convergence through UNFCCC Conferences. In I. Management Association (Ed.), *Natural Resources Management: Concepts, Methodologies, Tools, and Applications* (pp. 77-108). Hershey, PA: IGI Global. 10.4018/978-1-5225-0803-8.ch004

Dutta, U. (2017). Agro-Geoinformatics, Potato Cultivation, and Climate Change. In S. Londhe (Ed.), *Sustainable Potato Production and the Impact of Climate Change* (pp. 247–271). Hershey, PA: IGI Global. doi:10.4018/978-1-5225-1715-3.ch012

Edirisinghe, R., Stranieri, A., & Wickramasinghe, N. (2017). A Taxonomy for mHealth. In N. Wickramasinghe (Ed.), *Handbook of Research on Healthcare Administration and Management* (pp. 596–615). Hershey, PA: IGI Global. doi:10.4018/978-1-5225-0920-2.ch036

Ekpeni, N. M., & Ayeni, A. O. (2018). Global Natural Hazard and Disaster Vulnerability Management. In A. Eneanya (Ed.), *Handbook of Research on Environmental Policies for Emergency Management and Public Safety* (pp. 83–104). Hershey, PA: IGI Global. doi:10.4018/978-1-5225-3194-4.ch005

Ene, C., Voica, M. C., & Panait, M. (2017). Green Investments and Food Security: Opportunities and Future Directions in the Context of Sustainable Development. In M. Mieila (Ed.), *Measuring Sustainable Development and Green Investments in Contemporary Economies* (pp. 163–200). Hershey, PA: IGI Global. doi:10.4018/978-1-5225-2081-8.ch007

Escamilla, I., Ruíz, M. T., Ibarra, M. M., Soto, V. L., Quintero, R., & Guzmán, G. (2018). Geocoding Tweets Based on Semantic Web and Ontologies. In M. Lytras, N. Aljohani, E. Damiani, & K. Chui (Eds.), *Innovations, Developments, and Applications of Semantic Web and Information Systems* (pp. 372–392). Hershey, PA: IGI Global. doi:10.4018/978-1-5225-5042-6.ch014

Escribano, A. J. (2018). Marketing Strategies for Trendy Animal Products: Sustainability as a Core. In F. Quoquab, R. Thurasamy, & J. Mohammad (Eds.), *Driving Green Consumerism Through Strategic Sustainability Marketing* (pp. 169–203). Hershey, PA: IGI Global. doi:10.4018/978-1-5225-2912-5.ch010

Eudoxie, G., & Roopnarine, R. (2017). Climate Change Adaptation and Disaster Risk Management in the Caribbean. In W. Ganpat & W. Isaac (Eds.), *Environmental Sustainability and Climate Change Adaptation Strategies* (pp. 97–125). Hershey, PA: IGI Global. doi:10.4018/978-1-5225-1607-1.ch004

Farmer, L. S. (2017). Data Analytics for Strategic Management: Getting the Right Data. In V. Wang (Ed.), *Encyclopedia of Strategic Leadership and Management* (pp. 810–822). Hershey, PA: IGI Global. doi:10.4018/978-1-5225-1049-9.ch056

Fattal, L. R. (2017). Catastrophe: An Uncanny Catalyst for Creativity. In R. Shin (Ed.), *Convergence of Contemporary Art, Visual Culture, and Global Civic Engagement* (pp. 244–262). Hershey, PA: IGI Global. doi:10.4018/978-1-5225-1665-1.ch014

Forti, I. (2017). A Cross Reading of Landscape through Digital Landscape Models: The Case of Southern Garda. In A. Ippolito (Ed.), *Handbook of Research on Emerging Technologies for Architectural and Archaeological Heritage* (pp. 532–561). Hershey, PA: IGI Global. doi:10.4018/978-1-5225-0675-1.ch018

Gharbi, A., De Runz, C., & Akdag, H. (2017). Urban Development Modelling: A Survey. In S. Faiz & K. Mahmoudi (Eds.), *Handbook of Research on Geographic Information Systems Applications and Advancements* (pp. 96–124). Hershey, PA: IGI Global. doi:10.4018/978-1-5225-0937-0.ch004

Ghosh, I., & Ghoshal, I. (2018). Implications of Trade Liberalization for Food Security Under the ASEAN-India Strategic Partnership: A Gravity Model Approach. In V. Erokhin (Ed.), *Establishing Food Security and Alternatives to International Trade in Emerging Economies* (pp. 98–118). Hershey, PA: IGI Global. doi:10.4018/978-1-5225-2733-6.ch005

Glaves, H. M. (2017). Developing a Common Global Framework for Marine Data Management. In P. Diviacco, A. Leadbetter, & H. Glaves (Eds.), *Oceanographic and Marine Cross-Domain Data Management for Sustainable Development* (pp. 47–68). Hershey, PA: IGI Global. doi:10.4018/978-1-5225-0700-0.ch003

Godulla, A., & Wolf, C. (2018). Future of Food: Transmedia Strategies of National Geographic. In R. Gambarato & G. Alzamora (Eds.), *Exploring Transmedia Journalism in the Digital Age* (pp. 162–182). Hershey, PA: IGI Global. doi:10.4018/978-1-5225-3781-6.ch010

Gomes, P. P. (2018). Food and Environment: A Review on the Sustainability of Six Different Dietary Patterns. In A. Obayelu (Ed.), *Food Systems Sustainability and Environmental Policies in Modern Economies* (pp. 15–31). Hershey, PA: IGI Global. doi:10.4018/978-1-5225-3631-4.ch002

Gonzalez-Feliu, J. (2018). Sustainability Evaluation of Green Urban Logistics Systems: Literature Overview and Proposed Framework. In A. Paul, D. Bhattacharyya, & S. Anand (Eds.), *Green Initiatives for Business Sustainability and Value Creation* (pp. 103–134). Hershey, PA: IGI Global. doi:10.4018/978-1-5225-2662-9.ch005

Goodland, R. (2017). A Fresh Look at Livestock Greenhouse Gas Emissions and Mitigation. In I. Management Association (Ed.), Natural Resources Management: Concepts, Methodologies, Tools, and Applications (pp. 124-139). Hershey, PA: IGI Global. doi:10.4018/978-1-5225-0803-8.ch006

Goundar, S., & Appana, S. (2018). Mainstreaming Development Policies for Climate Change in Fiji: A Policy Gap Analysis and the Role of ICTs. In I. Management Association (Ed.), Sustainable Development: Concepts, Methodologies, Tools, and Applications (pp. 402-432). Hershey, PA: IGI Global. doi:10.4018/978-1-5225-3817-2.ch020

Granell-Canut, C., & Aguilar-Moreno, E. (2018). Geospatial Influence in Science Mapping. In M. Khosrow-Pour, D.B.A. (Ed.), Encyclopedia of Information Science and Technology, Fourth Edition (pp. 3473-3483). Hershey, PA: IGI Global. doi:10.4018/978-1-5225-2255-3.ch302

Grigelis, A., Blažauskas, N., Gelumbauskaitė, L. Ž., Gulbinskas, S., Suzdalev, S., & Ferrarin, C. (2017). Marine Environment Data Management Related to the Human Activity in the South-Eastern Baltic Sea (The Lithuanian Sector). In P. Diviacco, A. Leadbetter, & H. Glaves (Eds.), *Oceanographic and Marine Cross-Domain Data Management for Sustainable Development* (pp. 282–302). Hershey, PA: IGI Global. doi:10.4018/978-1-5225-0700-0.ch012

Guma, I. P., Rwashana, A. S., & Oyo, B. (2018). Food Security Policy Analysis Using System Dynamics: The Case of Uganda. *International Journal of Information Technologies and Systems Approach, 11*(1), 72–90. doi:10.4018/IJITSA.2018010104

Guma, I. P., Rwashana, A. S., & Oyo, B. (2018). Food Security Indicators for Subsistence Farmers Sustainability: A System Dynamics Approach. *International Journal of System Dynamics Applications, 7*(1), 45–64. doi:10.4018/IJSDA.2018010103

Gupta, P., & Goyal, S. (2017). Wildlife Habitat Evaluation. In A. Santra & S. Mitra (Eds.), *Remote Sensing Techniques and GIS Applications in Earth and Environmental Studies* (pp. 258–264). Hershey, PA: IGI Global. doi:10.4018/978-1-5225-1814-3.ch013

Hanson, T., & Hildebrand, E. (2018). GPS Travel Diaries in Rural Transportation Research: A Focus on Older Drivers. In I. Management Association (Ed.), Intelligent Transportation and Planning: Breakthroughs in Research and Practice (pp. 609-625). Hershey, PA: IGI Global. doi:10.4018/978-1-5225-5210-9.ch027

Hartman, M. B. (2017). Research-Based Climate Change Public Education Programs. In I. Management Association (Ed.), Natural Resources Management: Concepts, Methodologies, Tools, and Applications (pp. 992-1003). Hershey, PA: IGI Global. doi:10.4018/978-1-5225-0803-8.ch046

Hashim, N. (2017). Zanzibari Seaweed: Global Climate Change and the Promise of Adaptation. In I. Management Association (Ed.), Natural Resources Management: Concepts, Methodologies, Tools, and Applications (pp. 365-391). Hershey, PA: IGI Global. https://doi.org/ doi:10.4018/978-1-5225-0803-8.ch019

Herrera, J. E., Argüello, L. V., Gonzalez-Feliu, J., & Jaimes, W. A. (2017). Decision Support System Design Requirements, Information Management, and Urban Logistics Efficiency: Case Study of Bogotá, Colombia. In G. Jamil, A. Soares, & C. Pessoa (Eds.), *Handbook of Research on Information Management for Effective Logistics and Supply Chains* (pp. 223–238). Hershey, PA: IGI Global. doi:10.4018/978-1-5225-0973-8.ch012

Huizinga, T., Ayanso, A., Smoor, M., & Wronski, T. (2017). Exploring Insurance and Natural Disaster Tweets Using Text Analytics. *International Journal of Business Analytics, 4*(1), 1–17. doi:10.4018/IJBAN.2017010101

Hung, K., Kalantari, M., & Rajabifard, A. (2017). An Integrated Method for Assessing the Text Content Quality of Volunteered Geographic Information in Disaster Management. *International Journal of Information Systems for Crisis Response and Management, 9*(2), 1–17. doi:10.4018/IJISCRAM.2017040101

Husnain, A., & Avdic, A. (2018). Identifying the Contemporary Status of E-Service Sustainability Research. In I. Management Association (Ed.), Sustainable Development: Concepts, Methodologies, Tools, and Applications (pp. 467-485). Hershey, PA: IGI Global. https://doi.org/ doi:10.4018/978-1-5225-3817-2.ch022

Iarossi, M. P., & Ferro, L. (2017). "The Past is Never Dead. It's Not Even Past": Virtual Archaeological Promenade. In A. Ippolito & M. Cigola (Eds.), *Handbook of Research on Emerging Technologies for Digital Preservation and Information Modeling* (pp. 228–255). Hershey, PA: IGI Global. doi:10.4018/978-1-5225-0680-5.ch010

Ignjatijević, S., & Cvijanović, D. (2018). Analysis of Serbian Honey Production and Exports. In *Exploring the Global Competitiveness of Agri-Food Sectors and Serbia's Dominant Presence: Emerging Research and Opportunities* (pp. 109–139). Hershey, PA: IGI Global. doi:10.4018/978-1-5225-2762-6.ch005

Jana, S. K., & Karmakar, A. K. (2017). Globalization, Governance, and Food Security: The Case of BRICS. In I. Management Association (Ed.), Natural Resources Management: Concepts, Methodologies, Tools, and Applications (pp. 692-712). Hershey, PA: IGI Global. https://doi.org/ doi:10.4018/978-1-5225-0803-8.ch032

Jana, S. K., & Karmakar, A. K. (2017). Food Security in Asia: Is There Convergence? In I. Management Association (Ed.), Natural Resources Management: Concepts, Methodologies, Tools, and Applications (pp. 109-123). Hershey, PA: IGI Global. https://doi.org/ doi:10.4018/978-1-5225-0803-8.ch005

John, J., & Kumar, S. (2018). A Locational Decision Making Framework for Shipbreaking Under Multiple Criteria. In I. Management Association (Ed.), Operations and Service Management: Concepts, Methodologies, Tools, and Applications (pp. 504-527). Hershey, PA: IGI Global. doi:10.4018/978-1-5225-3909-4.ch024

John, J., & Srivastava, R. K. (2018). Decision Insights for Shipbreaking using Environmental Impact Assessment: Review and Perspectives. *International Journal of Strategic Decision Sciences*, *9*(1), 45–62. doi:10.4018/IJSDS.2018010104

Joshi, Y., & Rahman, Z. (2018). Determinants of Sustainable Consumption Behaviour: Review and Conceptual Framework. In A. Paul, D. Bhattacharyya, & S. Anand (Eds.), *Green Initiatives for Business Sustainability and Value Creation* (pp. 239–262). Hershey, PA: IGI Global. doi:10.4018/978-1-5225-2662-9.ch011

Juma, D. W., Reuben, M., Wang, H., & Li, F. (2018). Adaptive Coevolution: Realigning the Water Governance Regime to the Changing Climate. In I. Management Association (Ed.), Hydrology and Water Resource Management: Breakthroughs in Research and Practice (pp. 346-357). Hershey, PA: IGI Global. https://doi.org/ doi:10.4018/978-1-5225-3427-3.ch014

K., S., & Tripathy, B. K. (2018). Neighborhood Rough-Sets-Based Spatial Data Analytics. In M. Khosrow-Pour, D.B.A. (Ed.), *Encyclopedia of Information Science and Technology, Fourth Edition* (pp. 1835-1844). Hershey, PA: IGI Global. https://doi.org/ doi:10.4018/978-1-5225-2255-3.ch160

Kabir, F. (2018). Towards a More Gender-Inclusive Climate Change Policy. In N. Mahtab, T. Haque, I. Khan, M. Islam, & I. Wahid (Eds.), *Handbook of Research on Women's Issues and Rights in the Developing World* (pp. 354–369). Hershey, PA: IGI Global. doi:10.4018/978-1-5225-3018-3.ch022

Kabir, F. (2018). Towards a More Gender-Inclusive Climate Change Policy. In I. Management Association (Ed.), Climate Change and Environmental Concerns: Breakthroughs in Research and Practice (pp. 525-540). Hershey, PA: IGI Global. https://doi.org/ doi:10.4018/978-1-5225-5487-5.ch027

Kanyamuka, J. S., Jumbe, C. B., & Ricker-Gilbert, J. (2018). Making Agricultural Input Subsidies More Effective and Profitable in Africa: The Role of Complementary Interventions. In A. Obayelu (Ed.), *Food Systems Sustainability and Environmental Policies in Modern Economies* (pp. 172–187). Hershey, PA: IGI Global. doi:10.4018/978-1-5225-3631-4.ch008

Karimi, H., & Gholamrezafahimi, F. (2017). Study of Integrated Coastal Zone Management and Its Environmental Effects: A Case of Iran. In R. Singh, A. Singh, & V. Srivastava (Eds.), *Environmental Issues Surrounding Human Overpopulation* (pp. 64–88). Hershey, PA: IGI Global. doi:10.4018/978-1-5225-1683-5.ch004

Kaya, I. R., Hutabarat, J., & Bambang, A. N. (2018). "Sasi": A New Path to Sustain Seaweed Farming From Up-Stream to Down-Stream in Kotania Bay, Molucass. *International Journal of Social Ecology and Sustainable Development*, *9*(2), 28–36. doi:10.4018/IJSESD.2018040103

Khader, V. (2018). Technologies for Food, Health, Livelihood, and Nutrition Security. In I. Management Association (Ed.), Food Science and Nutrition: Breakthroughs in Research and Practice (pp. 94-112). Hershey, PA: IGI Global. https://doi.org/ doi:10.4018/978-1-5225-5207-9.ch005

Kocadağlı, A. Y. (2017). The Temporal and Spatial Development of Organic Agriculture in Turkey. In W. Ganpat, R. Dyer, & W. Isaac (Eds.), *Agricultural Development and Food Security in Developing Nations* (pp. 130–156). Hershey, PA: IGI Global. doi:10.4018/978-1-5225-0942-4.ch006

Koundouri, P., Giannouli, A., & Souliotis, I. (2017). An Integrated Approach for Sustainable Environmental and Socio-Economic Development Using Offshore Infrastructure. In I. Management Association (Ed.), Renewable and Alternative Energy: Concepts, Methodologies, Tools, and Applications (pp. 1581-1601). Hershey, PA: IGI Global. doi:10.4018/978-1-5225-1671-2.ch056

Kumar, A., & Dash, M. K. (2017). Sustainability and Future Generation Infrastructure on Digital Platform: A Study of Generation Y. In N. Ray (Ed.), *Business Infrastructure for Sustainability in Developing Economies* (pp. 124–142). Hershey, PA: IGI Global. doi:10.4018/978-1-5225-2041-2.ch007

Kumar, A., Mukherjee, A. B., & Krishna, A. P. (2017). Application of Conventional Data Mining Techniques and Web Mining to Aid Disaster Management. In A. Kumar (Ed.), *Web Usage Mining Techniques and Applications Across Industries* (pp. 138–167). Hershey, PA: IGI Global. doi:10.4018/978-1-5225-0613-3.ch006

Kumar, C. P. (2017). Impact of Climate Change on Groundwater Resources. In I. Management Association (Ed.), Natural Resources Management: Concepts, Methodologies, Tools, and Applications (pp. 1094-1120). Hershey, PA: IGI Global. doi:10.4018/978-1-5225-0803-8.ch052

Kumari, S., & Patil, Y. (2017). Achieving Climate Smart Agriculture with a Sustainable Use of Water: A Conceptual Framework for Sustaining the Use of Water for Agriculture in the Era of Climate Change. In P. Rao & Y. Patil (Eds.), *Reconsidering the Impact of Climate Change on Global Water Supply, Use, and Management* (pp. 122–143). Hershey, PA: IGI Global. doi:10.4018/978-1-5225-1046-8.ch008

Kumari, S., & Patil, Y. (2018). Achieving Climate Smart Agriculture With a Sustainable Use of Water: A Conceptual Framework for Sustaining the Use of Water for Agriculture in the Era of Climate Change. In I. Management Association (Ed.), Climate Change and Environmental Concerns: Breakthroughs in Research and Practice (pp. 111-133). Hershey, PA: IGI Global. doi:10.4018/978-1-5225-5487-5.ch006

Kursah, M. B. (2017). Least-Cost Pipeline using Geographic Information System: The Limit to Technicalities. *International Journal of Applied Geospatial Research*, *8*(3), 1–15. doi:10.4018/ijagr.2017070101

Lahiri, S., Ghosh, D., & Bhakta, J. N. (2017). Role of Microbes in Eco-Remediation of Perturbed Aquatic Ecosystem. In J. Bhakta (Ed.), *Handbook of Research on Inventive Bioremediation Techniques* (pp. 70–107). Hershey, PA: IGI Global. doi:10.4018/978-1-5225-2325-3.ch004

Lallo, C. H., Smalling, S., Facey, A., & Hughes, M. (2018). The Impact of Climate Change on Small Ruminant Performance in Caribbean Communities. In I. Management Association (Ed.), Climate Change and Environmental Concerns: Breakthroughs in Research and Practice (pp. 193-218). Hershey, PA: IGI Global. doi:10.4018/978-1-5225-5487-5.ch010

Laurini, R. (2017). Nature of Geographic Knowledge Bases. In S. Faiz & K. Mahmoudi (Eds.), *Handbook of Research on Geographic Information Systems Applications and Advancements* (pp. 29–60). Hershey, PA: IGI Global. doi:10.4018/978-1-5225-0937-0.ch002

Lawrence, J., Simpson, L., & Piggott, A. (2017). Protected Agriculture: A Climate Change Adaptation for Food and Nutrition Security. In I. Management Association (Ed.), Natural Resources Management: Concepts, Methodologies, Tools, and Applications (pp. 140-158). Hershey, PA: IGI Global. doi:10.4018/978-1-5225-0803-8.ch007

Leadbetter, A., Cheatham, M., Shepherd, A., & Thomas, R. (2017). Linked Ocean Data 2.0. In P. Diviacco, A. Leadbetter, & H. Glaves (Eds.), *Oceanographic and Marine Cross-Domain Data Management for Sustainable Development* (pp. 69–99). Hershey, PA: IGI Global. doi:10.4018/978-1-5225-0700-0.ch004

Lucas, M. R., Rego, C., Vieira, C., & Vieira, I. (2017). Proximity and Cooperation for Innovative Regional Development: The Case of the Science and Technology Park of Alentejo. In L. Carvalho (Ed.), *Handbook of Research on Entrepreneurial Development and Innovation Within Smart Cities* (pp. 199–228). Hershey, PA: IGI Global. doi:10.4018/978-1-5225-1978-2.ch010

Ma, X., Beaulieu, S. E., Fu, L., Fox, P., Di Stefano, M., & West, P. (2017). Documenting Provenance for Reproducible Marine Ecosystem Assessment in Open Science. In P. Diviacco, A. Leadbetter, & H. Glaves (Eds.), *Oceanographic and Marine Cross-Domain Data Management for Sustainable Development* (pp. 100–126). Hershey, PA: IGI Global. doi:10.4018/978-1-5225-0700-0.ch005

Mabe, L. K., & Oladele, O. I. (2017). Application of Information Communication Technologies for Agricultural Development through Extension Services: A Review. In T. Tossy (Ed.), *Information Technology Integration for Socio-Economic Development* (pp. 52–101). Hershey, PA: IGI Global. doi:10.4018/978-1-5225-0539-6.ch003

Malomo, B. I. (2018). A Review of Psychological Resilience as a Response to Natural Hazards in Nigeria. In A. Eneanya (Ed.), *Handbook of Research on Environmental Policies for Emergency Management and Public Safety* (pp. 147–165). Hershey, PA: IGI Global. doi:10.4018/978-1-5225-3194-4.ch008

Manchiraju, S. (2018). Predicting Behavioral Intentions Toward Sustainable Fashion Consumption: A Comparison of Attitude-Behavior and Value-Behavior Consistency Models. In I. Management Association (Ed.), Fashion and Textiles: Breakthroughs in Research and Practice (pp. 1-21). Hershey, PA: IGI Global. doi:10.4018/978-1-5225-3432-7.ch001

Manzella, G. M., Bartolini, R., Bustaffa, F., D'Angelo, P., De Mattei, M., Frontini, F., ... Spada, A. (2017). Semantic Search Engine for Data Management and Sustainable Development: Marine Planning Service Platform. In P. Diviacco, A. Leadbetter, & H. Glaves (Eds.), *Oceanographic and Marine Cross-Domain Data Management for Sustainable Development* (pp. 127–154). Hershey, PA: IGI Global. doi:10.4018/978-1-5225-0700-0.ch006

Mbonigaba, J. (2018). Comparing the Effects of Unsustainable Production and Consumption of Food on Health and Policy Across Developed and Less Developed Countries. In A. Obayelu (Ed.), *Food Systems Sustainability and Environmental Policies in Modern Economies* (pp. 124–158). Hershey, PA: IGI Global. doi:10.4018/978-1-5225-3631-4.ch006

McKeown, A. E. (2017). Nurses, Healthcare, and Environmental Pollution and Solutions: Breaking the Cycle of Harm. In I. Management Association (Ed.), Natural Resources Management: Concepts, Methodologies, Tools, and Applications (pp. 392-415). Hershey, PA: IGI Global. https://doi.org/ doi:10.4018/978-1-5225-0803-8.ch020

Mili, B., Barua, A., & Katyaini, S. (2017). Climate Change and Adaptation through the Lens of Capability Approach: A Case Study from Darjeeling, Eastern Himalaya. In I. Management Association (Ed.), Natural Resources Management: Concepts, Methodologies, Tools, and Applications (pp. 1351-1365). Hershey, PA: IGI Global. https://doi.org/ doi:10.4018/978-1-5225-0803-8.ch064

Mir, S. A., Shah, M. A., Mir, M. M., & Iqbal, U. (2017). New Horizons of Nanotechnology in Agriculture and Food Processing Industry. In B. Nayak, A. Nanda, & M. Bhat (Eds.), *Integrating Biologically-Inspired Nanotechnology into Medical Practice* (pp. 230–258). Hershey, PA: IGI Global. doi:10.4018/978-1-5225-0610-2.ch009

Moallem, M., Sterrett, W. L., Gordon, C. R., Sukhera, S. M., Mahmood, A., & Bashir, A. (2021). An Investigation of the Effects of Integrating Computing and Project- or Problem-Based Learning in the Context of Environmental Sciences: A Case of Pakistani STEM Teachers. In S. Schroth & J. Daniels (Eds.), *Building STEM Skills Through Environmental Education* (pp. 49–89). IGI Global. https://doi.org/10.4018/978-1-7998-2711-5.ch003

Mujere, N., & Moyce, W. (2017). Climate Change Impacts on Surface Water Quality. In W. Ganpat & W. Isaac (Eds.), *Environmental Sustainability and Climate Change Adaptation Strategies* (pp. 322–340). Hershey, PA: IGI Global. doi:10.4018/978-1-5225-1607-1.ch012

Mukherjee, A. B., Krishna, A. P., & Patel, N. (2018). Geospatial Technology for Urban Sciences. In *Geospatial Technologies in Urban System Development: Emerging Research and Opportunities* (pp. 99–120). Hershey, PA: IGI Global. doi:10.4018/978-1-5225-3683-3.ch005

Nagarajan, S. K., & Sangaiah, A. K. (2017). Vegetation Index: Ideas, Methods, Influences, and Trends. In N. Kumar, A. Sangaiah, M. Arun, & S. Anand (Eds.), *Advanced Image Processing Techniques and Applications* (pp. 347–386). Hershey, PA: IGI Global. doi:10.4018/978-1-5225-2053-5.ch016

Naraine, L., & Meehan, K. (2017). Strengthening Food Security with Sustainable Practices by Smallholder Farmers in Lesser Developed Economies. In W. Ganpat, R. Dyer, & W. Isaac (Eds.), *Agricultural Development and Food Security in Developing Nations* (pp. 57–81). Hershey, PA: IGI Global. doi:10.4018/978-1-5225-0942-4.ch003

Naraine, L., & Meehan, K. (2017). Strengthening Food Security with Sustainable Practices by Smallholder Farmers in Lesser Developed Economies. In W. Ganpat, R. Dyer, & W. Isaac (Eds.), *Agricultural Development and Food Security in Developing Nations* (pp. 57–81). Hershey, PA: IGI Global. doi:10.4018/978-1-5225-0942-4.ch003

Naraine, L., & Meehan, K. (2017). Strengthening Food Security with Sustainable Practices by Smallholder Farmers in Lesser Developed Economies. In W. Ganpat, R. Dyer, & W. Isaac (Eds.), *Agricultural Development and Food Security in Developing Nations* (pp. 57–81). Hershey, PA: IGI Global. doi:10.4018/978-1-5225-0942-4.ch003

Nikolaou, K., Tsakiridou, E., Anastasiadis, F., & Mattas, K. (2017). Exploring Alternative Distribution Channels of Agricultural Products. *International Journal of Food and Beverage Manufacturing and Business Models*, *2*(2), 36–66. doi:10.4018/IJFBMBM.2017070103

Nishat, K. J., & Rahman, M. S. (2018). Disaster, Vulnerability, and Violence Against Women: Global Findings and a Research Agenda for Bangladesh. In N. Mahtab, T. Haque, I. Khan, M. Islam, & I. Wahid (Eds.), *Handbook of Research on Women's Issues and Rights in the Developing World* (pp. 235–250). Hershey, PA: IGI Global. doi:10.4018/978-1-5225-3018-3.ch014

O'Hara, S., Jones, D., & Trobman, H. B. (2018). Building an Urban Food System Through UDC Food Hubs. In A. Burtin, J. Fleming, & P. Hampton-Garland (Eds.), *Changing Urban Landscapes Through Public Higher Education* (pp. 116–143). Hershey, PA: IGI Global. doi:10.4018/978-1-5225-3454-9.ch006

Obayelu, A. E. (2018). Integrating Environment, Food Systems, and Sustainability in Feeding the Growing Population in Developing Countries. In A. Obayelu (Ed.), *Food Systems Sustainability and Environmental Policies in Modern Economies* (pp. 1–14). Hershey, PA: IGI Global. doi:10.4018/978-1-5225-3631-4.ch001

Othman, R., Nath, N., & Laswad, F. (2018). Environmental Reporting and Accounting: Sustainability Hybridisation. In G. Azevedo, J. da Silva Oliveira, R. Marques, & A. Ferreira (Eds.), *Handbook of Research on Modernization and Accountability in Public Sector Management* (pp. 130–158). Hershey, PA: IGI Global. doi:10.4018/978-1-5225-3731-1.ch007

Ouadi, A., & Zitouni, A. (2021). Phasor Measurement Improvement Using Digital Filter in a Smart Grid. In A. Recioui & H. Bentarzi (Eds.), *Optimizing and Measuring Smart Grid Operation and Control* (pp. 100–117). IGI Global. https://doi.org/10.4018/978-1-7998-4027-5.ch005

Padigala, B. S. (2018). Traditional Water Management System for Climate Change Adaptation in Mountain Ecosystems. In I. Management Association (Ed.), Climate Change and Environmental Concerns: Breakthroughs in Research and Practice (pp. 630-655). Hershey, PA: IGI Global. doi:10.4018/978-1-5225-5487-5.ch033

Panda, C. K. (2018). Mobile Phone Usage in Agricultural Extension in India: The Current and Future Perspective. In F. Mtenzi, G. Oreku, D. Lupiana, & J. Yonazi (Eds.), *Mobile Technologies and Socio-Economic Development in Emerging Nations* (pp. 1–21). Hershey, PA: IGI Global. doi:10.4018/978-1-5225-4029-8.ch001

Pandian, S. L., Yarrakula, K., & Chaudhury, P. (2018). GIS-Based Decision Support System for Village Level: A Case Study in Andhra Pradesh. In S. Chaudhuri & N. Ray (Eds.), *GIS Applications in the Tourism and Hospitality Industry* (pp. 275–295). Hershey, PA: IGI Global. doi:10.4018/978-1-5225-5088-4.ch012

Quaranta, G., & Salvia, R. (2017). Social-Based Product Innovation and Governance in The Milk Sector: The Case of Carciocacio and Innonatura. In T. Tarnanidis, M. Vlachopoulou, & J. Papathanasiou (Eds.), *Driving Agribusiness With Technology Innovations* (pp. 293–310). Hershey, PA: IGI Global. doi:10.4018/978-1-5225-2107-5.ch015

Rahman, M. K., Schmidlin, T. W., Munro-Stasiuk, M. J., & Curtis, A. (2017). Geospatial Analysis of Land Loss, Land Cover Change, and Landuse Patterns of Kutubdia Island, Bangladesh. *International Journal of Applied Geospatial Research*, *8*(2), 45–60. doi:10.4018/IJAGR.2017040104

Rajack-Talley, T. A. (2017). Agriculture, Trade Liberalization and Poverty in the ACP Countries. In W. Ganpat, R. Dyer, & W. Isaac (Eds.), *Agricultural Development and Food Security in Developing Nations* (pp. 1–20). Hershey, PA: IGI Global. doi:10.4018/978-1-5225-0942-4.ch001

Rajamanickam, S. (2018). Exploring Landscapes in Regional Convergence: Environment and Sustainable Development in South Asia. In I. Management Association (Ed.), Sustainable Development: Concepts, Methodologies, Tools, and Applications (pp. 1051-1087). Hershey, PA: IGI Global. https://doi.org/ doi:10.4018/978-1-5225-3817-2.ch047

Rizvi, S. M., & Dearden, A. (2018). KHETI: ICT Solution for Agriculture Extension and Its Replication in Open and Distance Learning. In U. Pandey & V. Indrakanti (Eds.), *Open and Distance Learning Initiatives for Sustainable Development* (pp. 163–174). Hershey, PA: IGI Global. doi:10.4018/978-1-5225-2621-6.ch008

Roşu, L., & Macarov, L. I. (2017). Management of Drought and Floods in the Dobrogea Region. In I. Management Association (Ed.), Agri-Food Supply Chain Management: Breakthroughs in Research and Practice (pp. 372-403). Hershey, PA: IGI Global. https://doi.org/ doi:10.4018/978-1-5225-1629-3.ch016

Rouzbehani, K., & Rouzbehani, S. (2018). Mapping Women's World: GIS and the Case of Breast Cancer in the US. *International Journal of Public Health Management and Ethics*, *3*(1), 14–25. doi:10.4018/IJPHME.2018010102

Roy, D. (2018). Success Factors of Adoption of Mobile Applications in Rural India: Effect of Service Characteristics on Conceptual Model. In M. Khosrow-Pour, D.B.A. (Ed.), Green Computing Strategies for Competitive Advantage and Business Sustainability (pp. 211-238). Hershey, PA: IGI Global. https:// doi.org/ doi:10.4018/978-1-5225-5017-4.ch010

Sajeva, M., Lemon, M., & Sahota, P. S. (2017). Governance for Food Security: A Framework for Social Learning and Scenario Building. *International Journal of Food and Beverage Manufacturing and Business Models*, *2*(2), 67–84. doi:10.4018/IJFBMBM.2017070104

Sambhanthan, A., & Potdar, V. (2017). A Study of the Parameters Impacting Sustainability in Information Technology Organizations. *International Journal of Knowledge-Based Organizations*, *7*(3), 27–39. doi:10.4018/IJKBO.2017070103

Sanga, C., Kalungwizi, V. J., & Msuya, C. P. (2017). Bridging Gender Gaps in Provision of Agricultural Extension Service Using ICT: Experiences from Sokoine University of Agriculture (SUA) Farmer Voice Radio (FVR) Project in Tanzania. In I. Management Association (Ed.), Discrimination and Diversity: Concepts, Methodologies, Tools, and Applications (pp. 682-697). Hershey, PA: IGI Global. https://doi.org/ doi:10.4018/978-1-5225-1933-1.ch031

Santra, A., & Mitra, D. (2017). Role of Remote Sensing in Potential Fishing Zone Forecast. In A. Santra & S. Mitra (Eds.), *Remote Sensing Techniques and GIS Applications in Earth and Environmental Studies* (pp. 243–257). Hershey, PA: IGI Global. doi:10.4018/978-1-5225-1814-3.ch012

Schaap, D. (2017). SeaDataNet: Towards a Pan-European Infrastructure for Marine and Ocean Data Management. In P. Diviacco, A. Leadbetter, & H. Glaves (Eds.), *Oceanographic and Marine Cross-Domain Data Management for Sustainable Development* (pp. 155–177). Hershey, PA: IGI Global. doi:10.4018/978-1-5225-0700-0.ch007

Seckin-Celik, T. (2017). Sustainability Reporting and Sustainability in the Turkish Business Context. In U. Akkucuk (Ed.), *Ethics and Sustainability in Global Supply Chain Management* (pp. 115–132). Hershey, PA: IGI Global. doi:10.4018/978-1-5225-2036-8.ch006

Segbefia, A. Y., Barnes, V. R., Akpalu, L. A., & Mensah, M. (2018). Environmental Location Assessment for Seaweed Cultivation in Ghana: A Spatial Multi-Criteria Approach. *International Journal of Applied Geospatial Research*, *9*(1), 51–64. doi:10.4018/IJAGR.2018010104

Sen, Y. (2018). How to Manage Sustainability: A Framework for Corporate Sustainability Tools. In I. Management Association (Ed.), Sustainable Development: Concepts, Methodologies, Tools, and Applications (pp. 568-589). Hershey, PA: IGI Global. doi:10.4018/978-1-5225-3817-2.ch026

Shamshiry, E., Abdulai, A. M., Mokhtar, M. B., & Komoo, I. (2015). Regional Landfill Site Selection with GIS and Analytical Hierarchy Process Techniques: A Case Study of Langkawi Island, Malaysia. In P. Thomas, M. Srihari, & S. Kaur (Eds.), *Handbook of Research on Cultural and Economic Impacts of the Information Society* (pp. 248–282). Hershey, PA: IGI Global. doi:10.4018/978-1-4666-8598-7.ch011

Sharma, Y. K., Mangla, S. K., Patil, P. P., & Uniyal, S. (2018). Analyzing Sustainable Food Supply Chain Management Challenges in India. In M. Ram & J. Davim (Eds.), *Soft Computing Techniques and Applications in Mechanical Engineering* (pp. 162–180). Hershey, PA: IGI Global. doi:10.4018/978-1-5225-3035-0.ch008

Silvestrelli, P. (2018). The Impact of Events: To Which Extent Are They Sustainable for Tourist Destinations? Some Evidences From Expo Milano 2015. In M. Risso & S. Testarmata (Eds.), *Value Sharing for Sustainable and Inclusive Development* (pp. 185–204). Hershey, PA: IGI Global. doi:10.4018/978-1-5225-3147-0.ch008

Silvius, G. (2018). Sustainability Evaluation of IT/IS Projects. In I. Management Association (Ed.), Sustainable Development: Concepts, Methodologies, Tools, and Applications (pp. 26-40). Hershey, PA: IGI Global. https://doi.org/ doi:10.4018/978-1-5225-3817-2.ch002

Singh, R., Srivastava, P., Singh, P., Upadhyay, S., & Raghubanshi, A. S. (2017). Human Overpopulation and Food Security: Challenges for the Agriculture Sustainability. In R. Singh, A. Singh, & V. Srivastava (Eds.), *Environmental Issues Surrounding Human Overpopulation* (pp. 12–39). Hershey, PA: IGI Global. doi:10.4018/978-1-5225-1683-5.ch002

Singh, R., Srivastava, P., Singh, P., Upadhyay, S., & Raghubanshi, A. S. (2017). Human Overpopulation and Food Security: Challenges for the Agriculture Sustainability. In R. Singh, A. Singh, & V. Srivastava (Eds.), *Environmental Issues Surrounding Human Overpopulation* (pp. 12–39). Hershey, PA: IGI Global. doi:10.4018/978-1-5225-1683-5.ch002

Srivastava, N. (2017). Climate Change Mitigation: Collective Efforts and Responsibly. In I. Management Association (Ed.), Natural Resources Management: Concepts, Methodologies, Tools, and Applications (pp. 64-76). Hershey, PA: IGI Global. https://doi.org/ doi:10.4018/978-1-5225-0803-8.ch003

Stanganelli, M., & Gerundo, C. (2017). Understanding the Role of Urban Morphology and Green Areas Configuration During Heat Waves. *International Journal of Agricultural and Environmental Information Systems*, *8*(2), 50–64. doi:10.4018/IJAEIS.2017040104

Stewart, M. K., Hagood, D., & Ching, C. C. (2017). Virtual Games and Real-World Communities: Environments that Constrain and Enable Physical Activity in Games for Health. *International Journal of Game-Based Learning*, *7*(1), 1–19. doi:10.4018/IJGBL.2017010101

Stone, R. J. (2017). Modelling the Frequency of Tropical Cyclones in the Lower Caribbean Region. In W. Ganpat & W. Isaac (Eds.), *Environmental Sustainability and Climate Change Adaptation Strategies* (pp. 341–349). Hershey, PA: IGI Global. doi:10.4018/978-1-5225-1607-1.ch013

Syed, A., & Jabeen, U. A. (2018). Climate Change Impact on Agriculture and Food Security. In A. Eneanya (Ed.), *Handbook of Research on Environmental Policies for Emergency Management and Public Safety* (pp. 223–237). Hershey, PA: IGI Global. doi:10.4018/978-1-5225-3194-4.ch012

Tam, G. C. (2017). The Global View of Sustainability. In *Managerial Strategies and Green Solutions for Project Sustainability* (pp. 1–24). Hershey, PA: IGI Global. doi:10.4018/978-1-5225-2371-0.ch001

Tam, G. C. (2017). Understanding Project Sustainability. In *Managerial Strategies and Green Solutions for Project Sustainability* (pp. 110–139). Hershey, PA: IGI Global. doi:10.4018/978-1-5225-2371-0.ch005

Tam, G. C. (2017). Perspectives on Sustainability. In *Managerial Strategies and Green Solutions for Project Sustainability* (pp. 53–76). Hershey, PA: IGI Global. doi:10.4018/978-1-5225-2371-0.ch003

Tang, M., & Karunanithi, A. T. (2018). Visual Logic Maps (vLms). In *Advanced Concept Maps in STEM Education: Emerging Research and Opportunities* (pp. 108–149). Hershey, PA: IGI Global. doi:10.4018/978-1-5225-2184-6.ch005

Taşçıoğlu, M., & Yener, D. (2018). The Value and Scope of GIS in Marketing and Tourism Management. In S. Chaudhuri & N. Ray (Eds.), *GIS Applications in the Tourism and Hospitality Industry* (pp. 189–211). Hershey, PA: IGI Global. doi:10.4018/978-1-5225-5088-4.ch009

Thanh Tung, B. (2021). Pharmacology and Therapeutic Applications of Resveratrol. In A. Hussain & S. Behl (Eds.), *Treating Endocrine and Metabolic Disorders With Herbal Medicines* (pp. 321–333). IGI Global. https://doi.org/10.4018/978-1-7998-4808-0.ch014

Tianming, G. (2018). Food Security and Rural Development on Emerging Markets of Northeast Asia: Cases of Chinese North and Russian Far East. In V. Erokhin (Ed.), *Establishing Food Security and Alternatives to International Trade in Emerging Economies* (pp. 155–176). Hershey, PA: IGI Global. doi:10.4018/978-1-5225-2733-6.ch008

Tiftikçigil, B. Y., Yaşgül, Y. S., & Güriş, B. (2017). Sustainability of Foreign Trade Deficit in Energy: The Case of Turkey. In N. Ray (Ed.), *Business Infrastructure for Sustainability in Developing Economies* (pp. 94–109). Hershey, PA: IGI Global. doi:10.4018/978-1-5225-2041-2.ch005

Tiwari, S., Vaish, B., & Singh, P. (2018). Population and Global Food Security: Issues Related to Climate Change. In I. Management Association (Ed.), Climate Change and Environmental Concerns: Breakthroughs in Research and Practice (pp. 41-64). Hershey, PA: IGI Global. https://doi.org/ doi:10.4018/978-1-5225-5487-5.ch003

Toujani, A., & Achour, H. (2018). A Data Mining Framework for Forest Fire Mapping. In I. Management Association (Ed.), Information Retrieval and Management: Concepts, Methodologies, Tools, and Applications (pp. 771-794). Hershey, PA: IGI Global. doi:10.4018/978-1-5225-5191-1.ch033

Tripathy, B., & K., S. B. (2017). Rough Fuzzy Set Theory and Neighbourhood Approximation Based Modelling for Spatial Epidemiology. In I. Management Association (Ed.), *Public Health and Welfare: Concepts, Methodologies, Tools, and Applications* (pp. 1257-1268). Hershey, PA: IGI Global. https://doi.org/ doi:10.4018/978-1-5225-1674-3.ch058

Trukhachev, A. (2017). New Approaches to Regional Branding through Green Production and Utilization of Existing Natural Advantages. In I. Management Association (Ed.), Advertising and Branding: Concepts, Methodologies, Tools, and Applications (pp. 1758-1778). Hershey, PA: IGI Global. doi:10.4018/978-1-5225-1793-1.ch081

Tsobanoglou, G. O., & Vlachopoulou, E. I. (2017). Social-Ecological Systems in Local Fisheries Communities. In G. Korres, E. Kourliouros, & M. Michailidis (Eds.), *Handbook of Research on Policies and Practices for Sustainable Economic Growth and Regional Development* (pp. 306–316). Hershey, PA: IGI Global. doi:10.4018/978-1-5225-2458-8.ch026

Tuydes-Yaman, H., & Karatas, P. (2018). Evaluation of Walkability and Pedestrian Level of Service. In I. Management Association (Ed.), Intelligent Transportation and Planning: Breakthroughs in Research and Practice (pp. 264-291). Hershey, PA: IGI Global. doi:10.4018/978-1-5225-5210-9.ch012

Uddin, S., Chakravorty, S., Ray, A., & Sherpa, K. S. (2018). Optimal Location of Sub-Station Using Q-GIS and Multi-Criteria Decision Making Approach. *International Journal of Decision Support System Technology*, *10*(2), 65–79. doi:10.4018/IJDSST.2018040104

Uddin, S., Chakravorty, S., Sherpa, K. S., & Ray, A. (2018). Power Distribution System Planning Using Q-GIS. *International Journal of Energy Optimization and Engineering*, *7*(2), 61–75. doi:10.4018/IJEOE.2018040103

Uzun, F. V. (2018). Natural Resources Management. In A. Eneanya (Ed.), *Handbook of Research on Environmental Policies for Emergency Management and Public Safety* (pp. 1–21). Hershey, PA: IGI Global. doi:10.4018/978-1-5225-3194-4.ch001

V, M., Agrawal, R., Sharma, V., & T.N., K. (2018). Supply Chain Social Sustainability and Manufacturing. In I. Management Association (Ed.), *Technology Adoption and Social Issues: Concepts, Methodologies, Tools, and Applications* (pp. 226-252). Hershey, PA: IGI Global. https://doi.org/ doi:10.4018/978-1-5225-5201-7.ch011

van der Vliet-Bakker, J. M. (2017). Environmentally Forced Migration and Human Rights. In C. Akrivopoulou (Ed.), *Defending Human Rights and Democracy in the Era of Globalization* (pp. 146–180). Hershey, PA: IGI Global. doi:10.4018/978-1-5225-0723-9.ch007

Vargas-Hernández, J. G. (2022). A Comprehensive Entrepreneurship Model for the Internationalization of Green Innovation Businesses. In U. Akkucuk (Ed.), *Disruptive Technologies and Eco-Innovation for Sustainable Development* (pp. 131–149). IGI Global. https://doi.org/10.4018/978-1-7998-8900-7.ch008

Vaskov, A. G., Lin, Z. Y., Tyagunov, M. G., Shestopalova, T. A., & Deryugina, G. V. (2018). Design of Renewable Sources GIS for ASEAN Countries. In V. Kharchenko & P. Vasant (Eds.), *Handbook of Research on Renewable Energy and Electric Resources for Sustainable Rural Development* (pp. 1–25). Hershey, PA: IGI Global. doi:10.4018/978-1-5225-3867-7.ch001

Vázquez, D. G., & Gil, M. T. (2017). Sustainability in Smart Cities: The Case of Vitoria-Gasteiz (Spain) – A Commitment to a New Urban Paradigm. In L. Carvalho (Ed.), *Handbook of Research on Entrepreneurial Development and Innovation Within Smart Cities* (pp. 248–268). Hershey, PA: IGI Global. doi:10.4018/978-1-5225-1978-2.ch012

Wahab, I. N., & Soonthodu, S. (2018). Geographical Information System in Eco-Tourism. In S. Chaudhuri & N. Ray (Eds.), *GIS Applications in the Tourism and Hospitality Industry* (pp. 61–75). Hershey, PA: IGI Global. doi:10.4018/978-1-5225-5088-4.ch003

Weiss-Randall, D. (2018). Cultivating Environmental Justice. In *Utilizing Innovative Technologies to Address the Public Health Impact of Climate Change: Emerging Research and Opportunities* (pp. 110–143). Hershey, PA: IGI Global. doi:10.4018/978-1-5225-3414-3.ch004

Weiss-Randall, D. (2018). Cultivating Resilience. In *Utilizing Innovative Technologies to Address the Public Health Impact of Climate Change: Emerging Research and Opportunities* (pp. 204–235). Hershey, PA: IGI Global. doi:10.4018/978-1-5225-3414-3.ch007

Weiss-Randall, D. (2018). Climate Change Solutions: Where Do We Go From Here? In *Utilizing Innovative Technologies to Address the Public Health Impact of Climate Change: Emerging Research and Opportunities* (pp. 236–268). Hershey, PA: IGI Global. doi:10.4018/978-1-5225-3414-3.ch008

Whyte, K. P., List, M., Stone, J. V., Grooms, D., Gasteyer, S., Thompson, P. B., . . . Bouri, H. (2018). Uberveillance, Standards, and Anticipation: A Case Study on Nanobiosensors in U.S. Cattle. In I. Management Association (Ed.), Biomedical Engineering: Concepts, Methodologies, Tools, and Applications (pp. 577-596). Hershey, PA: IGI Global. https://doi.org/ doi:10.4018/978-1-5225-3158-6.ch025

Wulff, E. (2017). Data and Operational Oceanography: A Review in Support of Responsible Fisheries and Aquaculture. In P. Diviacco, A. Leadbetter, & H. Glaves (Eds.), *Oceanographic and Marine Cross-Domain Data Management for Sustainable Development* (pp. 303–324). Hershey, PA: IGI Global. doi:10.4018/978-1-5225-0700-0.ch013

Yener, D. (2017). Geographic Information Systems and Its Applications in Marketing Literature. In S. Faiz & K. Mahmoudi (Eds.), *Handbook of Research on Geographic Information Systems Applications and Advancements* (pp. 158–172). Hershey, PA: IGI Global. doi:10.4018/978-1-5225-0937-0.ch006

Yu, K., Liu, Y., & Sharma, A. (2021). Analyze the Effectiveness of the Algorithm for Agricultural Product Delivery Vehicle Routing Problem Based on Mathematical Model. *International Journal of Agricultural and Environmental Information Systems, 12*(3), 26–38. https://doi.org/10.4018/IJAEIS.2021070103

Zhou, X., Sharma, A., & Mohindru, V. (2021). Research on Linear Programming Algorithm for Mathematical Model of Agricultural Machinery Allocation. *International Journal of Agricultural and Environmental Information Systems, 12*(3), 1–12. https://doi.org/10.4018/IJAEIS.2021070101

About the Contributors

Perfecto Gatbonton Aquino Jr. is currently the Director of Council for Research and Creativity (CRC) and the Institute for Research and Advanced Studies, University of Cambodia. A former Research Associate cum Senior Lecturer at the Socio-Economics Research Institute and at the American Degree Program Division of The Duy Tan University in Danang City, Vietnam. He teaches Research Methods, Human Resources Management, Leadership and Organizational Behavior, CSR, and other Business Management related courses in higher education institutions. Prior to his assignment in Vietnam, Dr. Aquino was an Assistant Professor and Head of the Business Department at the Emirates College for Management and Information Technology (ECMIT) and was an Assistant Professor at the American University in the Emirates, United Arab Emirates. He was also an Assistant Professor at the Gulf College/Staffordshire University in Oman and at AMA International University-Bahrain. He is a holder of Doctor of Philosophy (Ph. D) in Public Administration and Management at the Pontifical and Royal University of Santo Tomas, Manila, Philippines. Dr. Aquino was awarded in 2019 a CBME Certificate of Professional Teaching by the Chartered Association of Business Schools in United Kingdom. He has published several papers in Scopus indexed-journals, Springer Elsevier and IGI Global International Publishing, among others. Furthermore, Dr. Aquino is an active peer reviewer on articles submitted to Sage, IGI Global Publishing, Routledge, Springer Elsevier, Emerald Publishing, USA to mention a few.

Revenio Cabanilla Jalagat Jr. is presently employed in Al-Zahra College for Women as Associate Professor for almost 11 years. He taught at Gulf College as a Business Lecturer from 2011-2013. Prior to that, he had worked in various disciplines in both the academic and business industry with almost 9 years of accumulated experience in the Philippines acting as Accountant, Chief Account, Finance Manager, Assistant Professor, and Professor. In 2023, he received an award as among the Top 50 Outstanding Philippine Awardee in the field of Research and Publication and Distinguished Researcher in his institution. He has published articles, edited books, and chapters indexed in Scopus and Web of Science indexes and served as an editorial member and reviewer in various international refereed journals in the field of business management. He was also awarded in 2021 the Certified Business Management Educator by the Chartered Association of Business Schools in the UK.

Mercia Selva Malar Justin, over the past six years, has served as Associate Professor at Xavier Institute of Management and Entrepreneurship, Bangalore. She has been teaching at the higher education institutions in India and abroad over the past thirty years. Some of the institutions she served are Wollongong University, ECMIT Dubai, SCMS-Cochin, LEAD College of Management, CMS College of Science and Commerce. She has authored the books Aspire to be a Great Teacher, Foreign Exchange

Exposure and Risk Management in Indian Hotel Industry and A Daughter's Tribute to a Godly Father. She won the Outstanding Women Educator and Scholar Award in 2016 and Distinguished Women Educator and Scholar Award in 2017, from National Foundation for Entrepreneurship Development (NFED). In 2023 she was awarded the National Trailblazers Triumph Award – Outstanding Social Sciences Author. She is a regular contributor to University News published by University Grants Commission, India. She served as the Editor of Journal of Management and Entrepreneurship published by XIME, Bangalore.

***

Sumi A. M. was born and brought up in Thiruvananthapuram, the capital city of Kerala, South India. Currently doing Ph.D. in Management at Institute of Management in Kerala (IMK), University of Kerala. Completed M. Phil in Management (2017) and M.B.A (2015) from the University of Kerala. A textbook published in Management, "History of Management Thoughts; Text and Cases" in 2019. This book traces the Principles of Management thoughts and their Evolution through history. Around ten articles were published in various reputed Journals and Edited books. More than 15 research papers were presented in national and international seminars. Also, she is interested in learning about countries and their culture, writing books and articles.

Ann Baby has a PhD in Computer Science and a University rank holder in her post-graduation and holds an M.Phil degree. With more than two decades of teaching experience in the Postgraduate level, she has to her credit several research papers, published in reputed National and International journals. Dr. Ann Baby was the conference coordinator of several workshops and seminars. She has also successfully completed funded Research Projects. She is a lifetime member of CSI (Computer Society of India). She is the resource person for several expert sessions on Swachtha, Environment Friendliness, Cyber Security, Data Analytics and Machine Learning.

Ummul Fatiha Binte Shafiq completed her BSc in Environmental Science and Engineering degree from Jatiya Kabi Kazi Nazrul Islam University Bangladesh. She did research in the field of land cover and land use change due to environmental extreme events using GIS technology, Landsat and other statistical software. She participated in the national international conference and presented paper. She published four articles in different journals and book chapters. Her interest is in the field of environmental restoration, innovation in the environmental technology to combat climate change impacts and ecofriendly planning for the green cities.

K. S. Chandrasekar currently working as Professor and Head of Institute of Management in Kerala (IMK) and is the former Director, School of Business Management and Legal studies, University of Kerala. He is the Chairman of the Board of Studies in Business Management of University of Kerala and Mahatma Gandhi University, Kottayam. He was the Vice-Chairman of the Centre for Management Development (A Government of Kerala institution), Trivandrum, Kerala. He held the post of Dean, Faculty of Management studies thrice. Since 9 th February 2002, he has been a visiting Professor at Commonwealth Open University in Spain and UK. He has written ten leading books and published 213 research articles in various reputed journals and newspapers. He presented 259 research papers in International and National seminars. He is the editorial board member of 18 reputed international and national journals. He won around 16 Awards for his academic excellence. He is currently the National council member of AIMA and ISTD, New Delhi.

Maybel Fernandez-Vallado is an alumna of the University of the Philippines (Diliman); and former Senior Lecturer 2 (September 2020 to June 2023) and former Faculty Member of the College of Arts and Sciences, University of the Philippines Visayas. Maybel was born in Quezon City (Philippines) but currently resides in the bustling city of Dubai. Her life is defined by several key passions and roles. First and foremost, Maybel is a devout follower of Jesus, and her faith is the guiding light of her life. She finds solace, strength, and purpose in her spirituality, exemplifying the teachings of love and compassion. As a loving wife to Emmanuel, Maybel shares a deep and enduring bond with her husband. Their relationship is built on the foundation of love, trust, and mutual respect. In addition, Maybel has a profound love for reading, with books being both a source of knowledge and a gateway to diverse worlds. She cherishes the written Word and uses it as a means of personal growth and connection with others. Another creative outlet for Maybel is her passion for crochet. She enjoys weaving intricate patterns and creating beautiful, handmade pieces that radiate warmth and care. Finally, Maybel finds joy in praying for people. She is always ready to offer her prayers and support to those in need, exemplifying her deep concern for the well-being of others. As a mother, Maybel and Emmanuel are blessed with six adult children who carry forward their values of faith, love, and kindness, making them proud parents. She also wrote Chapter 2: "Communication Dynamics of Women in MIcrofinance Projects as a Model for Women Empowerment: Microfinance for Poverty Alleviation in the Philippines" of the book entitled: Empowering Women in Microfinance in Developing Countries, published by IGI Global in October 2023.

Shafiqul Islam, Dr. Md., specializes in sustainable development issues including environmental hazard & disaster management, organic farming with particular expertise on tea garden, NGOs and civil society. Other interests include rural development, livelihood, food security, humanitarian assistance, maternal & child health and nutrition, organizational issues in development agencies, and bottom up approaches of development. An Environmentalist by education and doing, he has conducted extensive field research in cross range of field including climate change adaptation, disaster risk reduction, ecological resilience, and has advised and consulted for a wide range of national international development agencies, universities NGOs and private sectors. Shafiqul Islam is an Associate Professor of Environmental Science at the International University of Business Agriculture and Technology. He is a specialist on Disaster Management, with a regional focus on Drought in Barind Tract in Bangladesh. Dr. Islam has published widely on the Climate Change Adaptation. His recent work includes Cultivation and Drought Management in Agriculture (Springer Nature, 2022). After studying in Chittagong University, Shafiqul Islam was awarded a doctorate in Sustainable Approaches to Drought Management: Exploratory Study on selected regions of Northwest in Bangladesh' from the Institute of Disaster Management and Vulnerability Studies, University of Dhaka, Bangladesh. He worked as a Post-Doctoral Research Fellow at the London School of Economics and Political Science (LSE), UK and conducted a research on "Exploring Social Ecological Resilience of Fisher Folks to Coastal Disasters". He worked for 13 years with WB DFID, USAID, EU, GWAB, NETZ, DanChurchAid and ADD funded projects for research, collaboration, partnership, implementation, network linkages and dissemination. He returned to academia as an Assistant Professor at the Center for Sustainable Development in the University of Liberal Arts Bangladesh in 2010. Dr Islam is a Forester with broad interests in the ecology, silviculture, and conservation of biodiversity. He was a semester Faculty at the United International University in 2019. He was the

Associate Professor and Chairman at the Department of Environmental Science (Graduate Program) in State University of Bangladesh from 2019-2022. He supervised more than 80 Master's student theses as the major Professor from 2019 to 2022. After receiving his doctoral degree in Disaster Management from the University of Dhaka, Bangladesh. His expertise includes community-based disaster risk management, climate change adaptation, and disaster and environmental education. Prof. Islam has published more than 50 academic papers and book chapters. Dr. Islam has been involved in research in the areas of gender, disasters, environment, climate change, education, social inclusion, indigenous community and other areas of social development since late nineties.

Binoy Joseph has extensive expertise in the areas of Employee Resourcing, Performance Management, Reward Management and HR Accounting and Auditing and his core research interest lies in HRM Innovations. He has had professional experience in the hospitality industry. He has been a Visiting International Scholar at Western Michigan University, USA, and was also engaged in research on Employee Stock Option Plan and Corporate Performance. He has received his doctoral degree from Cochin University of Science and Technology in the topic "Innovative Human Resource Practices and HR Outcomes in Software Firms. He has received the prestigious AUDI Professorship 2013 Fellowship. He made a presentation in Audi Company in Ingolstadt Germany on 'People Management Challenges for German Managers in India'. He also took a course ' International Human Resource Management' for the students in Catholic University, Ingolstadt, Germany

Kiran Kakade has completed his PhD from the SRTM University, Nanded in Management Studies, and Three post-graduation degrees in LL.M., MBA (HR) and MCA (Computer Application) from the University of Pune with over one decade of professional experience serving the IT, education, Legal consultancy sectors in various capacities. Having devoted significant time engaging with many top management institutes as a full-time faculty and consultant, he recently working in the Faculty of HR as Associate Professor at Lala Lajpatrai Institute of Management, Mumbai, Maharashtra. His unconventional teaching style makes him a very popular mentor and sought-after educator. His interest area is Human Resource Management, Information Technology, Yoga and Meditation.

Somya Khare is an Assistant Professor of Law, associated with Faculty of Law Jagran Lakecity University since 2019 and teaching subjects like Administrative Law, Environmental Law, Legal Method, Jurisprudence and Intellectual Property Rights. She holds a Master's degree in Intellectual Property Rights. Her areas of interest are Intellectual Property Rights, Environmental Laws, Indigenous Community and Traditional Knowledge.

Anjali M. Kulkarni has done BE (IPE) from SGSITS, Indore and MBA in Finance from IMS, DAVV, Indore. She secured 99 percentile score in UGC NET in Management and qualified Maharashtra SET. She is also an Associate of Institution of Engineers and Licentiate from Insurance Institute of India.Prof. Anjali M. Kulkarni is having 16 years of teaching experience with various management institutes in Indore and Mumbai. She had also been associated with various management institutes in Navi Mumbai as a Visiting faculty. She has also taught BE, B.Pharm and M.Tech. Courses.

Swarnadeep Maity is an Assistant Professor of Finance at NSB Academy. Pursuing his Ph.D. in Finance from University of Mysore, he brings over seven years of teaching experience in the areas of

Finance. His research has been published in esteemed journals, reflecting his thought leadership in the field. He also participates in national and international conferences to contribute to the advancement of financial knowledge

Rohit Mishra is an Expert in Taxation, Economics, Management and Accountancy, and has an experience of more than 10 years in teaching. He is a post graduate in commerce from Barkatullah University, Bhopal and qualified CMA intermediate and pursuing Ph.D. on "Analysis of FIIs trend and Impact on Bombay Stock Exchange" He is recognized as a creative and practical problem solver. He is also actively involved in research in Indian Capital Market and Foreign Portfolio Investors. He was also involved in various Audit programmes of companies and organizations of higher repute.

Kasturi R. Naik is known for real-time, forthright, solutions to complex people's problems. Knows the difference between Passion, Profession, and Purpose and that has made all the difference to the way she lives her life. She has more than 12 years of academic experience and 5 years of consultancy experience in the fields of human resource management, leadership, and team building for various corporate, educational institutions and NGOs. She has been associated as a visiting faculty at JBIMS and GNIMS for MHRD and e-MBA courses and a resource person at XLRI and IIBF. She has also been associated with Vmentorai as a mentor and subject matter expert in the domain of human resource management. Dr Kasturi R Naik is the recipient of multiple national and international awards and accolades for her enriching work in field of emotional intelligence, personality enhancement, stress management and leadership development. Her credentials include PhD with Stress management as broad area of research, MMS, BPHARM, DLL, Internationally Certified & Licensed Heal Your Life Workshop Leader and Life Coach, Certified in Positive Psychology: Resilience Skills, Certified in Building Your Leadership Skills by HEC Paris, Certified Executive Life Coach, Certified Image Consultant, Certified Emotional Intelligence Practitioner, Certified NLP Practitioner and HRBP 2.0 Practitioner Her research capabilities are self-proven through the 30+ research papers that she has published nationally and internationally, out of which she got the best paper awards in a couple of conferences as well and published in the ABDC category journal. Dr. Kasturi R Naik has also been honored to play a role of the discussant and session chair in national and international conferences and has also won Best Paper track presenter award at ABRM International conference 2015, besides Best paper award at IMRC 2016.

Katyayani Singh, with a Ph.D. in Political Science, has been teaching as an Assistant Professor at Jagran Lakecity for over three years wherein she teaches courses such as that of political philosophy; political theory; and international relations. She also serves as a research member of Political Science Research Committee, Center for Global Nonkilling, Hawaii; Human Dignity and Humanitarian Studies and the General Secretary of the Center for Global Nonkilling India. She has authored a book The Nonkilling Paradigm: For World Peace and Enlightenment, published by Springer in 2020 and has written various articles relating to peace, conflict, and philosophy.

Joycia Thorat is Gender Policy and Thematic Advisor of CASA, India, and Post -Doctoral Fellow at Kwazulu-Natal University.

Index

U

W

www.ingramcontent.com/pod-product-compliance
Lightning Source LLC
LaVergne TN
LVHW080528300926
849162LV00007B/1564
* 9 7 8 1 6 6 8 4 5 9 8 6 7 *